Mycotoxin Prevention and Control in Agriculture

ACS SYMPOSIUM SERIES **1031**

Mycotoxin Prevention and Control in Agriculture

Michael Appell, Editor
USDA-ARS, National Center for Agricultural Utilization Research

David F. Kendra, Editor
USDA-ARS, National Center for Agricultural Utilization Research

Mary W. Trucksess, Editor
US Food and Drug Administration, Center for Food Safety and Applied Nutrition

Sponsored by the
ACS Division of Agriculture and Food Chemistry

American Chemical Society, Washington DC

Library of Congress Cataloging-in-Publication Data

Mycotoxin prevention and control in agriculture / [edited by] Michael Appell, David F. Kendra, Mary W. Trucksess ; sponsored by the ACS Division of Agriculture and Food Chemistry.

p. cm. -- (ACS symposium series ; 1031)

Includes bibliographical references and index.

ISBN 978-0-8412-6990-3 (alk. paper)

1. Mycotoxins. 2. Toxigenic fungi. 3. Food contamination. 4. Feeds--Contamination. 5. Food--Microbiology. 6. Mycotoxicoses. I. Appell, Michael. II. Kendra, David F. III. Trucksess, Mary W. IV. American Chemical Society. Division of Agricultural and Food Chemistry V. American Chemical Society. Meeting (235th : 20008 : New Orleans, La.)

RA1242.M94M924 2009

615.9'5295--dc22

2009042749

The paper used in this publication meets the minimum requirements of American National Standard for Information Sciences—Permanence of Paper for Printed Library Materials, ANSI Z39.48−1984.

Distributed by Oxford University Press

PRINTED IN THE UNITED STATES OF AMERICA

Cover Design By: David Weingaertner

Foreword

The ACS Symposium Series was first published in 1974 to provide a mechanism for publishing symposia quickly in book form. The purpose of the series is to publish timely, comprehensive books developed from the ACS sponsored symposia based on current scientific research. Occasionally, books are developed from symposia sponsored by other organizations when the topic is of keen interest to the chemistry audience.

Before agreeing to publish a book, the proposed table of contents is reviewed for appropriate and comprehensive coverage and for interest to the audience. Some papers may be excluded to better focus the book; others may be added to provide comprehensiveness. When appropriate, overview or introductory chapters are added. Drafts of chapters are peer-reviewed prior to final acceptance or rejection, and manuscripts are prepared in camera-ready format.

As a rule, only original research papers and original review papers are included in the volumes. Verbatim reproductions of previous published papers are not accepted.

ACS Books Department

Contents

Preface .... **xi**

Prevention and Exposure

1. **Biological Control of Molds and Mycotoxins in Foods** **1**
   Andreia Bianchini, Lloyd B. Bullerman

2. **Prevention of Mycotoxins in Dried Fruit, other Fruit Products, and Botanicals** **17**
   Peter M. Scott, Mary W. Trucksess

3. **Mycotoxin Contamination of Agricultural Products in the Southern United States and Approaches to Reducing it from Pre-harvest to Final Food Products** **37**
   Hamed K. Abbas, W. T. Shier

4. **The fate of ochratoxin A in soy milk and bean curd (tofu) productions** **59**
   Maria H. Iha, Mary W. Trucksess, Jeanne I. Rader

5. **Physical protection of apple products from patulin contamination** **69**
   Seiji Bandoh, Keisuke Higashihara, Masahiko Takeuchi, Katsumi Ohsawa, Tetsuhisa Goto

6. **Determination of *Aspergillus* section *Flavi* and their aflatoxin and cyclopiazonic acid production patterns in naturally dried figs** **77**
   Dilek Heperkan, Funda Karbancıoglu-Güler

Molecular Biology and Biosynthesis

7. **Elimination and control of aflatoxin contamination in agricultural crops through *Aspergillus flavus* genomics** **93**
   Thomas E. Cleveland, Deepak Bhatnagar, Jiujiang Yu

8. **Microorganisms and their enzymes for detoxifying mycotoxins posing a risk to livestock animals 107**
I. Rodrigues, E. M. Binder, G. Schatzmayr

9. **Transgenic Expression of Lytic Peptides in Food and Feed Crops to Control Phytopathogens and Preharvest Mycotoxin Contamination 119**
K. Rajasekaran, J.M. Jaynes, and J.W. Cary

10. **Phytotoxicity of trichothecenes 143**
S. P. McCormick

11. **Identification of Maize Breeding Markers through Investigations of Proteins Associated with Aflatoxin-Resistance 157**
Robert L. Brown, Zhi-Yuan Chen, Thomas E. Cleveland, Abebe Menkir, Ahmad Fakhoury

12. **Advances in Understanding the Biosynthesis of Fumonisins 167**
Ryan Gerber, Lili Lou, Justin Huffman, Xiangcheng Zhu, Ting Lin, Lucy Q Li, Isis Arreguin, Robert A.E. Butchko, Robert H. Proctor, Liangcheng Du

13. **Regulation of *Aspergillus flavus* aflatoxin biosynthesis and development183**
Jeffrey W. Cary, Leanne Szerszen, Ana M. Calvo

Analytical Approaches

14. **Sampling Plans of Mycotoxins in Foods and Dietary Supplements (Ginger Capsules) 207**
Mary W. Trucksess, Thomas B. Whitaker

15. **Extraction Efficiency Studies for Mycotoxins in Naturally Contaminated Commodities 223**
Ryan J. Malone

16. **Fumonsins in materials related to ethanol production from corn: LC-MS analysis of fumonsins in ethanol manufacture by-products 237**
Mark Busman, David F. Kendra

17. **LC/Ultraviolet/Fluorescence and LC/MS/MS Analyses of Multiple Mycotoxins in Spiked Beer Using Immunoaffinity Column Clean-Up 247**
Darsa P. Siantar, Mark Benvenuti, Jim Krol, Joe Romano, Sarita Cardozo, Norma R. Hill, Abdul Mabud

18. **Stable Isotope Labeled Mycotoxins as Standards for HPLC-MS/MS Analysis: Review and Evaluation of Published Procedures for the Introduction of Stable Isotopes into Mycotoxins.................................. 265**
Benedikt Cramer, Marita Beyer, Hans-Ulrich Humpf

19. **Molecularly Imprinted Polymer Nanomaterials for Mycotoxin Extraction 277**
Jorn C.C. Yu, Edward P.C. Lai

20. **A Closer Look at Cyclodextrins in Mycotoxin Analysis 293**
Michael Appell, Chris M. Maragos

Indexes

Author Index 309

Subject Index 311

Preface

Mycotoxins are naturally occurring secondary metabolites produced by molds, particularly under conditions that are suitable for mold growth but stressful for the susceptible host plants to resist mold invasion. These toxins are a great concern to the agricultural industry for the health risks posed to humans and animals. Furthermore, mycotoxin contamination negatively impacts livestock production, commodity values, and international trades. Factors contributing to mycotoxin production include, region of farming, stress on the plant and fungi, harvest, weather, post-harvest, transportation and storage conditions. Although there is a lack of consensus for the purpose of mycotoxin biosynthesis, the outcome of toxin production is mycotoxicoses, human and animal diseases. A wide-range of efforts is required to reduce exposure, including a multitude of pre-harvest and post-harvest approaches.

This book is the result of the symposium Mycotoxin Prevention and Control in Agriculture presented at the 235th American Chemical Society National Meeting and Exposition in New Orleans, LA. This symposium was sponsored by the Division of Agricultural and Food Chemistry of the American Chemical Society to bring together researchers from diverse fields in common effort to reduce exposure and gain insight into mycotoxin contaminated commodities. The symposium represented the comprehensive approaches developed to address the issue of mycotoxicology, study of mycotoxins and mycotoxicoses, which including Industrial Production and Management, Omics Technologies, Method Analysis, Analytical Approaches, and Toxicity and Decontamination.

The objective of this book is to disseminate to a more broad audience the research presented at the dynamic and stimulating symposium. The book opens with an overview of prevention of mycotoxin production by means of biological control and human exposure to contaminated foods, including tofu, apples, figs, and a broad range of fruits. The second section is focused on molecular biological approaches to control mycotoxin exposure. Topics covered include genomic approaches, detoxification, structure-activity studies of biosynthetic precursors, and mycotoxin biosynthesis. The success of these strategies is supported by the reduction of mycotoxin occurrence and incidence in foods. Accurate estimates of mycotoxin concentration in foods are important for food safety and wholesomeness. The final sections covers analytical approaches, including sampling plans for foods and dietary supplements, extraction efficiency of mycotoxins from naturally contaminated grains, traditional detection methods such as chromatography/mass spectrometry, and emerging techniques such as the use of molecularly imprinted polymers for purification and isolation and cyclodextrins for fluorescent enhancement of several mycotoxins.

Acknowledgements

We are very thankful to the authors and their employers for contributing their research, time and efforts to this book. We express our appreciation to the peer reviewers for their insightful comments on the book chapters. We thank David Weingaertner of the Food and Drug Administration for designing and providing the artwork for the cover of the book. We are very grateful to the Division of Agriculture and Food Chemistry and the American Chemical Society for providing an engaging environment for this symposium to be held.

Michael Appell
Mycotoxin Research Unit
USDA-ARS, National Center for Agricultural Utilization Research
1815 N. University St.
Peoria, IL, 61604
Phone: 309 681-6249
Fax: 309 681-6689
E-Mail: michael.appell@ars.usda.gov

David F. Kendra
Mycotoxin Research Unit
USDA-ARS, National Center for Agricultural Utilization Research
1815 N. University St.
Peoria, IL, 61604
Phone: 309 681-6579
Fax: 309 681-6689
E-mail: david.kendra@ars.usda.gov

Mary W. Trucksess
US Food and Drug Administration, Center for Food Safety and Applied Nutrition
5100 Paint Branch Parkway
College Park, MD 20740,
Phone: 301 436-1957
Fax: 301 436-2665
E-mail: mary.trucksess@fda.hhs.gov

Prevention and Exposure

Chapter 1

Biological Control of Molds and Mycotoxins in Foods

Andreia Bianchini and Lloyd B. Bullerman

Dept. of Food Science & Technology, University of Nebraska-Lincoln, Lincoln, NE 68583-0919

Growth and mycotoxin production by molds is influenced by the environment, including other microorganisms. Lactic acid bacteria (LAB), *Bacillus* species and sourdough bread cultures have been reported to inhibit mold growth which may result from competition for space and nutrients in general, competition for nutrients required for mycotoxin production but not for growth, and production of antimycotic and antimycotoxigenic metabolites. Changes in the pH of the substrate through production of organic acids, such as lactic, propionic and acetic, along with hydroxyl fatty acids, reuterin and low-molecular-weight peptides may account in part for this activity. LAB have also been reported as capable of binding mycotoxins, which shows promise for the possible use of these organisms as sequestering agents in fermented and other processed foods, as well as in the gut. In the literature there is also evidence of microbial degradation of mycotoxins, and while this area of research is promising, there are some risks associated with potential production of toxic by-products or only partial detoxification that need further study. In general, the "generally recognized as safe"(GRAS) status of LAB and sourdough cultures enhances the potential of this group of bacteria to be used in commercial applications as biological control agents in processed foods to prevent mold growth, to improve the shelf life of fermented products and to reduce potential health hazards associated with mycotoxins.

Introduction

In nature, molds share habitat with other microorganisms which can influence mold growth and mycotoxin production. This biocontrol may result from many factors, including competition for space and nutrients in general, competition for nutrients required for mycotoxin production but not for growth, production of antimycotic and antimycotoxigenic metabolites by co-existing microorganisms, changes in pH of the substratum or a combination of these factors.

The history of biocontrol applied to molds and mycotoxins began in the 1960s, when Ciegler *et al.* (*1*) screened over 1000 microorganisms for their ability to degrade aflatoxins, and identified *Flavobacterium aurantiacum* as capable of irreversibly removing aflatoxin from solutions. In later studies, it was observed that the activity of *Flav. aurantiacum* was influenced by pH and temperature (*2*).

In the literature, mold growth has been described as controlled by other molds, yeasts and bacteria. Examples of some mold species controlling others include *Trichoderma harzianum* and *T. viride* and their effect on *Fusarium moniliforme* and *Aspergillus flavus* (*3*). In this case the biocontrol involves a complementary action of antibiosis, nutrient competition and production of cell wall degrading enzymes. The biocontrol of molds by yeasts can be exemplified by the control of *Aspergillus* section Flavi by *Kluyveromyces* spp. (*4*). In this review the biocontrol of molds by bacteria, including *Bacillus* spp., propionic acid bacteria and lactic acid bacteria will be discussed more in depth.

Antifungal Activity

Bacillus

Bacillus pumilus has been described as a growth inhibitor of mold growth, including the genera *Aspergillus, Penicillium* and *Fusarium.* It also has been associated with inhibition of aflatoxins, cyclopiazonic acid, ochratoxin A and patulin production (*5*). In the presence of cell free supernatant of *B. pumilus*, Munimbazi & Bullerman (*6*) reported more than 99% inhibition in aflatoxin production by *Aspergillus parasiticus* and up to 53% inhibition of mold growth. *B. licheniformis* has been reported as a producer of fungicin M-4 (an anti-fungal polypeptide) and *B. subtilis* as a producer of rhizocticin A (also an anti-fungal polypeptide) and iturins (lipopetides) (*7, 8, 9*). Recently, Zhang and collaborators (*10*) described the production of a small thermo resistant peptide by a *Bacillus* strain isolated from tobacco stems. The peptide was named B-TL2 and was a strong inhibitor of mycelial growth of *Aspergillus niger* among other molds. *Bacillus* spp. have also been described as inhibitory to the development of peach brown rot caused by *Monilia fructicola*, with 92% suppression of brown rot incidence and 88% reduction of the diameter of the lesion (*11*).

Propionic Acid Bacteria

Propionic acid bacteria are all part of the genus *Propionibacterium* and they gain energy through fermentation of lactate and sugars to propionate, acetate and carbon dioxide (*12*). Their inhibitory effect is associated with reduction of pH and production of propionic and acetic acids, which are very efficient in reducing fungal growth. When testing a variety of molds, Lind *et al.* (*13*) found that the minimal inhibitory concentration of propionic and acetic acids to cause growth inhibition at pH 5.0 was between 10-120 mM, while lactic acid had a minimal inhibitory concentration of 500 mM. Lind and collaborators (*13*) also studied the antifungal activity of five dairy propionibacteria. Their results indicated that the most pronounced antifungal effect was obtained with *Prop. thoenii.* However, depending upon the culture media used for bacterial fermentation, the antifungal activity obtained was different, and the best results were obtained with deMan Rogosa and Sharpe (MRS) medium (with or without addition of acetate). The authors suggested that the good results obtained with MRS medium could be attributed to the stronger buffering capacity of SL medium (which is a lactate based medium) or to better production of antifungal compounds in a glucose base medium. Further studies (*14*) identified the active compounds produced by those bacteria as two diketopiperazines: cyclo(L-Phe-L-Pro) from *Prop. jensenii*, *Prop. acidipropionici* and *Prop. thoenii*, and cyclo (L-Ile-L-Pro) from *Prop. jensenii.* All five strains tested produced 3-phenyllactic acid in different amounts.

Lactic Acid Bacteria

Lactic acid bacteria (LAB) comprise mainly four genera: *Lactococcus, Lactobacillus, Leuconostoc,* and *Pediococcus*, which are used traditionally as starter cultures for dairy, vegetable, and meat fermentations because of their contribution to flavor development and preservative potential (*15*, *16, 17, 18, 19*). Foodborne and food grade LAB have been studied much more than *Bacillus* species or propionic bacteria and their antifungal activities have been reported by several authors throughout the years (*20, 21, 22, 23, 24, 25, 26, 27*). Table 1 shows some of the publications that have demonstrated the antifungal activity of LAB, which vary from broad to very specific among and within species.

Table 1: Publications reporting the antifungal activity of lactic acid bacteria.

LAB isolate	Activity spectrum	Reference
Streptococcus lactis C10	*Aspergillus parasiticus*	*28*
Lactobacillus casei var *rhamnosus*	Broad spectrum	*29*
Lactobacillus reuteri	Broad spectrum	*30*
Streptococcus lactis subsp. *diacetilactis* DRC1 and *S. thermophilus* 489	*Aspergillus fumigatus* *Aspergillus parasiticus* *Rhizopus stolonifer*	*31*
Lactobacillus spp.	*Aspergillus flavus* subsp. *parasiticus*	*20*
Lactobacillus casei subsp. *pseudoplantarum*	*Aspergillus flavus*	*32*
Lactococcus lactis subsp. *lactis*	*Aspergillus flavus* *A. parasiticus* *Fusarium* spp.	*33*
Lactobacillus casei	*Penicillium* spp.	*34*
Lactobacillus sanfrancisco CB1	*Fusarium* spp. *Penicillium* spp. *Aspergillus* spp. *Monilia* spp.	*35*
Lactobacillus plantarum	Broad spectrum	*25*
Lactobacillus rhamnosus VT1	Broad spectrum	*27*
Lactobacillus plantarum MiLAB 393	Broad spectrum	*36*
Lactobacillus plantarum VTT E-78076 and VTT E-79098	*Fusarium* spp.	*37*
Lactobacillus paracasei subsp. *paracasei*	Yeasts *Penicillium* spp.	*38*

Stiles *et al.* (*39*) demonstrated the inhibitory activity of *Lactobacillus rhamnosus* isolated from a Czech tartar sauce. They showed, in a simultaneous antagonism assay, the ability of *L. rhamnosus* to inhibit mycelial growth of *Penicillium commune* and *Aspergillus niger* when both organisms (mold and bacteria) were cultivated together for 10 days at 30ºC (Figures 1 and 2). In these experiments, growth was expressed as the weight of the mycelial mass after drying. *L. rhamnosus* showed inhibitory activity against *P. commune* mostly after 4 days of incubation, while *A. niger* was inhibited from the beginning of the experiment. The same group (*40*) further studied the inhibitory effect on mold growth by *L. rhamnosus* in commercial cottage cheese. In these

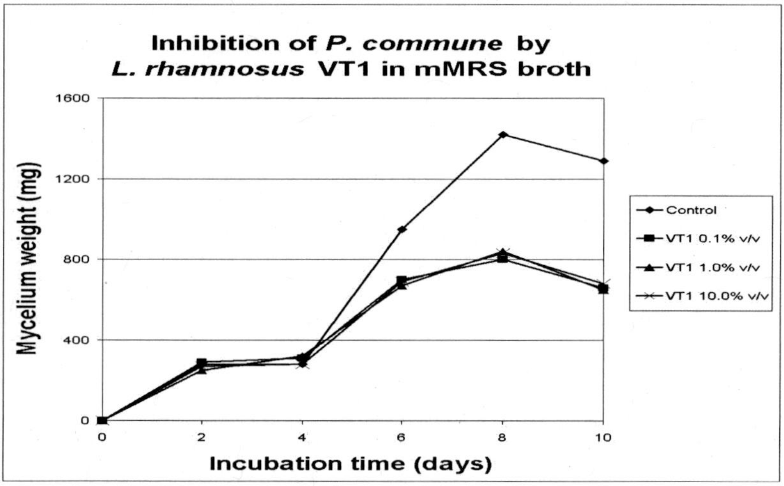

Figure 1: Inhibition of Penicillium commune NRRL 1889 by Lactobacillus rhamnosus VT1 (0.1, 1.0 and 10% v/v) in mMRS broth (39).

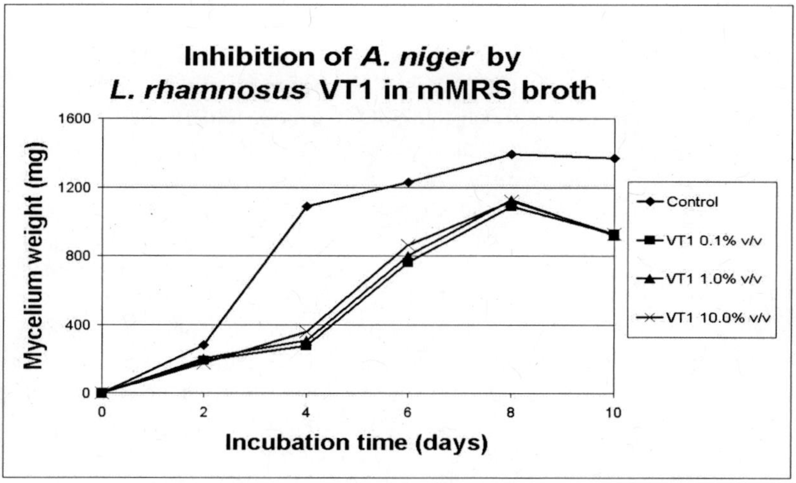

Figure 2: Inhibition of Aspergillus niger NRRL 326 by Lactobacillus rhamnosus VT1 (0.1, 1.0 and 10% v/v) in mMRS broth (39).

experiments, live bacterial cells were added to cheese that was simultaneously inoculated with *P. commune* and *A. niger*. After 10 days of incubation the growth of A. niger was inhibited more than 70% and *P. commune* more than 40%.

Sathe et al (*41*) studied the potential of LAB to prolong shelf-life of fresh vegetables. In their experiments cucumbers were wounded, inoculated with *L. plantarum*, and challenged with spoilage molds including *Aspergillus flavus*,

Fusarium graminearum, Rhizopus stolonifer and Botrytis cinerea. After 9 days of incubation at 20°C and 85% RH the size of the lesions were measured and reduction of up to 95% was observed for *R. stolonifer* (Table 2).

Table 2: Effect of L. plantarum treatment on lesion diameters (mm) produced by spoilage fungi on wounded cucumbers (41) after 9 days of incubation at 20°C.

Spoilage Fungi	Spoilage Fungus Only	Spoilage Fungus + *L. plantarum*
Aspergillus flavus	22.6	1.5
Fusarium graminearum	14.8	1.8
Rhizopus stolonifer	28.6	1.2
Botrytis cinerea	28.4	2.4

Also working with *L. plantarum*, Lavermicocca *et al.* (*25*) demonstrated that bread fermented with LAB along with *Saccharomyces cerevisiae* delayed *A. niger* growth for up to 7 days of storage at 20°C, when compared to control bread prepared only with *Saccharomyces.* In this experiment, breads made only with *S. cerevisae* showed growth of *A. niger* after 2 days of storage, while those made with *S. cerevisiae* and *L. plantarum* demonstrated mold growth after 7 days of storage.

It has been suggested that the inhibitory activity of LAB may result from production of organic acids (in particular, lactic, propionic and acetic acids), carbon dioxide, ethanol, hydrogen peroxide, diacetyl, reuterin and other small molecular weight metabolites, proteinaceous compounds or low-molecular weight peptides; competitive growth; decrease in the pH caused by acid production; or a combination of these factors.

Reports in the literature have shown that production of active organic acids vary with LAB species. *L. sanfranciscensis* has been reported to produce a mixture of acetic, formic, propionic, butyric, valeric, and caproic (key role) acids (*35*); *L. plantarum* has been reported to produce phenyllactic and 4-hydroxy-phenyllactic acids (*25*), as well as 3- phenyllactic acid (*36*); while *L. coryniformis, L. sakei,* and *Pediococcus pentosaceus* have been reported to produce phenyllactic acid (*42*). According to Schnürer (*43*), phenyllactic acid is only active against yeasts and molds at high concentrations ($mg{\cdot}ml^{-1}$), but may contribute to the overall antifungal effect.

Reuterin (β-hydroxypropionaldehyde) is a compound that was originally described from *L. reuteri* (*30*), and it is active against several different types of microorganisms, including fungi. Antifungal activity of this compound has been reported against species of *Candida, Torulopsis, Saccharomyces, Aspergillus,* and *Fusarium* (*44*). Reuterin is produced from glycerol by starving cells under anaerobic conditions and has been reported as produced by *L. reuteri, L. collinoides* (*45*), and *L. coryniformis* (*46, 47*).

Active compounds produced by LAB such as *Lactococcus lactis* and *Lactobacillus casei* have been suggested to be proteinaceous since the antimicrobial activity was lost after proteolytic treatment (*31, 33, 34*). Proteinaceous compounds with fungistatic effects against *Candida albicans* were actually purified from cultures of *Lact. pentosus* (*48*), but were not tested against molds. Magnusson and Schnürer (*49*) described proteinaceous compounds produced by *L. coryniformis* with activity against several molds and yeasts.

Other antifungal compounds include hydroxy fatty acids produced by *L. plantarum,* with minimum inhibitory concentration (MIC) in the ug·mL^{-1} range (*50*); and cyclic dipeptides produced by *L. plantarum* (*36*), and *L. coryniformis* (*42*), which are much less effective than the hydroxylated fatty acids, with a MIC in the mg mL^{-1} range.

Inhibition of Mycotoxin Production

Bacillus pumilus has been associated with inhibition of aflatoxin, cyclopiazonic acid, ochratoxin A and patulin production. Munimbazi and Bullerman (*6*) reported that more than 98% inhibition in aflatoxin production by *A. parasiticus* was caused by *B. pumilus*. Co-culturing *A. flavus* with *T. viride* in corn kernels was found to inhibit production of aflatoxin B_1 (73.5%) and aflatoxin G_1 (100%) (*51*).

The effect of small molecular weight metabolites produced by *Lactobacillus* species in the inhibition of aflatoxin production was demonstrated by Gourama & Bullerman (*21*). In their experiments a mixture of *Lactobacillus* species from a silage inoculant was cultivated inside of a dialysis sack, while *A. parasiticus* was cultivated on the outside. They found that after 14 days of incubation, the aflatoxin (B_1 and G_1) content of the mycelia grown in the presence of the *Lactobacillus* cultures was lower than the control, with little inhibition when the molecular weight cutoff (MWCO) was 1000, and great inhibition with MWCO of 6000-8000 and 12000-14000.

Stiles and Bullerman (*52*) studied the effect of *L. rhamnosus* on growth and mycotoxin production by *Fusarium* species, including *F. proliferatum, F. verticillioides* and *F. graminearum.* The results showed that production of fumonisin B_1 by *F. proliferatum* was reduced up to 63.2%, fumonisin B_2 produced by *F. verticillioides* up to 43.4%, and deoxynivalenol and zearalenone produced by *F. graminearum* up to 92% and 87.5%, respectively.

Gourama (*34*) reported the inhibitory activity of cell-free supernatants of *L. casei* strains on the growth of *Penicillium* spp. and the production of toxins (patulin and citrinin). The active compounds in the cell-free supernatant were found not related to the production of lactic acid or hydrogen peroxide, and were sensitive to some proteolytic enzymes (Table 3) and to high temperature (100°C).

Table 3: Effect of proteolytic enzymes on the antimycotoxigenic activity of L. casei (*34*).

	P. citrinum		*P. expansum*	
	Mycelia (mg)	Citrinin (μg/ mL)	Mycelia (mg)	Patulin (μg/ mL)
Control	25.1	4.1	27.1	3.5
L. casei cell-free supernatant	12.1	1.1	16.1	0.9
Trypsin	19.1	3.1	19.1	3.0
Pepsin	21.2	1.8	17.6	1.9

Karunaratne *et al.* (*20*) reported that aflatoxin production by *A. parasiticus* was considerably reduced in the presence of cell-free supernatants of *L. bulgaricus, L. acidophilus* and silage inoculant cultures. Also reported was a 100% inhibition of aflatoxin production in the presence of LAB cells. Even though the authors have not discussed, the difference observed between the inhibitory effect in the presence and absence of the LAB cells could be attributed to a binding or sequestration effect caused by the LAB cells. Binding of mycotoxins by LAB has been well documented in the literature (*53, 54, 55, 56, 57*).

Mycotoxin Binding

El-Nezami *et al.* (*58*) reported the ability of lactic acid bacteria to remove aflatoxin B_1 from artificially contaminated liquid media. The removal was strain dependent and very fast, with two strains of *L. rhamnosus* removing about 80% of the toxin at the beginning of the incubation time. The two best strains were further tested and their ability to remove aflatoxin was found to be dependent on temperature, with maximum removal at 37°C, and a bacterial concentration of at least 2×10^9 CFU/mL necessary for significant removal. Treating these strains with 1M HCl significantly enhanced their ability to remove aflatoxin (*59*). While screening dairy strains (LAB and bifidobacteria) for their ability to bind aflatoxin B_1, Peltonen *et al.* (*54*) found that between 5 and 60% of the aflatoxin in solution was bound by the bacteria, with *L. amylovorus* and *L. rhamnosus* being the most efficient species.

Fuchs *et al.* (*57*) screened thirty different LAB strains for their ability to remove ochratoxin A and patulin from solution, and they reported at least 95% removal of ochratoxin A in liquid medium by *L. acidophilus* and 80% removal of patulin by *Bifidobacterium animalis*. *L. rhamnosus* was evaluated for its potential to remove or degrade zearalenone and α-zearalenol, and both viable and non-viable cells (heat and acid treated) were able to remove about 50% of the toxin from solution, indicating that binding rather than metabolism was the mechanism in action (*55*). This finding was substantiated by the lack of detection of degradation products in solution. When *L. rhamnosus* was exposed to both toxins at the same time, its ability to remove zearalenone and α-zearalenol from solution was significantly reduced, indicating that these toxins may share the same binding site on the bacterial cell (*55*).

El-Nezami *et al.* (*56*) studied the ability of *Lactobacillus* and *Propionibacterium* strains to remove seven different *Fusarium* toxins (trichothecenes) from liquid media. *L. rhamnosus* GG and *Prop. freudenreichii* spp. *shermanii* JS were able to remove deoxynivalenol, diacetoxyscirpenol, and fusarenon, with the amount of removal varying from 18 to 93%. *L. rhamnosus* LC-705 removed only deoxynivalenol and diacetoxyscirpenol, with removal varying from 10 to 64%. Niderkorn *et al.* (*60*) screened lactic and propionic acid bacteria for their ability to remove deoxynivalenol and fumonisin from solution and they found that it was strain specific, with propionic acid bacteria being less efficient than LAB. *L. rhamnosus* removed up to 55% of deoxynivalenol, while *Leuconostoc mesenteroides* removed about 82% of fumonisin B_1, and *Lact. lactis* removed 100% of fumonisin B_2.

The effect of different variables on binding of aflatoxin B_1 to *L. rhamnosus* was studied by Haskard *et al.* (*53, 61*). Temperature, sonication and pH had no significant effect on the release of bound aflatoxin B_1 by *L. rhamnosus* (*53*), while salt (NaCl and $CaCl_2$) concentrations showed minor effects. Of the variables studied, urea had the greatest effect suggesting that hydrophobic interactions play a major role in binding (*61*).

The ability of lactic acid bacteria to remove toxins *in vitro* and *in vivo* was compared by El-Nezami *et al.* (*58, 62*). In *in vitro* experiments, *L. rhamnosus* was able to remove approximately 80% of the initial amount of aflatoxin B_1. In *in vivo* studies using the chicken duodenum loop technique, 54% (w/w) of the added aflatoxin B_1 was removed from the luminal fluid and intestinal absorption was reduced by 74% (w/w) (*62*).

Shetty *et al.* (*63*) found that the ability of *S. cerevisiae* to bind aflatoxin B_1 was strain specific with 7 strains binding 10-20%, 8 strains binding 20-40% and 3 strains binding more than 40% of the added aflatoxin B_1. El-Nezami *et al.* (*62*), testing the ability of *L. rhamnosus* to bind aflatoxin B_1 *in vivo,* also noted the specificity of the strains with one strain reducing absorption by 74% (strain GG) while another (strain LC-705) reduced absorption by only 37%.

In the literature it seems to be a consensus that the stability of the interactions involved in the binding of mycotoxins by bacterial or yeast cells depends on strain, treatment and environmental conditions (*53, 56, 58*). It is important to understand how different species and strains behave under different environmental conditions to evaluate their potential to be used as decontaminating agents in food processes and as binding agents in the GI tract of humans and animals.

Microbial Degradation

Organisms, enzymes and genes responsible for mycotoxin detoxification have been studied and this knowledge could be used for the development of inactivation procedures for mycotoxins. There are different approaches that could be used to screen microorganisms for their ability to degrade or metabolize mycotoxins. One involves the selection of strains based on their ability to use toxins as a source of carbon (*64*). The limitation of this method is that only a fraction of detoxifying organisms can utilize the respective toxins as a sole substrate. Another strategy tests randomly selected cultures for their ability to degrade toxins in different media (*65, 66*). This assay has no limitations but can be very laborious.

The ability of *Flavobacterium aurantiacum* to remove aflatoxin from solution has been well studied and toxicological examination of the bacteria-treated solution showed complete detoxification in a duckling assay (*1*). The rate of removal of aflatoxin by live cells (*Flav. aurantiacum*) was greater than with dead cells, but removal to some extent occurred with both viable and non-viable cells. Taking the experiments one step further, the investigators showed that use of C13-labeled aflatoxin resulted in release of labeled CO_2, indicating metabolism of the toxin by the live bacteria (*67*). Smiley and Draughon (*68*) studied the ability of crude protein extracts from *Flav. aurantiacum* to degrade aflatoxin B_1. Their results showed a 74.5% degradation of the toxin in the presence of crude protein extracts (800 μg of protein per mL). The crude protein extract treated with proteinase K was able to degrade 34.5% of the toxin in solution and when it was heat treated it degraded only 5.5% of the initial amount of aflatoxin B_1. These results strongly indicate that the degradation may be a result of enzymatic reactions.

Degradation of aflatoxins by other microorganisms, such as *Phoma* sp., *Rhizopus* sp., *Aspergillus niger, Corynebacterium rubrum, Candida lipolytica, Trichoderma viride* and *Mucor ambiguous,* has also been reported (*69, 70, 71, 72*).

Shima *et al.* (*65*) searching for organisms capable of metabolizing deoxynivalenol found, by enrichment culture, a bacterium from soil capable of transforming the toxin into 3-keto-4-deoxynivalenol. After isolation, the bacterium was classified as part of the *Agrobacterium-Rhizobium* group. The metabolite (3-keto-4-deoxynivalenol) was tested in a bioassay and it showed less than 10% of the immunosuppressive toxicity as deoxynivalenol. Another group of researchers (*66*) also found a mixed culture named D107 that was able to metabolize deoxynivalenol to 3-keto-4-deoxynivalenol; however, the group was not able to isolate a pure culture from the mixture that could consistently transform deoxynivalenol. The transformation was suppressed by glucose (> 0.07% w/v), tryptone and yeast extract.

El-Sharkawy and Abul-Hajj (*73*) have reported the detoxification of zearalenone by *Gliocladium roseum*. The product of the zearalenone metabolism was a mixture of two isomeric hydroxyketones that underwent spontaneous and irreversible decarboxylation.

Metabolism of ochratoxin A by atoxigenic *A. niger* has been described (*74*). First, ochratoxin A was converted to ochratoxin α, and then the concentration of the latter was gradually decreased to trace amounts after 10 days of incubation. More recently Varga *et al.* (*75*) reported the degradation of ochratoxin A by *Rhizopus* isolates, with some isolates degrading > 95% of the toxin in liquid medium within 16 days. One of the isolates (*R. stolonifer*) was also able to metabolize ochratoxin A on moistened wheat.

The idea of using a biological approach for mycotoxin degradation is very appealing; however, there are difficulties inherent to the use of microorganisms or bacterial enzymes for detoxification of mycotoxins. Active mixed cultures may lose activity after separation into pure cultures, as reported for zearalenone (*64*), because there is a co-metabolism and nutritional support for the degradation of the toxin or because the detoxification process is part of a reaction chain. Another potential problem is this approach may result in the production of compounds that have greater toxicity than the original mycotoxin.

Conclusions

There is evidence that *Bacillus*, propionic acid bacteria and LAB, among other microorganisms, are able to inhibit mold growth and mycotoxin production. The "generally recognized as safe" (GRAS) status of some of these organisms, such as LAB, offers the potential to use these bacteria in commercial applications as biological control agents in foods to prevent mold growth, improve the shelf life of fermented products and reduce the health hazards associated with mycotoxins. As for the ability to bind mycotoxins, bacterial cells (active or inactive) show promise as sequestering agents for mycotoxins in fermented and other processed foods. The field of mycotoxin detoxification by microorganisms is promising even though it seems that only very few organisms can effectively degrade toxins. However, risks associated with potential production of toxic by-products or only partial detoxification need further study.

Future Trends

So far most of the research focus has been concentrated on the biological control of molds and mycotoxin in simple systems and more attention should be devoted to a broader and complete evaluation of the practical use of these organisms or their enzymes in food and feeds. It is also very important to assess the potential of specific strains of LAB for binding mycotoxins and the factors that affect the stability of the toxin sequestration. The importance of this knowledge is two-fold. The first involves the potential for these bacteria (active and inactive cells) to be applied as part of decontamination procedures for food products where contamination with a mycotoxin is a problem, and the second is the potential for these bacteria to act as binding agents in the gastro-intestinal tract of humans and animals, preventing absorption of mycotoxins. Another area of research that has potential for expansion is the search for microbial detoxification systems based on genetic information. DNA ("metagenome") obtained from the environment shared by toxigenic molds can be used for the construction of genomic libraries that could be searched for the genes of specific detoxification enzymes.

References

1. Ciegler, A.; Lillehoj, E. B.; Peterson, R. E.; Hall, H. H. Microbial detoxification of aflatoxin. *Appl. Microbiol.* **1966**, *14*, 934-939.
2. Lillehoj, E. B.; Ciegler, A.; Hall, H. H. Aflatoxin B_1 uptake by *Flavobacterium aurantiacum* and resulting toxic effects. *J. Bacteriol.* **1967,** *93*, 464-471.
3. Calistru, C.; McLean, M.; Berjak, P. *In vitro* studies on the potential for biological control of *Aspergillus flavus* and *Fusarium moniliforme* by *Trichoderma* species. *Mycopathologia* **1997**, *137*, 115-124.
4. La Penna, M.; Nesci, A.; Etcheverry, M. *In vitro* studies on the potential for biological control on *Aspergillus* section *Flavi* by *Kluyveromyces* spp. *Lett Appl. Microbiol.* **2004**, *38*, 257-264.
5. Munimbazi, C.; Bullerman, L.B. Isolation and partial characterization of antifungal metabolites of *Bacillus pumilus. J. Appl. Microbiol.* **1998**, *84*, 959-968.
6. Munimbazi, C.; Bullerman, L.B. Inhibition of aflatoxin production of *Aspergillus parasiticus* NRRL 2999 by *Bacillus pumilus. Mycopathologia* **1998,** *140*, 163-169.
7. Kugler, M.; Loeffler, W.; Rapp, C.; Kern, A.; Jung, G. Rhizocticin A, an antifungal phosphono-oligopeptide of *Bacillus subtilis* ATCC 6633: biological properties. *Arch. Microbiol.* **1990**, *153*, 276-281.
8. Maget-Dana, R.; Peypoux, F. Iturins, a special class of pore-forming lipopeptides: biological and physicochemical properties. *Toxicol.* **1994**, *87*, 151-174.
9. Lebbadi, M.; Gálvez, A.; Maqueda, M.; Martínez-Bueno, M.; Valdivia, E. Fungicin M4: a narrow spectrum peptide antibiotic from *Bacillus licheniformis* M-4. *J. Appl. Microbiol.* **2008**, *77*, 49-53.
10. Zhang, B.; Xie, C.;Yang, X. A novel small antifungal peptide from *Bacillus* strain B-TL2 isolated from tobacco stems. *Peptides* **2008,** *29*, 350-355.
11. Zhou, T.; Schneider, K. E.; Li, X.-Z. Development of biocontrol agents from food microbial isolates for controlling post-harvest peach brown rot caused by *Monilinia fructicola. Int. J. Food Microbiol.* **2008**, *126*, 180-185.
12. Piveteau, P. Metabolism of lactate and sugars by dairy propionibacteria: a review. *Lait.* **1999**, *75*, 331-343.
13. Lind, H.; Jonsson, H.; Schnürer, J. Antifungal effect of dairy propionibacteria – contribution of organic acids. *Int J Food Microbiol* **2005,** 98, 157-165.
14. Lind, H.; Sjögren, J.; Gohil, S.; Kenne, L.; Schnürer, J.; Broberg, A. Antifungal compounds from cultures of dairy propionibacteria type strains. *FEMS Microbiol. Lett.* **2007**, *271*, 310-315.
15. Holzapfel, W.H.; Geisen, R.; Schillinger, U. Biological preservation of foods with reference to protective cultures, bacteriocins and food grade enzymes. *Int. J. Food Microbiol.* **1995**, *24*, 343-362.
16. Urbach, G. Contribution of lactic acid bacteria to flavour compound formation in dairy products. *Int. Dairy J.* **1995**, *5*, 877-903.

17. Lindgren, S.E.; Dobrogosz, W. J. Antagonistic activities of lactic acid bacteria in food and feed fermentations. *FEMS Microbiol. Lett.* **2006**, *87*, 149-164.
18. Olson, N. F. The impact of lactic acid bacteria on cheese flavor. *FEMS Microbiol. Lett.* **2006**, *87*, 131-147.
19. Buckenhüskes, H.J. Selection criteria for lactic acid bacteria to be used as starter cultures for various food commodities *FEMS Microbiol. Rev.* **2006**, *12*, 253-271.
20. Karunaratne, A.; Wezenberg, E.; Bullerman, L. B. Inhibition of mold growth and aflatoxin production by *Lactobacillus* spp. *J. Food Prot.* **1990**, *53*, 230-236.
21. Gourama, H.; Bullerman, L. B. Inhibition of growth and aflatoxin production of *Aspergillus flavus* by a *Lactobacillus* species. *J. Food Prot.* **1995**, *58*, 1249-1256.
22. Gourama, H.; Bullerman, L. B. Antimycotic and antiaflatoxigenic effect of lactic acid bacteria. A Review. *J. Food Prot.* **1995**, *57*, 1275-1280.
23. Stiles, J. Antifungal activity of lactobacilli. *Ph.D. Thesis*. Institute of Chemical Technology, Prague, Czech Republic, **1999**.
24. Stiles, J.; Plocková, M.; Toth, V.; Chumchalová, J. Inhibition of *Fusarium* sp. DMF 0101 by *Lactobacillus* strains grown in MRS and Elliker broths. *Adv. Food Sci.* **1999**, *21*, 117-121.
25. Lavermicocca, P.; Valerio, F.; Evidente, A.; Lazzaroni, S.; Corsetti, A.; Gobbetti, M. Purification and characterization of novel antifungal compounds from the sourdough *Lactobacillus plantarum* strain 21 B. *Appl. Environ. Microbiol.* **2000**, *66*, 4084-4090.
26. Plocková, M.; Stiles, J.; Chemchalova, J.; Halfarova, R. Control of mould growth by *Lactobacillus rhamnosus* VTI and *Lactobacillus reuteri* CCM 3525 on milk agar plates. *Czech J. Food Sci.* **2001**, *19*, 46-50.
27. Stiles, J.; Penkar, S.; Plocková, M.; Chumchalová, J.; Bullerman, L.B. Research Note: Antifungal Activity of Sodium Acetate and *Lactobacillus rhamnosus*. *J. Food Prot.* **2002**, *65*, 1188-1191.
28. Wiseman, D. W.; Marth, E. H. Growth and aflatoxin production by *Aspergillus parasiticus* when in the presence of *Streptococcus lactis. Mycopathologia* **1981,** *73*, 49-56.
29. King, S.W.; Fowler, G. G.; Vandenbergh, P. A. Method for inhibiting fungi. *European Patent Application EP 0,221,499,A2* **1986**.
30. Talarico, T.L.; Casas, I.A.; Chung, T.C.; Dobrogosz, W.J. Production and isolation of reuterin, a growth inhibitor produced by *Lactobacillus reuteri. Antimicrob Agents Chemother* **1988**, *32*, 1854-1858.
31. Batish, V.K.; Grover, S.; Lal, R. Screening lactic starter cultures for antifungal activity. *Cultur Dairy Prod. J.* **1989,** *24*, 21-25.
32. Gourama, H.; Bullerman, L.B. Anti-aflatoxigenic activity of *Lactobacillus casei pseudoplantarum. Int J Food Microbiol* **1997**, 34, 131-143.
33. Roy, U.; Batish, V.K.; Grover, S.; Neelakantan, S. Production of antifungal substance by *Lactococcus lactis* subsp. *lactis* CHD-28.3. *Int. J. Food Microbiol.* **1996**, *32*, 27-34.
34. Gourama, H. Inhibition of growth and mycotoxin production of *Penicillium* by *Lactobacillus* species. *Lebenson Wiss. Technol.* **1997**, *30*, 279-283.

35. Corsetti, A.; Gobbetti, M.; Rossi, J.; Damiani, P. Antimould activity of sourdough lactic acid bacteria: identification of a mixture of organic acids produced by *Lactobacillus sanfrancisco* CB1. *Appl. Microbiol. Biotechnol.* **1998**, *50*, 253-256.
36. Ström, K.; Sjögren, J.; Broberg, A.; Schnürer, J. *Lactobacillus plantarum* MiLAB 393 produces the antifungal cyclic dipeptides Cyclo(L-Phe-L-Pro) and Cyclo(L-Phe-trans-4-OH-L-Pro) and 3-Phenyllactic Acid. *Appl. Environ. Microbiol* .**2002**, *68*, 4322-4327.
37. Laitila, A.; Alakomi, H-L.; Raaska, L.; Mattila-Sandholm, T.; Haikara, A. Antifungal activities of two *Lactobacillus plantarum* strains against *Fusarium* moulds *in vitro* and in malting of barley. *J. Appl. Microbiol.* **2002**, *93*, 566-576.
38. Schwenninger, S. M.; Ah, U. V.; Niederer, B.; Teuber, M.; Meile, L. Detection of antifungal properties in *Lactobacillus paracasei* subsp. *paracasei* SM20, SM29, and SM63 and molecular typing of the strains. *J. Food Prot.* **2005**, *68*, 111-119.
39. Stiles, J.; Carter, V.; Bullerman, L.B. Effect of *Lactobacillus rhamnosus* VT1 on the growth of *Penicillium* and *Aspergillus* species. Presented at the *7th International Mycological Congress*. Oslo, Norway, August 11-17, **2002**.
40. Stiles, J.; Carter, V.; Bullerman, L.B. Inhibition of *Penicillium commune* and *Aspergillus niger* by *Lactobacillus rhamnosus* VT1 in commercial cottage cheese. Presented at the *89th Annual Meeting of International Association for Food Protection*. San Diego, CA, June 30 – July 3, **2002**.
41. Sathe, S. J.; Nawani, N. N.; Dhakephalkar, P. K.; Kapadnis, B.P. Antifungal lactic acid bacteria with potential to prolong shelf-life of fresh vegetables. *J. Appl. Microbiol.* **2007,** *103*, 2622-2628.
42. Magnusson, J.; Ström, K.; Roos, S.; Sjögren, J.; Schnürer, J. Broad and complex antifungal activity among environmental isolates of lactic acid bacteria. *FEMS Microbiol. Lett.* **2003**, *219*, 129-135.
43. Schnürer, J. Antifungal lactic acid bacteria and propionic acid bacteria as new food biopreservatives. Proceedings of *Food Mycology 2007: Emerging Mold Problems and Spoilage in Food and Beverages*. Eds: Samson, R. A.; Dijksterhuis, J.; Houbraken, J.; Rico, E.; Johnson, S. Key West USA, June 7-8, **2007**.
44. Chung, T.C.; Axelsson, L.; Lindgren, S.E.; Dobrogosz, W.J. *In vitro* studies on reuterin synthesis by *Lactobacillus reuteri. Microb. Ecol. Health Dis.* **1989**, *2*, 137-144.
45. Claisse, O; Lonvaud-Funel, A. Assimilation of glycerol by a strain of *Lactobacillus collinoides* isolated from cider. *Food Microbiol.* **2000**, *17*, 513-519.
46. Nakanishi, K.; Tokuda, H.; Ando, T.; Yajima, M.; Nakajima, T.; Tanaka, O.; Ohmomo, S. Screening of lactic acid bacteria having the ability to produce reuterin. *Jpn. J. Lactic Acid Bact.* **2002**, *13*, 37-45.
47. Magnusson, J. Antifungal Activity of Lactic Acid Bacteria. *PhD Thesis*, Swedish University of Agricultural Sciences, Uppsala, Sweden, **2003**.
48. Okkers, D.J.; Dicks, L.M.T.; Silvester, M.; Joubert, J.J.; Odendaal, H.J. Characterization of pentocin TV35b, a bacteriocin-like peptide isolated

from *Lactobacillus pentosus* with a fungistatic effect on *Candida albicans*, *J. Appl. Microbiol.* **1999**, *87*, 726–734.

49. Magnusson, J.; Schnürer, J. *Lactobacillus coryniformis* subsp. *coryniformis* strain Si3 produces a broad-spectrum proteinaceus antifungal compound. *Appl. Environ. Microbiol.* **2001**, *67*, 1-5.
50. Sjögren, J.; Magnusson, J.; Broberg, A.; Schnürer, J.; Kenne, L. Antifungal 3-hydroxy fatty acids from *Lactobacillus plantarum* MiLAB 14. *Appl. Environ. Microbiol.* **2003**, *69*, 7554-7557.
51. Choudhary, A.K. Influence of microbial co-inhabitants on aflatoxin synthesis of *Aspergillus flavus* on maize kernels. *Lett. Appl. Microbiol.* **1992**, *14*, 143-147.
52. Stiles, J.; Bullerman, L.B. Inhibition of *Fusarium* species and mycotoxin production by *Bacillus pumilus* NE B1 and *Lactobacillus rhamnosus* VT1. Proceedings of *13th International Reinhardsbrunn Symposium.* In: Modern Fungicides and Antifungal Compounds III AgroConcept GmbH. Eds: Dehne, H.W.; Gisi, U.; Kuck, K.H.; Russell, P.E.; Lyr, H. Bonn, Germany, May 14-18, **2002**.
53. Haskard, C.A.; El-Nezami, H.S.; Kankaanpää, P.E.; Salminen, S.; Ahokas, J.T. 2001. Surface binding of aflatoxin B_1 by lactic acid bacteria. *Appl .Environ. Microbiol.* **2001**, *67*, 3086-3091.
54. Peltonen, K.; El-Nezami, H.; Haskard, C.; Ahokas, J.; Salminen, S. Aflatoxin B_1 binding by dairy strains of lactic acid bacteria and bifidobacteria. *J. Dairy Sci.* **2001**, *84*, 2152-2156.
55. El-Nezami, H.; Polychronaki, N.; Salminen, S.; Mykkänen, H. Binding rather than metabolism may explain the interaction of two food-grade *Lactobacillus* strains with zearalenone and its derivative α-zearalenol. *Appl. Environ. Microbiol.* **2002**, *68*, 3545-3549.
56. El-Nezami, H.S.; Chrevatidis, A.; Auriola, S.; Salminen, S.; Mykkänen, H. Removal of common *Fusarium* toxins *in vitro* by strains of *Lactobacillus* and *Propionibacterium. Food. Addit. Contam.* **2002,** *19*, 680-686.
57. Fuchs, S.; Sontag, G.; Stidl, R.; Ehrlich, V.; Kundi, M.; Knasmüller, S. Detoxification of patulin and ochratoxin A, two abundant mycotoxins, by lactic acid bacteria. *Food Chem. Toxicol.* **2008**, *46*, 1398-1407.
58. El-Nezami, H.; Kankaanpaa, P.; Salminen, S.; Ahokas, J. Ability of dairy strains of lactic acid bacteria to bind a common food carcinogen, aflatoxin B_1. *Food Chem. Toxicol.* **1998,** *36*, 321-326.
59. El-Nezami, H.; Kankaanpaa, P.; Salminen, S.; Ahokas, J. Physicochemical alterations enhance the ability of dairy strains of lactic acid bacteria to remove aflatoxin from contaminated media. *J. Food Prot.* **1998**, *61*, 466-468.
60. Niderkorn, V.; Boudra, H.; Morgavi, D.P. Binding of *Fusarium* mycotoxins by fermentative bacteria *in vitro*. *J. Appl. Microbiol.* **2006**, *101*, 849-856.
61. Haskard, C.; Binnion, C.; Ahokas, J. Factors affecting the sequestration of aflatoxin by *Lactobacillus rhamnosus* strain GG. *Chem. Biol. Interact.* **2000**, *128*, 39-49.
62. El-Nezami, H.; Mykkänen, H.; Kankaanpää, H.; Salminen, S.; Ahokas, J. Ability of *Lactobacillus* and *Propionibacterium* strains to remove aflatoxin B_1 from the chicken duodenum. *J. Food Prot.* **2000**, *63*, 549-552.

63. Shetty, P.H.; Hald, B.; Jespersen, L. Surface binding of aflatoxin B_1 by *Saccharomyces cerevisiae* strains with potential decontaminating abilities in indigenous fermented foods. *Int. J. Food Microbiol.* **2007**, *113*, 41-46.
64. Megharaj, M.; Garthwaite, I.; Thiele, J.H. Total biodegradation of the oestrogenic mycotoxin zearalenone by a bacterial culture. *Lett. Appl. Microbiol.* **1997**, *24*, 329-333.
65. Shima, J.; Takase, S.; Takahashi, Y.; Iwai, Y.; Fujimoto, H.; Yamazaki, M.; Ochi, K. Novel detoxification of the trichothecene mycotoxin deoxynivalenol by a soil bacterium isolated by enrichment culture. *Appl. Environ. Microbiol.* **1997**, *63*, 3825-3830.
66. Völkl, A.; Vogler, B.; Schollenberger, M.; Karlovsky, P. Microbial detoxification of mycotoxin deoxynivalenol. *J. Basic Microbiol.* **2004**, *44*, 147-156.
67. Line, J.E.; Brackett, R.E.; Wilkinson, R.E. Evidence for degradation of aflatoxin B_1 by *Flavobacterium aurantiacum. J. Food Prot.* **1994**, *57*, 788-791.
68. Smiley, R.D.; Draughon, F.A. Preliminary evidence that degradation of aflatoxin B_1 by *Flavobacterium aurantiacum* is enzymatic. *J. Food Prot.* **2000**, *63*, 415-418.
69. Cole, R.J.; Kirksey, J.W. Aflatoxin G_1 metabolism by *Rhizopus* species. *J. Agric. Food Chem.* **1971,** *19*, 222-223.
70. Mann R.; Rehm, H.J. Degradation products from aflatoxin B_1 by *Corynebacterium rubrum, Aspergillus niger, Trichoderma viride,* and *Mucor ambiguous. Eur. J. Appl. Microbiol.* **1976**, *2*, 297-306.
71. Faraj, M.K.; Smith, J.E.; Harran, G. Aflatoxin biodegradation: effects of temperature and microbes. *Mycol. Res.* **1993**, *97*, 1388-1392.
72. Shantha, T. Fungal degradation of aflatoxin B_1. *Nat. Toxins* **1999**, *7*, 175-178.
73. El-Sharkawy, S.; Abul-Hajj, Y.J. Microbial cleavage of zearalenone. *Xenobiotica* **1988**, *18*, 365-371.
74. Varga, J.; Rigó, K.; Téren, J. Degradation of ochratoxin A by *Aspergillus* species. *Int. J. Food Microbiol.* **2000**, *59*, 1-7.
75. Varga, J.; Péteri, Z.; Tábori, K.; Téren, J.; Vágvölgyi, C. Degradation of ochratoxin A and other mycotoxins by *Rhizopus* isolates. *Int. J. Food Microbiol.* **2005**, *99*, 321-328.

Chapter 2

Prevention of Mycotoxins in Dried Fruit, other Fruit Products, and Botanicals

Peter M. Scott[1] and Mary W. Trucksess[2]

[1]**Bureau of Chemical Safety, Health Canada, 2203D, Ottawa, Ontario, K1A 0K9, Canada**
[2]**Food and Drug Administration, 5100 Paint Branch Parkway, College Park, MD 20740, USA**

The occurrence of ochratoxin A (OTA) in vine products and some botanicals are health concerns. Strategies to minimize fungal contamination and the subsequent OTA reduction for both commodities are discussed in this chapter. The use of insecticides, fungicides, good vineyard management (soil cultivation and soil amendments), sanitized containers, and proper sorting and drying are essential for lowering OTA in wine and raisins. The problem of aflatoxins in figs needs good orchard management, including removal of damaged fruits, additional solar drying and fluorescence sorting and treatment with sulfur dioxide and metabisulfites. Prevention of patulin occurrence in apple juice and cider can be achieved by washing apples with water, culling, and trimming. Good agricultural practices should also be applied to harvested botanicals. Cultural practices such as wetting capsicum (chilli) pods should be avoided. Modification of product processing and extraction solvents resolved many mycotoxin problems in herbal drugs. Levels of OTA in licorice roots can be reduced by peeling off the outer skin. Irradiation of fruits and botanicals has been shown to decrease fungal counts significantly.

Introduction

General strategies to prevent mycotoxin contamination of food and animal feed were reviewed by Kabak *et al.* (*1*). They are divided into pre-harvest strategies, harvest management, and post-harvest strategies and are based on Good Agricultural Practice (GAP) and Good Manufacturing Practice (GMP). If prevention of formation in the agricultural product cannot be achieved, then physical, chemical, and biological decontamination processes may be used to prevent mycotoxins reaching the consumer. The review covers mainly grains and nuts, although patulin in apples is mentioned. Many of the preventive principles may apply to fruits, fruit products, and botanicals, although much less work has been done on them. Breeding fruit trees for resistance to the causative fungi has not yet been proposed as a viable preventive method and would entail very time consuming research. One recent approach to prevent mycotoxicosis in farm animals is to inhibit mycotoxin absorption in the gastrointestinal tract by addition of an adsorbent to the feed; at present this strategy has no relevance for mycotoxins in fruits, fruit products, and botanicals, which are, of course, not generally used for animal feeds.

Ochratoxin A in Fruit and Fruit products

Occurrence

The most important mycotoxin occurring in vine fruit (grapes) and in grape products such as wine and dried vine fruits is ochratoxin A (OTA) (*2-6*). The European Union has set maximum limits for OTA of 2.0 μg/kg, 2.0 μg/kg, and 10.0 μg/kg in wine, grape juice, and dried vine fruit (currants, raisins and sultanas), respectively (*7*). Wine has been identified as a major source of OTA in the human diet, originating from black aspergilli, principally *Aspergillus carbonarius*, which infect grapes in the vineyard. A European multi-laboratory project, WINE-OCHRA RISK, was initiated in 2001 for the the risk assessment and integrated management of OTA in grapes and wine (*8*). The objectives included identification and characterisation of OTA producing fungi in grapes grown in the Mediterranean basin; definition of factors able to influence OTA content in grapes and wine; determination of the critical control points (CCP) for OTA synthesis during grape production; specification of critical limits for all CCP; preventive and corrective actions; establishment of monitoring systems; and risk assessment and elaboration of a decision support system (DSS).

OTA has been found also in other dried fruits such as figs, apricots, plums, dates, and quince (*2*).

Prevention of OTA in Grapes and Grape Products

Recent reviews pertaining to reduction of OTA in grapes have been published by Leong *et al.* (*9*) and Hocking *et al.* (*10*).

The Organisation Internationale de la Vigne et du Vin (OIV) developed a *Code of Sound Vitivinicultural Practices in Order to Minimise Levels of Ochratoxin A in Vine-Based Products* (Resolution Viti-Oeno 1/2005) (*11*). The *Code* covered grape cultivation in the vineyard, harvest practices, and treatment in the winery. The Codex Alimentarius Commission subsequently developed, along the same lines, a *Proposed Draft Code of Practice for the Prevention and Reduction of Ochratoxin A Contamination in Wine* (*12*).

Cultivation Practices in the Vineyards

The recommendations of the OIV and Codex (*11*, *12*) included training of the producers (which should address the correct identification of ochratoxigenic fungi in the vineyard); avoiding direct contact of grapes bunches with the soil (the espalier cropping system produced the most OTA in the grape (*13*)); choosing less rootstock vigor and varieties which are less prone to developing mold and grape rot; avoiding use of marc containing toxigenic fungi as a fertilizer; and pest and disease control - avoidance of lesions on the berries caused by diseases, insects, and sun burn; control of dangerous fungal diseases affecting grape quality; and control of grape berry moths, grape mealy bugs, and grape leafhoppers. Berry damage must be minimized (*14*). Fungicides effective against black aspergilli include cyprodinil, fludioxonil, and carbendazim (*9*, *15-19*). The grape moth is an important factor in occurrence of OTA in wine (*16*) and surveys in southern Italy have shown an interaction between damage of grapes from larvae of the grape berry moth (*Lobesia botrana*) and OTA levels (*20*). Not mentioned in these recommendations are birds, which can cause considerable fruit damage in most vineyards and can be controlled with noisemakers, flashy streamers, and netting (*21*). Deer and other animals also damage grapes, if not controlled with electric fences.

In contrast to biocontrol of aflatoxins in peanuts by competitive exclusion with non-aflatoxigenic strains of *A. flavus* (*1*, *22*), there is very little information on biocontrol of *A. carbonarius* and *A. niger* on grapes. Biocontrol has been achieved experimentally using antagonisic yeasts, especially *Issatchenkia orientalis* (*23*).

Practices at Harvest

The codes of practice for the prevention and reduction of ochratoxin A in vine-based products (*11*, *12*) continue with recommendations that include ensuring the hygiene of containers to be used at harvest and/or the drying of grapes; sorting grapes and using only grapes not damaged by insects and not contaminated by mold. If grapes are moderately contaminated with toxigenic mold and are to be used in wine production, grapes with black mold must be discarded and the grapes transported as quickly as possible to the winery.

Although not part of a code of practice, it has been shown experimentally that gamma-radiation at 3.5 kGy significantly reduced fungal counts in grapes

(*24*). Irradiation of foods to prevent mold growth and formation of mycotoxins has been reviewed in general (*1*).

Treatment at the Winery

Further OIV and Codex recommendations include determining the concentration of OTA in the must, if risk of its contamination appears likely. Prevention of ochratoxin A formation is preferable to its removal. However, if musts are found to be contaminated, the wine can be treated with enological charcoal as recommended in the codes of practice (*11*, *12*), although this is inadvisable for red grape musts (*25*); quick clarification of the must by filtration, centrifugation, and flotation are preferable to using pectolytic enzymes; fermentation and maturing should be carried out in smooth walled containers (these are more easily cleaned to avoid contamination from previous fermentations and maturings); maturing on lees can help in reducing the OTA level; and the vintner should become informed about the efficiency of fining agents for reducing the level of OTA (enological charcoal is the most effective).

Considerable research has been carried out on pre-fermentation treatments, fermentation, maturing, and clarification of wine. Grazioli *et al.* (*26*), Leong *et al.* (*27*), Fernandes *et al.* (*28, 29*), Caridi *et al.* (*30*), and Lataste *et al.* (*31*) studied the fate of OTA during the complete winemaking process using naturally contaminated grapes, vine-inoculated grapes, or spiked squashed grapes. No new OTA was produced, but each operation after grape crushing reduced the concentration. The main losses of OTA occur in the solid–liquid separation after fermentation (*27-29*); there is also significant reduction during the malo-lactic fermentation. Port wine vinification was studied by Ratola *et al.* (*32*) and again losses of OTA (up to 92%) were observed. Removal of OTA during fermentation was probably carried out by the yeast cell wall (*30*) and was dependent on the yeast strain (*33*). From radiolabeling studies with tritium, OTA did not appear to be changed to other products by the yeast (*31*).

Fining (clarification) is a common winery practice. Potassium caseinate and activated charcoal were found experimentally to be the best fining agents to remove OTA from wine (*34-40*). Other effective materials were silica gel, gelatin, and bentonite (*36-39, 41*). Microbiological adsorbents of OTA from wine or grape juice are active dry yeast (*41*, *42*), yeast lees (*42*), heat treated yeast (*Saccharomyces cerevisiae*) (*43*), the yeast *Phaffia rhodozyma* (*44*), lactic acid bacteria such as *Lactobacillus plantarum* (*38, 45*), and living and dead conidia of non-ochratoxigenic black Aspergilli (*46*, *47*).

Oak chips and powders can be used as substitutes for oak barrels so have been studied for reduction of OTA levels in wine. The highest reductions were over 50%; effectiveness depended on the amounts of oak chips and powder and the storage time (*48*).

Reduction of Ochratoxin A Contamination of Dried Vine Fruit

Vineyard management and harvest practices should be the same as for wine grapes. The OIV and Codex recommendations (*11*, *12*) include some that apply to raisins, e.g. placing grapes to be dried or raisined (dried) in a single layer and avoiding overstacking; avoiding fruit fly infestation; drying in well ventilated conditions; and covering at night. Fungi of the *Aspergillus* section *Nigri* (which includes *A. carbonarius*) develop post-harvest and during sun-drying (*49*), so grapes should be dried rapidly at > 30°C to a safe water activity. Electron beam radiation of raisins can reduce the fungal and yeast bioburden (*50*).

Prevention of Mycotoxins in Dried Figs

Mycotoxins known to occur in dried figs are aflatoxins, ochratoxin A, patulin, kojic acid, and fumonisins (*2-4*, *51*). The most important are the aflatoxins, which can occur at high levels. The Codex Committee on Contaminants in Foods has proposed a *Draft Code of Practice for the Prevention and Reduction of Aflatoxin Contamination in Dried Figs* (*52*, *53*). Some of the recommendations are included below. The implementation of HACCP principles will minimize aflatoxin contamination through application of preventive controls to the extent feasible in the production, handling, storage, and processing of the dried fig crop.

Orchard Management

Cultivation practices that might disperse *A. flavus/A. parasiticus*, and other fungal spores in the soil to aerial parts of trees should be avoided. Decaying organic matter should be buried (*54*). An integrated pest management programme must be applied; fruits or vegetables that promote infestation with dried fruit beetles or vinegar flies should be removed from the fig orchards, since these pests act as vectors for the transmission of fungi especially into the fruit cavity. All equipment and machinery to be used for harvesting, storage and transportation of crops should be clean. Prevention of aflatoxin contamination of figs is important at the critical firm ripe stage (*55*). Figs to be dried are not harvested when they mature but kept on the trees for over-ripening. If fruits are picked from the tree this should be done by hand (*56*). Fig fruits that fall to the ground need to be collected daily to reduce aflatoxin formation, and other losses caused by diseases or pests.

Drying, Storage and Processing

Fig fruits can be dried artificially in driers or under the sun with the help of solar energy. A simple solar drier for use in orchards was described by Le Bars (*54*). Low quality figs which are separated as cull and have the risk of contamination, should be dried and stored separately to prevent cross contamination. The optimum storage conditions for dried figs are at temperatures of 5-10 °C and relative humidity less than 65 % (*52*). Dried figs are fumigated, stored, sized, washed, cleaned, sorted, and packed. Dried fig lots entering into the processing plant must be sampled and analyzed as an initial screening for quality moisture content and presence of bright greenish yellow fluorescent (BGYF) figs, which are removed by hand sorting. Dried figs contaminated with aflatoxins have a significant correlation with BGYF under long wave (360 nm) UV light – this is true for Turkish figs (*57*) but the BGYF is not a useful preventive measure for aflatoxins in California figs (not dried) (*58*) as the fluorescence is internal and not usually external. The moisture content and water activity level of dried fig fruits must be below the critical level (moisture content can be set at 24 %). If demanded by the buyer, figs are washed and dried again before packaging.

Treatment of figs with sodium or potassium metabisulfite solutions followed by solar drying decreased mold counts (*56*). Microflora of dried figs can be inactivated by ozone (*59*). As with grapes, it has been shown that gamma-radiation at 3.5 kGy significantly reduces fungal counts in figs (*24*).

Aflatoxin in Coconut Products

Aflatoxins are not found in unbroken coconuts (*60*) but the occurrence of aflatoxins in dry stored copra is well known and levels of up to 4000 ppb aflatoxin B_1 have been reported (*60, 61*). Citrinin as well as aflatoxins has been found in dried coconut (*62*). In 2003 it was reported that while all traders in Andhra Pradesh, India were aware that microflora played a vital role in rotting of copra, 72% were not aware that copra can be contaminated with aflatoxin (*63*), so there is a need for training and preventive measures to be undertaken. A study of drying methods found that direct smoke drying in a tapahan gave lower aflatoxin levels than sun drying or indirect drying in a kukum dryer; drying to a moisture content of 9 % should yield copra cake containing less than 20 ppb aflatoxin B_1 (*64*). As with figs, a good correlation has been observed between the presence of aflatoxin in copra and bright greenish yellow fluorescence (BGYF) (*65*).

Coconut oil containing aflatoxin B_1 can be decontaminated by exposure to sunlight (*66*).

Prevention of Patulin in Fruit Products

Patulin is found in moldy fruits, in particular apples, where it is produced by *Penicillium expansum* (*67, 68*). The European Union maximum level is 50 μg/kg in apple juice and apple juice ingredients in other beverages (*69*). The Codex Alimentarius *Code of Practice for the Prevention and Reduction of Patulin Contamination in Apple Juice* comprises recommended practices based on Good Agricultural Practice (GAP) and recommended practices based on Good Manufacturing Practices (GMP) (*70*). Preharvest practices based on GAP include controlling pests and diseases such as canker, eye rot (*Botrytis* spp. and

Nectria spp.), certain moths, sawfly, and dock sawfly that allow entry sites for patulin-producing moulds; and application of fungicide to prevent spore germination and fungal growth. Practices during harvesting and transportation of fruit include picking the fruit in dry weather; removing deteriorated fallen fruit from the ground before shaking the trees; using containers that are clean, dry, and free of soil for transporting harvested fruit; and placing the apples in cold storage within 18 hours of harvest. Post-harvest handling and storage practices include controlled atmosphere storage. Some recommended practices based on GMP are: careful sorting during processing and prior to pressing to remove any visually moldy fruit, followed by thorough washing; cleaning and sanitizing juice presses and other manufacturing equipment; analysis of the juice for patulin and only sending it for packing after the analysis shows patulin is below the maximum limit; and reducing the temperature of the juice to 5° C or less, if the juice is to be held for a period prior to use. Factors to help control patulin in apple juice have also been listed by the Ontario Government (*71*).

Pre-processing Reduction of Penicillium expansum

Chemicals that have been used to reduce postharvest decay caused by *Penicillium expansum* in apples include diphenylamine, thiabendazoles, calcium salts, and short chain organic acid fumigants (*67*).

Recent research has studied *Candida sake* and *Pantoea agglomerans* as biocontrol agents for *Penicillium expansum* growth and patulin accumulation in apples (*72*). *P. expansum* growth was controlled in *C. sake* treated apples while *P. agglomerans* better controlled patulin accumulation. However, these biocontrol agents could not control blue rot and patulin accumulation during further storage at room temperature. Further examples of biological control of post-harvest decay fungi, including *P. expansum*, are reviewed by Jackson and Dombrink-Kurtzman (*67*); yeast- and bacteria-based prophylactic products are available commercially.

Cold storage of apples inoculated with *Penicillium expansum* at 1°C for 6 weeks did not lead to patulin accumulation (*73*). Controlled atmosphere storage of inoculated apples at 1°C for longer time periods has also been studied (*74*, *75*). The oxygen level has to be as low as possible to suppress patulin production during storage (*75*). Polyethylene packaging material, compared to polypropylene, inhibited the growth of *P. expansum* at 25°C, regardless of the gaseous environment (*76*).

Washing the apples using high-pressure water spraying to remove rotten apple flesh, and patulin, from the fruit has been investigated as a control measure. Studies have shown that washing in this way can remove more than half of the patulin present in the fruit (*77*, *78*). Patulin levels in apple juice can also be decreased by water washing without high pressure (*79*, *80*). However, sound tissue can still contain low levels of patulin and would not be washed away (*81*); *P. expansum* can still be present in the apple juice (*80*). Important control measures which can be used before pressing the apples to make juice are sorting to remove rotten fruit and trimming the rotten parts (*67*, *68*, *79-81*).

The effects of gamma-radiation on mycotoxigenic *Aspergillus* and *Penicillium* spp. naturally occurring in various fruits, including apple, strawberry, apricot, plum, peach, and mulberry was studied by Aziz and Moussa (*24*). The total fungal counts decreased significantly after irradiation at 3.5 kGy and much lower levels of patulin remained in apples (also apricots and pears) following storage under refrigeration for 28 days.

Removal of Patulin from Apple Juice

Storage of apple juice concentrates for one month at 22°C reduced patulin levels by 45-64% (*82*). There was greater reduction in the presence of L-ascorbic acid (*83*). There are some contradictory results for the stability of patulin during experimental heat treatment of apple juice, as in pasteurization (*84*); in general it has been shown that there is no significant reduction. Clarification processes - fining with bentonite, treatment with pectinase, filtration, or centrifugation – reduced patulin levels by 70-89 % (*85*). Ultrafiltration was less efficient for reducing patulin in apple juice than conventional clarification and filtration (*78*).

Activated charcoal has been studied. This was a successful way for patulin removal from apple juice concentrate and apple juice (*86-90*); however, there could be a reduction in color.

Irradiation of apple juice concentrate containing patulin at 0.35 kGy destroyed half the patulin and there were no changes in its concentration on storage for 8 weeks (*91*).

Fermentation of apple juice with *Saccharomyces cerevisiae* is a practical way of reducing patulin levels if alcoholic cider is to be the end product (*92-94*). As much as 99% reduction of patulin has been achieved (*93*). E- and Z-ascladiol, among other metabolites, were formed; it should be noted that E-ascladiol is itself a mycotoxin, but less toxic than patulin (*95*).

Prevention of Mycotoxins in Botanicals

Since the history of mankind, botanicals or herbal plants have been consumed as food or used as medicines. With the advance of modern medicines, botanicals become alternative medicines or dietary supplements and remain as traditional medicines in many parts of the world. Medicinal plants after harvesting are usually dried and preserved. Humidity during harvesting, drying procedures, and prolonged storage if not under proper control can increase mold growth (*96*). Several surveys of toxigenic molds in botanicals have found high levels of *Aspergillus*, *Penicillium*, and *Fusarium* spp. (*97-99*). In general these molds produce little or no mycotoxins in the plants. Some plants contain naturally occurring antimicrobial compounds that may limit mold growth and mycotoxin production. However, under certain adverse conditions mycotoxins could be found and at levels in excess of most of the regulatory limits. Incidence and occurrence of mycotoxins in botanicals such as medicinal plants, ginseng, ginger, capsicum, licorice, senna pods, garlic, and red fermented rice, have been reviewed in a recently published article (*2*). In this part of our review we cover some of the important preventive strategies of mycotoxin production in these commonly used botanicals.

Use of Proper Harvesting, Drying, and Storage Practices

Botanicals are harvested, dried, and preserved. Reduction of fungal population is the key to mycotoxin prevention. The weather condition during harvesting, drying procedures, and prolonged storage can provide suitable conditions for increasing fungal growth. Adequate gathering, drying, and storage conditions for herbal drugs have been recommended to reduce mycological contamination (*100*). These conditions were evaluated in a study of aflatoxins in pods of the medicinal plant *Cassia senna angustifolia* in India (*101*). Dried leaves and pods of *C. senna* contain up to 7% sennosides A and B. Senna leaves are usually free of aflatoxins. But senna pods are often found to contain aflatoxins. In this study two harvesting methods were compared; one cutting the whole plant, the other picking the pods. Aflatoxin levels were much lower using the first harvesting method. It could be that damage to the pods was avoided and so mold growth was reduced.

Conventional sun drying of the senna pods for two or three days significantly increased aflatoxin concentrations after drying (*101*). The use of a solar tunnel dryer to reach a moisture content of 8% within 24 h gave aflatoxin levels that were 50% lower compared with sun drying. There was no significant difference in aflatoxin levels whether senna pods were stored loose in jute bags or in pressed bales.

Mycotoxins in ginseng roots have not been a problem. Aflatoxins were found only on rare occasions (*102*). The drying process for ginseng may be responsible for fungal infection. Roots were dried for 4-6 weeks by spreading on screen racks in the shade with frequent turning. No aflatoxins were found in roots that were flash dried in forced air electric dryers for 3-4 days. Probably drying and watering the plants could be the mean to control mycotoxin contamination. The use of good agricultural practices pre- and post- harvesting might reduce the contaminations of fumonisins, aflatoxins, and ochratoxin A in medicinal plants.

Gamma Irradiation

Microbiological decontamination of botanicals can be achieved by irradiation. A study of the use of radiation treatment of dried chamomile flowers to reduce microbiological contamination found that the survival curves of the irradiated microflora were a function of radiation dose (*103*). After treatment with 10 kGy ionizing radiation medical herbs have been found to maintain pharmacological activity and show reduction of microbiological contamination (*104*). The use of gamma irradiation for sterilization of traditional Chinese medicines (TCMs) was investigated (*105*). It was found that under certain conditions sterilization could be achieved without inducing biological and physicochemical changes of the TCMs. Before irradiation treatment, the microbiological contamination should be reduced, the plant should be dried, and powdered TCMs should be mixed with honey forming a bolus, which can minimize the decomposition before irradiation. The appropriate dosages are 5 kGy for some special herbal medicines and under 5, 7, or 10 kGy for dry herbs.

The effect of gamma irradiation on microbiological, chemical, and sensory characteristics of licorice root products was reported (*106*). Licorice root products were irradiated at doses of 0, 5, 10, 15, and 20 kGy. After irradiation, products were stored at room temperature for 12 months. Results indicated that 5 and 10 kGy were the effective doses for microbial decontamination. In comparison with non-irradiated products the counts of microorganisms were much less. The mineral ion concentrations in extracts of irradiated products were lower than for non-irradiated products. The quality of a solution produced from licorice root product was improved after irradiation of the roots as evidenced by increased glycyrrhizinic acid and maltose concentrations and there was no loss of sensory characteristics after 12 months of storage.

Modification of Product Processing

Methods of extraction and microbial populations in crude herbal drug were investigated (*107*). Fungal decontamination procedures were used prior to water extraction of the crude herbal drug. Alcohol or sodium hypochlorite as surface decontaminants followed by rinsing with distilled water reduced microbiological counts. Hot water and ethanol extractions showed great reduction of microbials (100).

The procedures used in preparing licorice tea can also affect the amount of OTA transfer from licorice roots to the beverages (*108*). About 5% of the OTA present in dry licorice roots was transferred to the corresponding decoction tea and only 1% of ochratoxin A in infusion tea. Levels of OTA can be reduced by peeling off the outer skin of licorice roots (*109*). OTA was detected in 100% of unpeeled licorice roots and 7% in peeled roots. The median OTA levels were <0.1 and 4.3 μg/kg for peeled and unpeeled roots, respectively. A report of OTA in licorice as affected by processing methods was published recently (*110*). Fresh roots with uniform shape and bright-yellow color were separated from damaged roots with uneven shape and brownish color. All samples contained OTA and the levels were similar. Therefore sorting did not reduce OTA contamination. Washing the roots was also ineffective. Peeling the roots reduced OTA contents by more than 50% in this study. Processing into extract and block licorice further reduced OTA concentrations. It is worthwhile to investigate if this processing procedure is useful in reducing the OTA and AF levels in other roots, such as in ginger.

The stability of aflatoxins in herbal sauces subjected to microwave cooking and gas cooking was examined (*111*). Aflatoxin in herbal sauce remained constant. A comprehensive review included the effect of processing on mycotoxin contamination in final capsicum products (*112*).

Changes in Cultural Practices

High levels of aflatoxin and OTA in chilies could be minimized by changing cultural practices (*113*, *114*). Chillies are grown in countries with tropical climates. Chillies are often spread on the ground for sun-drying in the open air. Prolonged sun drying of chillies increases mold growth and subsequently mycotoxin production. Chilli pods are often wetted by sprinkling with water prior to marketing in order to increase their weight and to enhance their color. The additional moisture promotes toxin contamination and this practice should be discouraged.

Treatment of *Monascus* Strains with Mutagens

Monascus is a food fungus used for fermentation of steamed rice. The fermentation product, red fermented rice (RFR), has been consumed as a dietary supplement. RFR contains monacolin K, which is a cholesterol-lowering agent (*115*). Some *Monascus purpureus* and its fermentation products were found contaminated with citrinin, a hepato-nephrotoxic toxin. In order to avoid citrinin contamination, screening strains of *Monascus* for non-production or low production of citrinin is necessary. A mutant strain, *Monascus* sp. M12-69, was acquired by treatment with mutagenic agents of a wild strain of *Monascus* (M12) screened from FRF samples gathered in China (*116*). This mutant strain under optimum conditions could produce RFR with a high concentration of monacolin K (2.52 mg/g) and low concentration of citrinin (0.13 ng/g) in RFR.

Summary

Research is needed on prevention of the contamination of mycotoxin-producing fungi in the orchards and fields where fruits and botanicals are grown, during drying, and storage of the plants. Irradiation of both fruits and botanicals has been shown to decrease fungal counts significantly. If control over the prevention of mycotoxin contamination in the field, during handling, and during storage is not possible, then effective processing and decontamination methods should be developed. Surveillance of toxin susceptible fruit products and botanicals for mycotoxins and administration of regulatory limits could reduce the number of toxin contaminated fruit products and botanicals entering the market.

References

1. Kabak, B.; Dobson, A. D. W.; Var, I. Strategies to prevent mycotoxin contamination of food and animal feed: a review. *Crit. Rev. Food Sci. Nutr.* **2006**, *46*, 593-619.
2. Trucksess, M. W.; Scott, P. M. Mycotoxins in botanicals and dried fruit: a review. *Food Addit. Contam.* **2008**, *25*, 181-192.

3. Drusch, S.; Aumann, J. Mycotoxins in fruits: microbiology, occurrence, and changes during fruit processing. *Adv. Food Nutr. Res.* **2005**, *50*, 33-78.
4. Drusch, S.; Ragab, W. Mycotoxins in fruits, fruit juices, and dried fruits. *J. Food Protect.* **2003**, *66*, 1514-1527.
5. Scott, P. M. Mycotoxins in alcoholic beverages and fruit juices: occurrence and analysis; Siantar, D. P.; Trucksess, M. W.; Scott, P. M.; Herman, E. M., Eds.; *Food Contaminants. Mycotoxins and Food Allergies; ACS Symposium Series 1001*; American Chemical Society: Washington, DC, 2008; pp. 170-191.
6. Mateo, R.; Medina, Á.; Mateo, E.M.; Mateo, F.; Jiménez, M. An overview of ochratoxin A in beer and wine. *Int. J. Food Microbiol.* **2007**, *119*, 79-83.
7. COMMISSION REGULATION (EC) No 1881/2006 of 19 December 2006 setting maximum levels for certain contaminants in foodstuffs. *Offic. J. Eur. Union* L 364/5-24.
8. Battilani, P. (co-ordinator). Risk assessment and integrated ochratoxin A (OTA) management in grapes and wine. Catholic University Piacenza, Faculty of Agriculture, Italy. http://www.ochra-wine.com/
9. Leong, S.-L.; Hocking, A. D.; Pitt, J. I.; Kazi, B. A.; Emmett, R. W.; Scott, E. S. Australian research on ochratoxigenic fungi and ochratoxin A. *Int. J. Food Microbiol.* **2006**, *111*, *Suppl.1*, S10-S17.
10. Hocking, A. D.; Leong, S.-L. L.; Kazi, B. A.; Emmett, R. W.; Scott, E. S. Fungi and mycotoxins in vineyards and grape products. *Int. J. Food Microbiol.* **2007**, *119*, 84-88.
11. Castellucci, F. Resolution Viti-Oeno 1/2005. Code of sound vitivinicultural practices in order to minimise levels of ochratoxin A in vine-based products. Office International de la Vigne et du Vin: Paris, 2005; 5 pp. http://news.reseau-concept.net/images/oiv_uk/Client/VITI-OENO_1-2005_EN.pdf
12. Codex Committee on Contaminants in Foods. Proposed draft code of practice for the prevention of ochratoxin A contamination in wine, Beijing, China, 16-20 April 2007. ftp://ftp.fao.org/codex/cccf1/cf01_11e.pdf
13. Visconti, V.; Perrone, G.; Cozzi, G.; Solfrizzo M. Managing ochratoxin A risk in the grape-wine food chain. *Food Addit. Contam.* **2008**, *25*, 193-202.
14. Bellí, N.; Marin, S.; Coronas, I.; Sanchis, V.; Ramos, A. J. Skin damage, high temperature and relative humidity as detrimental factors for *Aspergillus carbonarius* infection and ochratoxin A production in grapes. *Food Control* **2007**, *11*, 1343-1349.
15. Mínguez, S.; Cantus, J. M.; Pons, A.; Margot, P.; Cabañes, F. X.; Masque, C.; Accensi, F.; Elorduy, X.; Giralt, L.; Vilavella, M.; Rico, S.; Domíngo, C.; Blasco, M.; Capdevila, J. Influence of the fungus control strategy in the vineyard on the presence of ochratoxin A in the wine. *Bull. O. I. V.* **2004**, *77*, 821-831.
16. Drouillard, J.-B.; Sage, L.; Pladeau, V.; Dubernet, M. Ochratoxine A dans les vins. Un partenariat filière pour des solutions pratiques au vignoble. *Phytoma* **2003**, *565*, 30-35. (*Chem. Abstr.* 140:365250p, 2004).
17. Bellí, N.; Marín, S.; Sanchis, V.; Ramos, A. J. Impact of fungicides on *Aspergillus carbonarius* growth and ochratoxin A production on synthetic

grape-like medium and on grapes. *Food Addit. Contam.* **2006**, *23*, 1021-1029.

18. Bellí, N.; Marin, S.; Argilés, E.; Ramos, A. J.; Sanchis, V. Effect of chemical treatments on ochratoxigenic fungi and common mycobiota of grapes (*Vitis vinifera*). *J. Food Protect.* **2007**, *70*, 157-163.
19. Medina, Á.; Mateo, R.; Valle-Algarra, F. M.; Mateo, E. M.; Jiménez, M. Effect of carbendazim and physicochemical factors on the growth and ochratoxin A production of *Aspergillus carbonarius* isolated from grapes. *Int. J. Food Microbiol.* **2007**, *119*, 230-235.
20. Cozzi, G.; Pascale, M.; Perrone, G.; Visconti, A.; Logrieco, A. Effect of *Lobesia botrana* damages on black aspergilli rot and ochratoxin A content in grapes. *Int. J. Food Microbiol.* **2006**, *111, Suppl.* 1, 88–92.
21. Fraser, H. W.; Fisher K.H.; Frensch, I. Bird control on grape and tender fruit farms. Ontario Ministry of Agriculture & Rural Affairs Factsheet, 1998. http://www.omafra.gov.on.ca/english/engineer/facts/98-035.htm
22. Pitt, J. I.; Hocking, A. D. Mycotoxins in Australia: biocontrol of aflatoxins in peanuts. *Mycopathologia* **2006**, *162*, 233-243.
23. Bleve, G.; Grieco, F.; Cozzi, G.; Logrieco, A.; Visconti, A. Isolation of epiphytic yeasts with potential for biocontrol of *Aspergillus carbonarius* and *A. niger* on grape. *Int. J. Food Microbiol.* **2006**, *108*, 204-209.
24. Aziz, N. H.; Moussa, L. A. A. Influence of gamma-radiation on mycotoxin producing moulds and mycotoxins in fruits. *Food Control* **2002**, *13*, 281-288.
25. Roset, M. Survey on ochratoxin A in grape juice. A review of analytical data and research realised by the French grape industry since 1999. *Fruit Process.* **2003**, *13*, 167-172.
26. Grazioli, B.; Fumi, M. D.; Silva, A. The role of processing on ochratoxin A content in Italian must and wine: a study on naturally contaminated grapes. *Int. J. Food Microbiol.* **2006**, *111 Suppl. 1*, S93-S96.
27. Leong, S. L.; Hocking, A. D.; Varelis, P.; Giannikopoulos, G.; Scott, E. S. Fate of ochratoxin A during vinification of Semillon and Shiraz grapes. *J. Agric. Food Chem.* **2006**, *54*, 6460-6464.
28. Fernandes, A.; Venâncio, A.; Moura, F.; Garrido, J.; Cerdeira, A. Fate of ochratoxin A during a vinification trial. *Asp. Appl. Biol.* **2003**, *68*, 73-80.
29. Fernandes, A.; Ratola, N.; Cerdeira, A.; Venancio, A. Changes in ochratoxin A concentration during winemaking. *Am. J. Enol. Vitic.* **2007**, *58*, 92-96.
30. Caridi, A.; Galvano, F.; Tafuri, A.; Ritieni, A. Ochratoxin A removal during winemaking. *Enzyme Microb. Technol.* **2006**, *40*, 122-126.
31. Lataste, C.; Guérin, L.; Solanet, D.; Berger, J. L.; Cottereau, P.; Molot, B. Évolution de la contamination en ochratoxine A: du vignoble français au vin. *Progr. Agric. Vitic.* **2004**, *121*, 57-64.
32. Ratola, N.; Abade, E.; Simões, T.; Venâncio, A.; Alves, A. Evolution of ochratoxin A content from must to wine in Port Wine microvinification. *Anal. Bioanal. Chem.* **2005**, *382*, 405-411.
33. Cecchini, F.; Morassut, M.; Moruno, E. G.; Di Stefano, R. Influence of yeast strain on ochratoxin A content during fermentation of white and red must. *Food Microbiol.* **2006**, *23*, 411-417.

34. Castellari, M.; Versari, A.; Fabiani, A.; Parpinello, G. P.; Galassi, S. Removal of ochratoxin A in red wines by means of adsorption with chemical fining agents. *J. Agric. Food Chem.* **2001**, *49*, 3917-3921.
35. Var, I.; Kabak, B.; Erginkaya, Z. Reduction in ochratoxin A levels in white wine, following treatment with activated carbon and sodium bentonite. *Food Control* **2008**, *19*, 592-598.
36. Kurtbay, H. M.; Bekçi, Z.; Merdivan, M.; Yurdakoç, K. Reduction of ochratoxin A levels in red wine by bentonite, modified bentonites, and chitosan. *J. Agric. Food Chem.* **2008**, *56*, 2541–2545.
37. Leong, S-L. L.; Hocking, A. D.; Scott, E. S. The effect of juice clarification, static or rotary fermentation and fining on ochratoxin A in wine. *Aust. J. Grape Wine Res.* **2006**, *12*, 245-252.
38. Silva, A.; Galli, R.; Grazioli, B.; Fumi, M. D. Metodi di reduzione di ocratossine A nei vini (Methods for ochratoxin A reduction in wines). *Ind. Bev.* **2003**, *32*, 467-472.
39. Dumeau, F.; Trione, D. Wine processing in relation to ochratoxin A. *Vignevini* **2000** *27*, 79-81 (in Italian) (*Chem. Abstr.* 134:177696t, 2001).
40. Gambuti, A.; Strollo. D.; Genovese, A.; Ugliano, M.; Ritieni, A.; Moio, L. Influence of enological practices on ochratoxin A concentration in wine. *Am. J. Enol. Vitic.* **2005**, *56*, 155-162.
41. Salaha, M. I.; Metafa, M.; Lanaridis, P. Ochratoxin A occurrence in Greek dry and sweet wines. *J. Int. Sci. Vigne Vin.* **2007**, *41*, 225-230.
42. Moruno, E. G.; Sanlorenzo, C.; Boccaccino, B.; Di Stefano, R. Treatment with yeast to reduce the concentration of ochratoxin A in red wine. *Am. J. Enol. Vitic.* **2005**, *56*, 73-76.
43. Bejaoui, H.; Mathieu, F.; Taillandier, P.; Lebrihi, A. Ochratoxin A removal in synthetic and natural grape juices by selected oenological *Saccharomyces* strains. *J. Appl. Microbiol.* **2004**, *97*, 1038-1044.
44. Péteri, Z.; Téren, J.; Vágvölgyi, C.; Varga, J. Ochratoxin degradation and adsorption caused by astaxanthin-producing yeasts. *Food Microbiol.* **2007**, *24*, 205-210.
45. Del Prete, V.; Rodriguez, H.; Carrascosa. A. V.; de las Rivas, B.; Garcia-Moruno, E.; Muñoz, R. *In vitro* removal of ochratoxin A by lactic acid bacteria. *J. Food Protect.* **2007**, *70*, 2155-2160.
46. Bejaoui, H.; Mathieu, F.; Taillandier, P.; Lebrihi, A. Conidia of black aspergilli as new biological adsorbents for ochratoxin A in grape juices and musts. *J. Agric. Food Chem.* **2005**, *53*, 8224-8229.
47. Bejaoui, H.; Mathieu, F.; Taillandier, P.; Lebrihi, A. Biodegradation of ochratoxin A by *Aspergillus* section Nigri species isolated from French grapes: a potential means of ochratoxin A decontamination in grape juices and musts. *FEMS Microbiol. Lett.* **2006**, *255*, 203-208.
48. Savino, M.; Limosani, P.; Garcia-Moruno, E. Reduction of ochratoxin A contamination in red wines by oak wood fragments. *Am. J. Enol. Vitic.* **2007**, *58*, 97-101.
49. Valero, A.; Oliván, S.; Marin, S.; Sanchis, V.; Ramos, A. J. Effect of intra and interspecific interaction on OTA production by *A.* section *Nigri* in grapes during dehydration. *Food Microbiol.* **2007**, *24*, 254-259.

50. Ic, E.; Kottapalli, B.; Maxim, J.; Pillai, S. D. Electron beam radiation of dried fruits and nuts to reduce yeast and mold bioburden. *J. Food Protect.* **2007**, *70*, 981-985.
51. Güler, F. K.; Heperkan, D. Natural occurence of fumonisins in dried figs. XIIth IUPAC International Symposium on Mycotoxins and Phycotoxins, 21-25 May, 2007, Istanbul, Turkey, Abstr. 1247.
52. Codex Committee on Contaminants in Foods. Joint FAO/WHO Food Standards Programme. Proposed draft code of practice for the prevention and reduction of aflatoxin contamination in dried figs. CX/CF 08/2/13. The Hague, the Netherlands, March 31-April 4, 2008. ftp://ftp.fao.org/codex/cccf2/cf02_13e.pdf
53. Codex Committee on Contaminants in Foods. Joint FAO/WHO Food Standards Programme. Proposed draft code of practice for the prevention and reduction of aflatoxin contamination in dried figs. CX/CF 08/2/13 Add.1. The Hague, the Netherlands, March 31-April 4, 2008. ftp://ftp.fao.org/codex/cccf2/cf0213ae.pdf
54. Le Bars, J. Contribution to a practical strategy for preventing aflatoxin contamination of dried figs. *Microbiol. Alim. Nutr.* **1990**, *8*, 265-270.
55. Boudra, H.; Le Bars, J.; Le Bars, P.; Dupuy, J. Time of *Aspergillus flavus* infection and aflatoxin formation in ripening of figs. *Mycopathologia* **1994**, *127*, 29-33.
56. Özay, G.; Aran, N.; Pala, M. Influence of harvesting and drying techniques on microflora and mycotoxin contamination of figs. *Nahrung* **1995**, *39*, 156-165.
57. Steiner W.; Rieker, R. H.; Battaglia, R. Aflatoxin contamination in dried figs: distribution and association with fluorescence. *J. Agric. Food Chem.* **1988**, *36*, 88–91.
58. Doster, M. A.; Michailides, T. J. Production of bright greenish yellow fluorescence in figs infected by *Aspergillus* species in California orchards. *Plant Dis.* **1998**, *82*, 669-673.
59. Öztekin, S.; Zorlungenç, B.; Zorlungenç, F. K. Effects of ozone treatment on microflora of dried figs. *J. Food Eng.* **2006**, *75*, 396-399.
60. Kumari, C. K.; Nusrath, M; Reddy, B. N. Aflatoxin contamination of coconut products from Andhra Pradesh. *Indian Phytopathol.* **1984**, *37*, 284-287.
61. Samarajeewa, U.; Arseculeratne, S. N. A survey of aflatoxin contamination of coconut products in Sri Lanka; incidence, origins and recommendations. *J. Nat. Sci. Counc. Sri Lanka* **1983**, *11*, 225-235.
62. Saxena, J.; Mehrotra, B. S. The occurrence of mycotoxins in some dry fruits retail marketed in Nainital district of India. *Acta Alim.* **1990**, *19*, 221-224.
63. Srinivasulu, B.; Aruna, K.; Kumar, K. V. K.; Rao, D. V. R. Investigations on post harvest aflatoxin contamination of copra. *Indian Coconut J.* **2003**, *34*, 8-9.
64. Head, S. W.; Swetman, T. A.; Nagler, M. J. Studies on deterioration and aflatoxin contamination in copra during storage. *Oléagineux Corps Gras, Lipides* **1999**, *6*, 349-359 (*Chem. Abstr.* 133:3929k, 2000).

65. Kumari, C. K.; Nusrath, M. BGY fluorescence - its application in rapid screening of aflatoxin-containing copra. *Indian J. Bot.* **1988**, *11*, 84-86.
66. Samarajeewa, U.; Arseculeratne, S. N. Degradation of aflatoxins in coconut oil and copra meal (poonac). *J. Nat. Sci. Counc. Sri Lanka* **1977**, *5*, 1-12.
67. Jackson, L.; Dombrink-Kurtzman, M. A. Patulin. In *Microbiology of Fruits and Vegetables*; Sapers, G.; Gorny, J. R.; Yousef, A. E., Eds.; CRC Press: Boca Raton, FL, 2005; pp 281-311.
68. Moake, M. M.; Padilla-Zakour, O. I.; Worobo, R. W. Comprehensive review of patulin control methods in foods. *Compr. Rev. Food Sci. Food Safety* **2005**, *1*, 8-20.
69. European Commission (EC). Commission Regulation EC no.1425/2003 of 11 August 2003 amending regulation (EC) No 466/2001 as regards patulin. *Offic. J. Eur. Union* **2003**, *L 203*, 1-3.
70. Joint FAO/WHO Food Standards Programme. Codex Alimentarius Commission Report of the 34th Session of the Codex Committee on Food Additives and Contaminants, Rotterdam, The Netherlands 11-15 March 2002. Proposed draft code of practice for the prevention of patulin contamination in apple juice and apple juice ingredients in other beverages. Alinorm 03/12, Appendix XI, pp. 92-98. www.codexalimentarius.net/download/report/28/Al03_12e.pdf
71. Jones, S. Patulin - a chemical concern for apple producers & processors. Ontario Ministry of Agriculture, Food & Rural Affairs. Factsheet 211/57, 2004. www.omafra.gov.on.ca/english/crops/facts/04-043.htm
72. Morales, H.; Sanchis, V.; Usall, J.; Ramos, A. J.; Marín, S. Effect of biocontrol agents *Candida sake* and *Pantoea agglomerans* on *Penicillium expansum* growth and patulin accumulation in apples. *Int. J. Food Microbiol.* **2008**, *122*, 61-67.
73. Morales, H.; Marin, S.; Rovira, A.; Ramos, A. J.; Sanchis, V. Patulin accumulation in apples by *Penicillium expansum* during postharvest stages. *Lett. Appl. Microbiol.* **2007**, *44*, 30-35.
74. Morales, H.; Sanchis, V.; Rovira, A.; Ramos, A. J.; Marin, S. Patulin accumulation in apples during postharvest: Effect of controlled atmosphere storage and fungicide treatments. *Food Control* **2007**, *18*, 1443-1448.
75. Baert, K.; Devlieghere, F.; Flyps, H.; Oosterlinck, M.; Ahmed, M.M.; Rajkovic, A.; Verlinden, B.; Nicolai, B.; Debevere, J.; De Meulenaer, B. Influence of storage conditions of apples on growth and patulin production by *Penicillium expansum*. *Int. J. Food Microbiol.* **2007**, *119*, 170-181.
76. Moodley, R. S.; Govinden, R.; Odhav, B. The effect of modified atmospheres and packaging on patulin production in apples. *J. Food Protect.* **2002**, *65*, 867-871.
77. Wu, R.; Qiu, N.; Li, J.; Wang, P.; Yang, X. Dynamic analysis of patulin during apple concentrate processing. *Shipin Gonye Keji* **2005**, *26*, 162-164 (in Chinese) (*Chem. Abstr.* 145:6821c, 2006).
78. Acar, J.; Gökmen, V.; Taydas, E. E. The effects of processing technology on the patulin content of juice during commercial apple juice concentrate production. *Z. Lebensm. Unters. Forsch.* **1998**, *207*, 328-331.
79. Jackson, L. S.; Beacham-Bowden, T.; Keller, S. E.; Adhikari, C.; Taylor, K. T.; Chirtel, S. J.; Merker, R. I. Apple quality, storage, and washing

treatments affect patulin levels in apple cider. *J. Food Protect.* **2003**, *66*, 618-624.

80. Sydenham, E. W.; Vismer, H. F.; Marasas, W. F. O.; Brown, N. L.; Schlechter, M.; Rheeder, J. P. The influence of deck storage and initial processing on patulin levels in apple juice. *Food Addit. Contam.* **1997**, *14*, 429-434.
81. Marín, S.; Morales, H.; Hasan, H. A. H.; Ramos, A. J.; Sanchis, V. Patulin distribution in Fuji and Golden apples contaminated with *Penicillium expansum*. *Food Addit. Contam.* **2006**, *23*, 1316-1322.
82. Koca, N.; Ekşi, A. Reduction of patulin in apple juice concentrates during storage. *J. Food Safety* **2005**, *25*, 1-8.
83. Brackett, R. E.; Marth, E. H. Ascorbic acid and ascorbate cause disappearance of patulin from buffer solutions and apple juice. *J. Food Protect.* **1979**, *42*, 864-866.
84. Kadakal, Ç.; Nas, S. Effect of heat treatment and evaporation on patulin and some other properties of apple juice *J. Sci. Food Agric.* **2003**, *83*, 987-990.
85. Bissesseur, J.; Permaul, K.; Odhav, B. Reduction of patulin during apple juice clarification. *J. Food Prot.* **2001**, *64*, 1216-1219.
86. Fathi-Achachlouei, B.; Ahmadi-Zenouz, A.; Assadi, Y.; Hesari, J. Reduction of patulin content in apple juice concentrate using activated carbon and its effects on several chemical constituents. *J. Food Agric. Environ.* **2007**, *5*, 12-16.
87. Artik, N.; Cemeroğlu, B.; Aydar, G.; Sağlam, N. Use of activated carbon for patulin control in apple juice concentrate (in Turkish). *Turk. J. Agric. Forestr.* **1995**, *19*, 259-265.
88. Kadakal, C.; Nas, S. Effect of activated charcoal on patulin, fumaric acid and some other properties of apple juice. *Nahrung* **2002**, *46*, 31-33.
89. Sands, D. C.; McIntyre, J. L.; Walton, G. S. Use of activated charcoal for the removal of patulin from cider. *Appl. Environ. Microbiol.* **1976**, *32*, 388-391.
90. Leggott, N. L.; Shephard, G. S., Stockenström, S.; Staal, E.; van Schalkwyk, D. J. The reduction of patulin in apple juice by three different types of activated carbon *Food Addit. Contam.* **2001**,*18*, 825-829.
91. Żegota, H.; Żegota, A.; Bachmann, S. Effect of irradiation and storage on patulin disappearance and some chemical constituents of apple juice concentrate. *Z. Lebensm. Unters. Forsch.* **1988**, *187*, 321-324.
92. Harwig, J.; Scott, P. M.; Kennedy, B. P. C.; Chen, Y.-K. Disappearance of patulin from apple juice fermented by *Saccharomyces* spp. *Can. Inst. Food Sci. Technol. J.* **1973**, *6*, 45-46.
93. Stinson, E. E.; Osman, S. F.; Huhtanen, C. N.; Bills, D. D. Disappearance of patulin during alcoholic fermentation of apple juice. *Appl. Environ. Microbiol.* **1978**, *36*, 620-622.
94. Moss, M. O.; Long, M. T. Fate of patulin in the presence of the yeast *Saccharomyces cerevisiae*. *Food Addit. Contam.* **2002**, *1*, 387-399.
95. Suzuki, T.; Takeda, M.; Tanabe, H. A new mycotoxin produced by *Aspergillus clavatus*. *Chem. Pharm. Bull.* **1971**, *19*, 1786-1788.
96. Cvetnić, Z.; Pepeljnjak, S. Mycological contamination of stored herbal drugs. *Acta Pharmaceut.* **1999**, *49*, 201-209.

97. Abeywickrama, K.; Bean, G. A. Toxigenic *Aspergillus flavus* and aflatoxins in Sri Lankan medicinal plant material. *Mycopathologia* **1991**, *113*, 187-190.
98. Halt, M. Moulds and mycotoxins in herb tea and medicinal plants. *Eur. J. Epidemiol.* **1998**, *14*, 269-274.
99. Rizzo, I.; Vedova, G.; Maurutto, S.; Haidukowski, M.; Varsavsky, E. Assessment of toxigenic fungi on Argentinean medicinal herbs. *Microbiol. Res.* **2004**, *159*, 113-120.
100. Kneifel, W.; Czech, E.; Kopp, B. Microbial contamination of medicinal plants – a review. *Planta Med.* **2002**, *68*, 5-15.
101. Müller, P.; Basedow, T. Aflatoxin contamination of pods of Indian *Cassia senna* L. (*Caesalpinaceae*) before harvest, during drying and in storage: Reasons and possible methods of reduction. *J. Stored Prod. Res.* **2007**, *43*, 323-329.
102. D'Ovidio, K.; Trucksess, M.; Weaver, C.; Horn, E.; McIntosh, M.; Bean, G. Aflatoxins in ginseng roots. *Food Addit. Contam.* **2006**, *23*, 174-180.
103. Owczarczyk, H. B.; Migdal, W.; Kedzia, B. The pharmacological activity of medical herbs after microbiological decontamination by irradiation. *Radiat. Phys. Chem.* **2000**, *57*, 331-335.
104. Katušin-Ražem, B.; Ražem, D.; Dvornik, I.; Matic, S. Radiation treatment of herb tea for the reduction of microbial contamination (*Flores chamomillae*). *Radiat. Phys. Chem.* **1983**, *22*, 707-713.
105. Xingwang, F.; Jilan, W. Feasibility of sterilizing traditional Chinese medicines by gamma-irradiation. *Radiat. Phys. Chem.* **1998**, *52*, 53-58.
106. Al-Bachir, M.; Al-Adawi, M. A.; Al-Kaid, A. Effect of gamma irradiation on microbiological, chemical and sensory characteristics of licorice root product. *Radiat. Phys. Chem.* **2004**, *69*, 333-338.
107. Efuntoye, M.O. Mycotoxins of fungal strains from stored herbal plants and mycotoxin contents of Nigerian crude herbal drugs. *Mycopathologia* **1999**, *147*, 43-48.
108. Ariño, A.; Herrera, M.; Estopañan, G.; Juan, T. High levels of ochratoxin A in licorice and derived products. *Int. J. Food Microbiol.* **2007**, *114*, 366-369.
109. Majerus, P.; Max, M.; Klaffke, H.; Plavinskas, R. Ochratoxin A in Sűssholz, Lakritze und daraus hergestellten Erzeugnissen. *Deut. Lebensm. Rundsch.* **2000**, *96*, 451-454.
110. Ariño, A.; Herrera, M.; Langa, E.; Raso, J.; Herrera, A. Ochratoxin A in liquorice as affected by processing methods. *Food Addit. Contam.* **2007**, *24*, 987-992.
111. MacDonald, S.; Castle, L. A UK retail survey of aflatoxins in herbs and spices and their fate during cooking. *Food Addit. Contam.* **1996**, *13*, 121-128.
112. Santos, L.; Marin, S.; Sanchis, V.; Ramos, A. J. Capsicum and mycotoxin contamination: State of the art in a global context. *Food Sci. Technol. Int.* **2008**, *14* , 5-20.
113. Thirumala-Devi, K.; Mayo, M. A.; Reddy, G.; Reddy, S. V.; Delfosse, P.; Reddy, D. V. R. Production of polyclonal antibodies against ochratoxin A

and its detection in chilies by ELISA. *J. Agric. Food Chem.* **2000**, *48*, 5079-5082.

114. Reddy, S. V.; Mayi, D. K.; Reddy, M. U.; Thirumala-Devi, K.; Reddy, D. V. R. Aflatoxins B_1 in different grades of chillies (*Capsicum annum* L.) in India as determined by indirect competitive-ELISA. *Food Addit. Contam.* **2001**, *18*, 553-558.

115. Heber, D.; Lembertas, A.; Lu, Q. Y.; Bowerman, S.; Go, V. L. An analysis of nine proprietary Chinese red yeast rice dietary supplements: implications of variability in chemical profile and contents. *J. Altern. Complement. Med.* **2001**, *7*, 133-139.

116. Chen, F.; Hu, X. Study on red fermented rice with high concentration of monacolin K and low concentration of citrinin. *Int. J. Food Microbiol.* **2005**, *103*, 331-337

Chapter 3

Mycotoxin Contamination of Agricultural Products in the Southern United States and Approaches to Reducing it from Pre-harvest to Final Food Products

Hamed K. Abbas[1] and W. T. Shier[2]

[1]Crop Genetics & Production Research Unit, Agricultural Research Service, United States Department of Agriculture, Stoneville, MS 38776; and [2]Dept. of Medicinal Chemistry, College of Pharmacy, University of Minnesota, Minneapolis, MN 55455, USA

E-mail: Hamed.Abbas@ars.usda.gov

Mycotoxins represent >300 fungal natural products, including aflatoxins, trichothecenes, zearalenones, ochratoxins, patulin and fumonisins. Mycotoxins contaminate cereal grains, causing acute and chronic illnesses in livestock and humans, including teratogenesis, carcinogenesis, endocrine disruption and immunosuppression. Mycotoxin-producing fungi infect plants from soil reservoirs or as endophytes, which infect seeds and plants they growing from them. Mycotoxins can be produced pre-harvest or post-harvest, but near-ideal crop handling and storage conditions in developed countries allow regulatory agencies to assume zero post-harvest production. Cereal grains range in mycotoxin contamination susceptibility from corn, the most susceptible, to rice, the least susceptible. In this paper we provide an overview of mycotoxin contamination problems in the southern United States, and give examples of efforts to reduce mycotoxin contamination pre-harvest, post-harvest and during food processing.

Introduction

Mycotoxins are chemically-stable, toxic secondary metabolites produced by a more than 100 diverse species of fungi. More than 300 mycotoxins are now readily identified due to advances in a host of technologies including fermentation, detection, biotechnology, molecular genetics and others (*1-3*). Among the most important toxigenic fungi for agriculture and food are species of *Aspergillus*, *Fusarium* and *Penicillium*. Many other phytopathogenic fungi (e.g., *Alternaria* spp.) produce mycotoxins which are stable, toxic in experimental systems, and believed to act as virulence factors for the producing fungi. However, these mycotoxins have not been identified as food safety hazards, possibly because they do not accumulate to toxic levels in the harvested agricultural product. Some fungal species produce a single mycotoxin, while most produce multiple toxins that may or may not be chemically related. Only a handful of the 300 mycotoxins are of concern to food safety (*1*,*3*). These mycotoxins include aflatoxins, trichothecenes, fumonisins, ochratoxins, and zearalenone (Figure 1). Ingestion of these mycotoxins by animals and/or humans may result in toxicological problems, including teratogenic, carcinogenic, estrogenic or immunosuppressive effects (*3*). These toxicological problems can ultimately result in economic losses in agriculture (*3*, *4-12*). Because of the risks these mycotoxins pose, they have been subjected to government regulation in about 100 countries (*7*, *13*, *14*).

In principle, mycotoxins can be produced both pre-harvest and post-harvest, but in developed countries rapid drying of crops to <20% moisture (*15*) immediately after harvest, and subsequent near-ideal crop handling and storage conditions, which maintain low moisture levels, allows regulatory agencies to assume zero post-harvest production (*14*). In contrast, food and feed in developing countries are frequently contaminated with mycotoxins, which is often due to the usage of diseased plant tissues infected with a toxigenic fungus pre-harvest (*16-22*). Major crops vary widely in susceptibility to mycotoxin contamination. Corn (maize) is widely considered to be among the most susceptible of major crops to mycotoxins, while rice is among the least susceptible crops (*18-20*, *23-28*). Mycotoxins are very persistent in food and once they are present, there are no general mechanisms that will completely remove them from any type of food. Consequently, there is a need for research on improved methods to mitigate the mycotoxin problem in three areas: (i) pre-harvest to minimize mycotoxin formation in grain through effective crop management and environmental manipulation; (ii) post-harvest to prevent formation of additional mycotoxin in storage by rapid and effective drying, and to develop effective remediation methods to reduce mycotoxin levels; and (iii) during certain types of food processing to destroy mycotoxins in contaminated foods meant for direct human consumption (*29-31*). In this paper, we provide an overview of mycotoxin contamination in the crops of the southern United States; and present examples of efforts to reduce mycotoxin contamination of agricultural products by preventing pre-harvest production; by minimizing post-harvest production and by degradation during food processing.

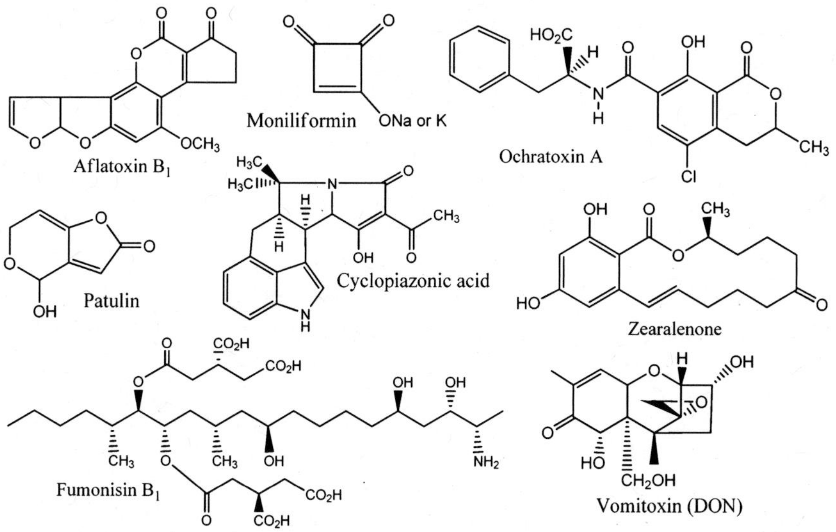

Figure 1. Structures of some mycotoxins of interest to agriculture and food safety.

Mycotoxin Contamination of Corn Grown in the Southern USA

Corn (*Zea mays*) is frequently infested with fungi which produce mycotoxins and thus affect the quality and safety of food and animal feeds. Two of the most costly contamination problems are due to *Aspergillus flavus* and *Fusarium verticillioides* (syn. *F. moniliforme*) which produce aflatoxins and fumonisins, respectively (*20-22*). In 1998, both aflatoxin and fumonisin were responsible for major mycotoxin contamination problems in corn in the southern USA. The increase in mycotoxin production presumably is related to the observed weather conditions, particularly drought and high temperatures (*20*, *29*, *32-41*). In 1998 and 1999 corn hybrids in Mississippi and Arkansas were grown under typical agricultural practices used for the region which allowed them to be naturally infected with *Aspergillus* species and *Fusarium* spp. At harvest, corn kernel samples were evaluated for the presence of aflatoxins and fumonisins.

In 1998 all hybrids exceeded regulatory action levels of 20 ppb aflatoxin (102 - 8,100 ppb) and 5 ppm fumonisin (6 - 47 ppm). In 1999 weather conditions more closely approached 30-year norms and toxin levels were much lower than in 1998. The observed aflatoxin levels were 0 - 30 ppb and fumonisin levels were 0 - 5.8 ppm; deoxynivalenol was also detected in some samples, but no zearalenone was detected. This study supported the widely held belief that production of high levels of mycotoxins in crops by *A. flavus* and *Fusarium* spp. is associated with stress conditions, such as had occurred in 1998. The 1998 toxin levels were high enough to be of concern for food safety. Therefore, we initiated a research program focused on monitoring the production of the mycotoxins produced in harvested corn by *Aspergillus flavus* and *Fusarium verticillioides* (Figure 2).

Although the last major aflatoxin epidemic observed in US corn was in 1998 (*20*, *21*), aflatoxin has been detected throughout our corn kernel survey, and concentrations have varied by year, location, and climatic conditions in which the samples were collected (Table 1). Similarly, in the nine years since 1998, we detected fumonisins in almost all corn samples that were analyzed (>90%) regardless of the source of the sample (Table 2). It is more common during cool, wet weather for corn to become contaminated with *Fusarium* species, notably *F. graminearum*. In cooler climates such as Minnesota, this contamination can result in a wide range of trichothecenes (particularly DON) and zearalenone (*16*, *17*). Due to the hotter, dry climate these fungi do not commonly cause disease in corn in the southern US. However, corn residues left on soil surfaces after harvest, which then over-winter in the field, contained readily-detected levels of zearalenone (0.08 to 0.80 ppm), but no deoxynivalenol (DON) (*42*).

Additional fungal species can contribute to drastically increased mycotoxin levels. In 2006 a high degree of common smut caused by *Ustilago maydis* was observed in southern US corn with ~17% of ears being infected. There was a lower incidence in 2007 (~3% infected ears), and expression of the Bt gene in the corn had no significant effect. In 2006, grain from smut-infected corn contained an average of 2437 ppm total aflatoxin, compared to ~50 ppm total aflatoxin in corn not infected with smut. In 2007, both Bt and non-Bt corn that were infected with smut had higher levels of aflatoxin and fumonisin compared to smut-free corn. Fumonisin levels were five to ten-fold higher in smut-infected corn during both years with no significant effect of the Bt status of the corn hybrids (*43*). Additional research is needed to determine how the occurrence of smut leads to increased mycotoxin contamination.

Table 1. Aflatoxin levels in corn samples by year (μg/kg)*

Source	1998	1999	2000	2001	2002	2003	2004	2005	2006	2007
Arkansas										
Number	298	120	nd	121	161	122	69	64	nd	nd
Range	5-4544	5-114	nd	6-525	7-224	5-1390	11-550	5-122	nd	nd
% Positive	91	30	nd	5	2	31	14	20	nd	nd
Georgia										
Number	nd	nd	nd	nd	nd	nd	nd	nd	646	nd
Range	nd	nd	nd	nd	nd	nd	nd	nd	5-572	nd
% Positive	nd	nd	nd	nd	nd	nd	nd	nd	13	nd
Illinois										
Number	nd	nd	nd	nd	nd	nd	nd	nd	83	87
Range	nd	nd	nd	nd	nd	nd	nd	nd	20-1385	5-1425
% Positive	nd	nd	nd	nd	nd	nd	nd	nd	100	99
Louisiana										
Number	nd	nd	nd	552	843	882	211	95	96	176
Range	nd	nd	nd	5-34950	5-10600	5-3200	5-2305	5-2435	6-55	17-1470
% Positive	nd	nd	nd	63	74	56	32	86	8	83
Mississippi										
Number	25	103	489	506	249	104	92	46	46	450
Range	41-590	5-30	5-2665	6-605	5-3120	5-195	5-427	7-143	5-1710	14-175
% Positive	72	32	52	12	53	12	55	37	65	2
Tennessee										
Number	nd	nd	nd	nd	nd	72	nd	nd	nd	nd
Range	nd	nd	nd	nd	nd	5-7	nd	nd	nd	nd
% Positive	nd	nd	nd	nd	nd	3	nd	nd	nd	nd
Texas										
Number	nd	nd	nd	nd	nd	nd	80	nd	nd	nd
Range	nd	nd	nd	nd	nd	nd	10-1692	nd	nd	nd
% Positive	nd	nd	nd	nd	nd	nd	95	nd	nd	nd

Abbreviation: nd = not done

* Aflatoxin levels were measured by commercial ELISA assays with limit of detection = 5 μg/kg. Data for Mississippi in 1998 was supplied by Dr. Gary Windham, Starksville, MS.

Table 2. Fumonisin levels in corn samples by year (μg/g)*

Source	1998	1999	2000	2001	2002	2003	2004	2005	2006	2007
Arkansas										
Number	256	116	nd	120	161	122	69	72	nd	nd
Range	4-128	0.5-12	nd	4-170	0.5-58	0.5-17	0.6-52	0.6-23	nd	nd
% Positive	100	78	nd	100	81	84	99	88	nd	nd
Georgia										
Number	nd	nd	nd	nd	nd	nd	nd	nd	168	673
Range	nd	nd	nd	nd	nd	nd	nd	nd	0.5-24	0.5-155
% Positive	nd	nd	nd	nd	nd	nd	nd	nd	43	98
Illinois										
Number	nd	nd	nd	nd	nd	nd	nd	nd	83	88
Range	nd	nd	nd	nd	nd	nd	nd	nd	0.5-13	0.5-42
% Positive	nd	nd	nd	nd	nd	nd	nd	nd	73	78
Louisiana										
Number	nd	nd	706	550	1,676	1,728	935	969	595	176
Range	nd	nd	0.5-32	0.5-14	0.5-68	0.5-460	0.5-176	0.5-51	0.5-20	0.5-18
% Positive	nd	nd	93	77	95	95	92	87	60	78
Mississippi										
Number	312	511	838	599	790	887	714	313	384	683
Range	2-74	0.5-13	0.5-32	0.5-50	0.5-43	0.5-50	0.5-60	0.5-64	0.5-28	0.5-32
% Positive	100	89	82	98	95	96	89	90	82	39
North Carolina										
Number	nd	nd	nd	nd	nd	nd	nd	nd	120	nd
Range	nd	nd	nd	nd	nd	nd	nd	nd	0.5-11	nd
% Positive	nd	nd	nd	nd	nd	nd	nd	nd	67	nd
Tennessee										
Number	nd	nd	nd	nd	nd	72	nd	nd	nd	nd
Range	nd	nd	nd	nd	nd	0.5-36	nd	nd	nd	nd
% Positive	nd	nd	nd	nd	nd	99	nd	nd	nd	nd
Texas										
Number	nd	nd	nd	nd	nd	18	80	nd	nd	nd
Range	nd	nd	nd	nd	nd	4-24	0.8-69	nd	nd	nd
% Positive	nd	nd	nd	nd	nd	100	98	nd	nd	nd

Abbreviation: nd = not done

* Fumonisin levels were measured by commercial ELISA assays with limit of detection = 0.5 μg/g.

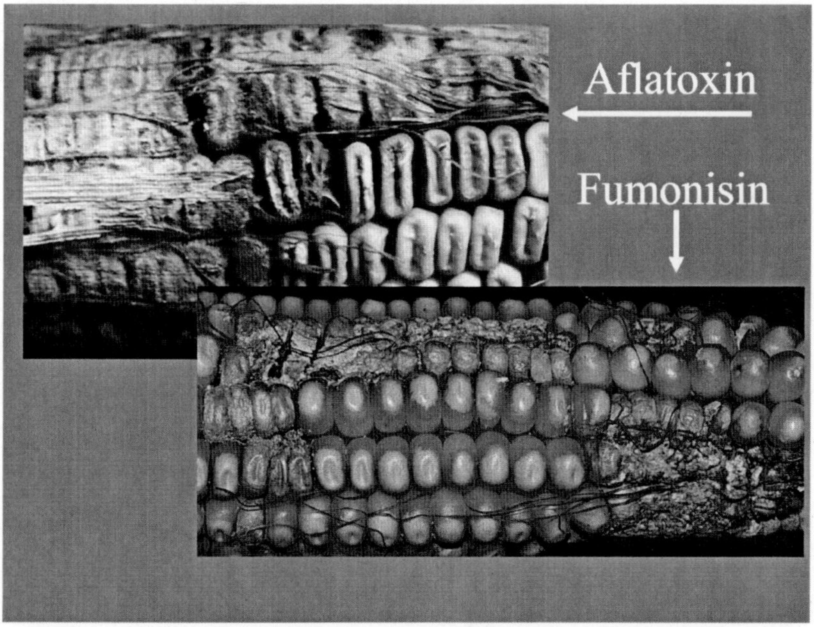

Figure 2. Corn (maize) cobs infected with fungi that produce the mycotoxins aflatoxin (Aspergillus flavus) and fumonisin (Fusarium verticillioides) (syn. F. moniliforme).

Mycotoxin Contamination of Rice Grown in the Southern USA

Rice is widely regarded as unusually resistant to mycotoxin contamination, despite having kernels that provide an exceptionally favorable culture medium for fungal growth and mycotoxin production. We are interested in learning why rice is mycotoxin-resistant, and if any of the resistance mechanisms can be transferred to mycotoxin-susceptible crops such as corn. Among the plausible mechanisms for mycotoxin resistance in rice are the following: (i) rice kernels develop and mature in a structure anatomically resistant to fungal infection, due at least in part to the kernels being physically separated so a fungal infection can not spread directly from kernel-to-kernel, as it can in corn; and (ii) all upland rice and most paddy rice is harvested in the dry season in Asia, which ensures that the crop is effectively dried and that it spends most of its storage time under near ideal conditions. In contrast, most corn is grown in regions where there is no reliable cycle of wet and dry seasons to synchronize with.

In order to gain information about plausible explanations for the resistance of rice to mycotoxin contamination, we have examined rice harvested in situations in which one of the proposed resistance mechanisms was not operative. To investigate the importance of the first mycotoxin resistance mechanism proposed above (i.e., mycotoxin resistance is due to the resistance of rice kernels to fungal infection), we examined rough rice samples (67 in Arkansas in 1998; 33 in Mississippi in 1999 and 2000) harvested from fields exhibiting *Fusarium* sheath rot disease, in which *F. proliferatum* had succeeded in infecting lower parts of the rice plants. Of the samples tested, 27% contained *F. proliferatum* (84% toxigenic); 11% contained fumonisins (0.5-6.2 ppm). A sample of rough rice with 5.13 ppm fumonisin yielded hulls with 16.8 ppm fumonisin; brown rice with 0.87 ppm; and bran with 3.53 ppm; but no fumonisin was detected in polished rice (*19, 37*). To examine the importance of the second mycotoxin resistance mechanism proposed above (i.e., rice is stored under near ideal conditions), we measured mycotoxin levels in rice samples that had gotten wet during storage. No fumonisin was detected, but aflatoxin was present in all samples of rough rice studied (average, 143 ppb; range: 50-272 ppb). A sample of rough rice with 61.8 ppb aflatoxin yielded hulls with 16.3 ppb aflatoxin; brown rice with 88.3 ppb; bran with 587 ppb; polished rice with 63.5, and it yielded cooked rice with 28.4 ppb aflatoxin. Thus, the results obtained so far are consistent with mycotoxin resistance in rice being due at least in part to resistance to fungal infection and near ideal processing and storage conditions. However, the observations of mycotoxins in rice hulls and bran at much higher levels than in the endosperm, the part people consume, is consistent with the presence of an unidentified barrier(s) to endosperm contamination, which would represent an additional protective mechanism different from the separation of kernels mechanism discussed above.

Rice samples which showed symptoms of *Fusarium* sheath rot were collected during the 1995 harvest season from Arkansas (7 samples) and Texas (13 samples). Fumonisins (FB_1 and FB_2) were isolated from 40% of the rice samples tested that showed symptoms of *Fusarium* sheath rot. The level of fumonisins was ≤4.3 ppm. *Fusarium proliferatum* isolates were obtained from these rice samples and shown to produce fumonisins at levels of 14 to 230 ppm. Also, these isolates produced moniliformin in a range of 7 to 6018 ppm and beauvericin at 109 to 1350 ppm. Shelling and milling of unpolished rice showed that fumonisin levels were very high in hulls (≥17 ppm), low in brown rice (≤0.9 ppm), moderate in bran (≤4 ppm), but were below the level of detection in polished rice.

More than 100 rice samples were selected out of thousands collected by a farmer's cooperative in the Mid-South area during an outbreak of scab disease caused by *Fusarium graminearum*. These samples were analyzed for scab disease, *Fusarium* spp., deoxynivalenol (DON, vomitoxin), and zearalenone. DON was detected in 94 of 100 samples at 0.1 to 1.6 ppm (ELISA) when naturally-contaminated rice kernels were extracted with water. DON and zearalenone levels measured by ELISA were affected by the type of extraction solvent used. DON was detected by GC/MS in 38% of samples (0.1 to 0.4 ppm) and zearalenone was detected by GC/MS in 65% of samples (0.3 to 2.2 ppm). Samples were analyzed for aflatoxin and fumonisins by ELISA as well. Aflatoxins were found in 10 out of 100 samples at levels of 20.2 to 28.6 ppb and fumonisins were found in 5 out of 80 samples at levels of 0.1 to 0.5 ppm. Representative isolates of *F. graminearum* produced DON and its derivatives (7 to 430 ppm) and zearalenone (583 to 9,883 ppm) (*45*).

Mycotoxin Stability in Stored Agricultural Products

There has been extensive research on remediation methods to reduce mycotoxin contamination in stored grain for use in animal feeds, or at least to reduce the toxicity. Mycotoxin stability in agricultural products depends heavily on the chemical and physical properties of the mycotoxins. Aflatoxins in commodity products are considered very stable compounds based on their physical properties. Aflatoxins are highly reactive with alkaline oxidizing agents, such as sodium hypochlorite. Therefore these agents are usually used in many laboratories to inactivate aflatoxin on glassware and laboratory bench surfaces. Most research on removal of aflatoxin from agricultural products has focused on "ammoniation" (treatment of dry grains or milled products with anhydrous ammonia) (*46-48*), screening out fines (fungus-infected kernels are more likely to break than sound ones, so removing broken kernels reduces the aflatoxin level) and incorporation of additives such as montmorillite clay or charcoal to bind toxin in the gastrointestinal tract (*49*). Aflatoxin can be degraded by urea and sunlight (*50*, *51*). Aflatoxins in foods can be reduced by cooking and extruding (*33*, *34*), techniques in which relatively high temperatures and pressures were applied. Roasting temperatures ranging from 143 to 149°C reduced aflatoxin concentrations in corn to about 50% of original values (*52*). Fumonisins were found to be reduced by 50% to other detectable fumonisins derivatives such as hydrolyzed fumonisin analogs in tortillas using the traditional nixtamalization method (*53*, *54*). Recently, Voss et al. (*55*) reported that extrusion processing of corn grits in the presence of a reducing sugar (*56*), glucose, significantly reduced the toxicity of contaminating fumonisin in rats. Cramer and Humpf reported that ochratoxin A can be reduced by 30% to 80% depending on temperatures used in dry heating or during roasting processes. They found that ochratoxin A degradation products were formed, which were less toxic than the parent compound (*57*). Patulin can be removed from apple and apple products by using the correct temperature for storage before processing and also by removing damaged portions of the fruit before processing (*58*).

Aflatoxins are sufficiently stable in food and food products that the heat in cooking, baking, frying, and autoclaving (Table 3), and the enzymes in brewing reduced but did not destroy all the aflatoxin (Abbas, unpublished results). We have found that shelling, milling, baking, and cooking did not remove aflatoxin, DON, zaralenone, and fumonisin from rice, wheat, or corn products (*59*). These results showed that aflatoxin persists in rice even when it is boiled and cooked for about 40 minutes (*59*), whereas cooked rice became free from extractable fumonisin (*18*).

It is widely assumed that the mycotoxin content of agricultural products remains constant as long as they are stored under relatively dry conditions (i.e., less than 15% moisture content). Thus, the expense of determining mycotoxin content at one point post-harvest need not be repeated. We have tested this assumption by tracking the aflatoxin (Figure 3) and fumonisin (Figure 4) contents of naturally-contaminated corn over a period of one year under various storage conditions. We observed that the levels of aflatoxins and to a lesser extent, levels of fumonisins dropped significantly during the last six months of storage in the dark at both 35°C and -20°C. The reduction in mycotoxin levels occurred whether or not the corn meal was autoclaved first to inactivate degradation enzymes from the corn or any microbial contaminants. The destruction of the mycotoxins in the dark without effects of enzymes or storage temperature is most consistent with destruction of the mycotoxins by an oxidative or peroxidative process(es) following the depletion of endogenous antioxidants.

Table 3. Effect of heat treatment on aflatoxin contamination levels in harvested corn kernels.

	Aflatoxin levels (ppb)				% reduction of total aflatoxin by heat
	Not autoclaved		Autoclaved*		
Corn sample	AFB1	AFB2	AFB1	AFB2	
Sample 1	414	7.4	172	3.8	42
Sample 2	578	18	165	4.5	29

* Autoclaved = heated at 120°C for 15 minutes

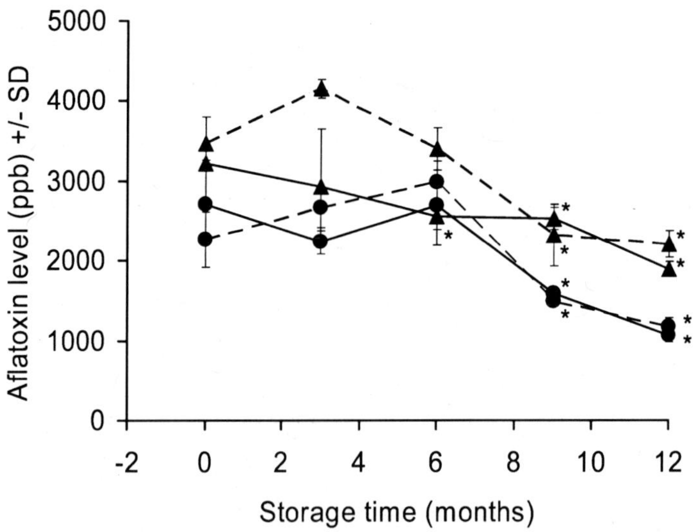

Figure 3. The effect of storage time on aflatoxin levels in ground corn which was either autoclaved at 120°C for 15 minutes to inactivate enzymes in the corn and any microbial contaminants which might degrade the toxin (solid triangles), or left unheated (solid circles). Triplicate samples were stored in either a warm environment, 35°C (solid lines), or in a freezer at -20°C (dashed lines). Aflatoxin levels were measured with commercial ELISA assays. (= significantly less than at zero months, Student's t-test, P<0.05; SD = standard deviation).*

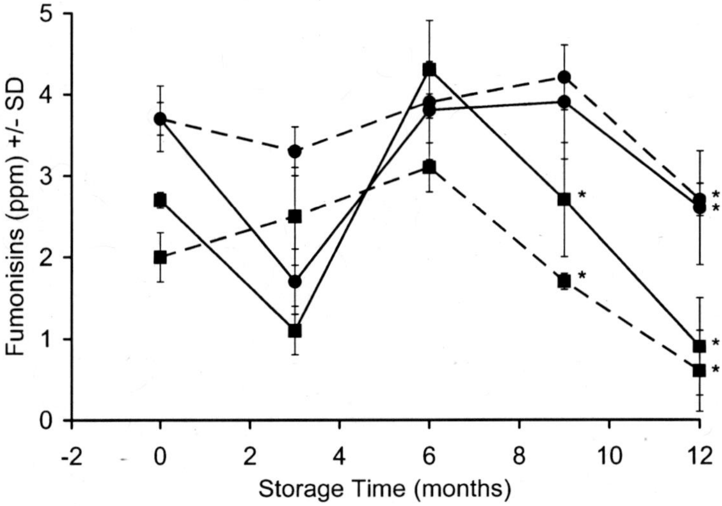

Figure 4. The effect of storage time on fumonisin levels in ground corn which was either autoclaved at 120°C for 15 minutes to inactivate enzymes in the corn and any microbial contaminants which might degrade the toxin (solid circles), or left unheated (solid squares). Triplicate samples were stored in either a warm environment, 35°C (solid lines), or in a freezer at -20°C (dashed lines). Fumonisin levels were measured with commercial ELISA assays. (= significantly less than at six months, Student's t-test, P<0.05; SD = standard deviation).*

Effect of Food Processing on Mycotoxin Content

Government regulations in developed countries require foods intended for direct human consumption to be made from grains with low levels of mycotoxin contamination (e.g., aflatoxins less than 20 ppb in the United States and less than 5 ppb in Europe), even when some standard food processing techniques are known to reduce mycotoxin contamination levels. Thus, the effects of food processing-induced changes in the structures or content of regulated mycotoxins is generally not a concern in the US and other developed countries, as long as food processing doesn't convert mycotoxins to forms with greater toxicity or higher bioavailability. Changes in the structures of biologically active compounds usually result in destruction or reduction of activity. This rule of thumb appears to apply to most mycotoxins as well, but several lines of research from numerous laboratories indicate that fumonisins are an exception to this rule.

Food Processing-Induced Changes in Fumonisins

The first indication that fumonisins might be unusual in their ability to retain activity despite molecular modification was the observation (*60*) that hydrolyzing the side chains off of fumonisin B_1 gave hydrolyzed fumonisin B_1 (HFB_1) which retained substantial cytotoxic activity with cultured mammalian cells. Intact fumonisin B_1 (FB_1) exhibited strong cytotoxic activity with some cell lines that exhibit significant levels of differentiated properties in culture, including epithelial cell lines (e.g., MDCK, PK-15, Caco-2) and liver cell lines (e.g., H4TG), but the more commonly used undifferentiated cell lines were resistant. In contrast, HFB_1 exhibited a lower level of cytotoxic activity, but was cytotoxic with all cell lines. It has not been established whether the selective toxicity of FB_1 in the differentiated cell lines is related to higher levels of transport into the cells, or higher levels of metabolism, or both. HFB_1 has been studied by many investigators in a variety of contexts including its formation during food and feed processing, its toxicity demonstrated in some feeding studies, and its detection in commercial food products (*56, 61-66*).

Biogenic and abiogenic changes in the structures of fumonisins during food processing can, in principle, lead to increased toxicity by various mechanisms including increased bioavailability, altered metabolism, and inherently increased toxicity by tighter binding to its site of action.

a) Food Processing-induced Changes to Fumonisins Which May Increase Bioavailability

There are many unanswered questions about the bioavailability of fumonisins (*66*). Studies by numerous research groups on the toxicokinetics of radiolabeled and unlabeled fumonisins in numerous animal species have indicated that they are poorly absorbed after oral administration. Bolger et al. (*67*) have summarized these studies by concluding that absorption of FB_1 is negligible after oral administration (4% of dose). Studies have also provided evidence that fumonisins undergo very limited, if any, functional or non-functional metabolism (reviewed in *66*), except HFB_1 (*68*). Nevertheless, fumonisins exhibit readily demonstrated toxic effects in some species, particularly horses, but toxicity has been more difficult to demonstrate in other species, including humans for whom the evidence is limited to correlations between exposures and higher rates of cancer and neural tube birth defects in certain areas. The existence of robust toxic effects of oral FB_1 in some species, but very low oral absorption has been called the "fumonisin paradox" (*66*). There are various possible explanations for this apparent paradox, including the following: (i) fumonisins act by a cascading mechanism of action, which amplifies the small response to the 4% which is absorbed, into readily measured toxicity; (ii) fumonisins exhibit greatly increased bioavailability/metabolism at environmental concentrations, which are generally much lower than those used in the reported toxicokinetic studies; or (iii) abiogenic conversion of fumonisins occurs during processing of foods and feeds, which converts fumonisins to a form(s) with much greater bioavailability than the unaltered fumonisins used in the reported toxicokinetic studies. Abiogenic conversion products would have to either retain biological activity or be efficiently converted back to active fumonisin forms once inside the body.

The "fumonisin paradox" has yet to be resolved, but experiments carried out in our laboratories (*69*, *70*) and others (*71*, *72*) have provided additional information about the possible species that can be formed abiogenically during normal food processing and might provide a resolution similar to option (iii) given above. FB_1 was biosynthetically radiolabeled by feeding *Fusarium verticillioides* cultures methionine radiolabeled with tritium on the S-methyl group. Tritium in this position of methionine becomes incorporated into the FB_1 backbone on methyl side chains. A tritium label at this point in the fumonisin molecule is as chemically stable as a carbon-14 label would be. Incorporating radiolabeled FB_1 into corn meal and roasting it in a laboratory model of corn flake manufacturing resulted in greater than 90% of the radiolabel being covalently bound to a water-insoluble, detergent solution-soluble, non-dialyzable, trichloroacetic acid-precipitable substance, consistent with the toxin being covalently bound to protein. As a result the radiolabeled FB_1 was in a form not extractable by the aqueous methanol or acetonitrile solvent systems used to determine the FB_1 content of processed foods and feeds. Sodium dodecylsulfate-polyacrylamide gel electrophoresis analysis with autoradiographic detection indicated most of the label was bound to non-zein proteins, with very little bound to zeins, the major storage proteins in corn kernels. This observation is consistent with radiolabeled FB_1 binding predominantly to the ε-amino groups of lysine residues in protein, which are absent from zeins. Alkaline hydrolysis released the radiolabel from protein to give intact radiolabeled HFB_1 in high yield, consistent with radiolabeled FB_1 binding to protein through one of its side chains. Heating solid FB_1 alone at temperatures encountered in roasting yielded non-dialyzable, presumably polymeric material that gave intact HFB_1 in high yield on alkaline hydrolysis. FA_1, which has the free amino group of FB_1 blocked with an acetyl group, heated alone at temperatures encountered in roasting yielded mixtures containing a substance with NMR and mass spectral properties consistent with N-acetyl-FB_1-anhydride, but it was too unstable for isolation and rigorous structural confirmation. These results are consistent with one of the tricarballyl side chains of FB_1 loosing a molecule of water to generate an succinic anhydride moiety, which can then react with whatever free amino group is most abundant, which are ε-amino groups of lysine residues in non-zein proteins in food products derived from corn (see Figure 5). In the course of the reaction the ring of the cyclic anhydride side chain is opened and the FB_1 becomes covalently bound to protein through an amide linkage.

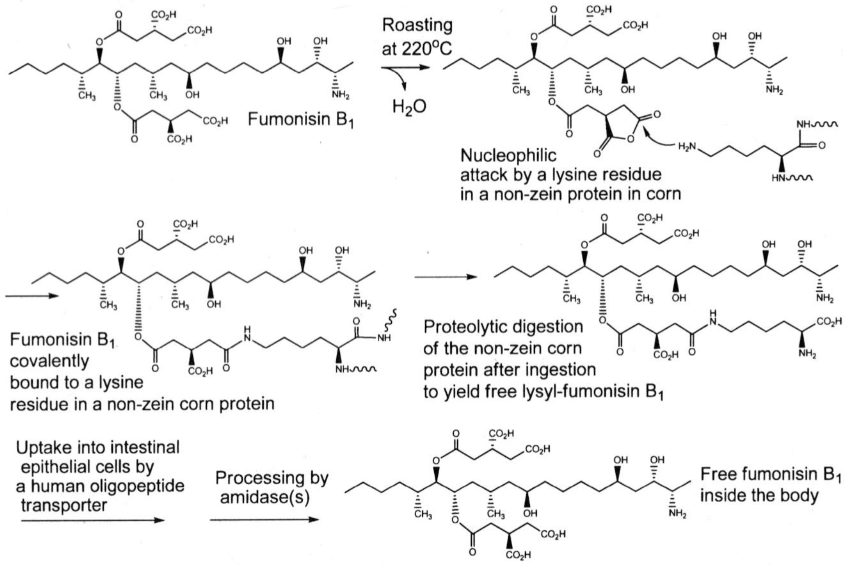

Figure 5. A proposed mechanism for covalent binding of tritium labeled fumonisin B_1 to e-amino groups of lysine residues in non-zein corn proteins in a laboratory model of corn flake manufacture.

When first discovered (*63*), the disappearance of extractable, HPLC-detectable FB_1 from corn-derived foods during heating involved in food processing was encouraging, and taken as an indicator that toxicity also disappeared. However, in subsequent research Scott and associates (*71*, *72*) showed that the amount of HFB_1 released from corn flakes, tortilla chips and corn chips by alkaline hydrolysis corresponded to about twice as much FB_1 being present in bound form, presumably through a side chain, as was present in free form. Identifying the reaction that forms a side chain anhydride with subsequent covalent coupling to the ε-amino groups of lysines in proteins provides a plausible mechanism for the binding of FB_1 to corn components. However, it is also a source of concern, because attachment to an amino acid provides a mechanism for efficient uptake of bound FB_1 from the gastrointestinal tract. There exist various peptide transporters in intestinal epithelial cells, the best characterized of which is oligopeptide-T, which has been shown to be relatively promiscuous in its transporting specificity (*73*). Not only does oligopeptide-T transport into the body miscellaneous di- and tri-peptides generated by the partial digestion of proteins, it also transports various drugs, including valcyclovir and penicillins (for examples, see Figure 6), giving them much higher bioavailability than would otherwise be expected. The oligopeptide transporter appears to transport particularly well drugs that are about the size of a di- or tri-peptide with an amino acid attached to a moiety with a wide range of possible structures. Lysyl-FB_1 fits within the range of structures which might be transported by oligopeptide-T and some other transporters, but the transport characteristics of lysyl-FB_1 have not yet been studied.

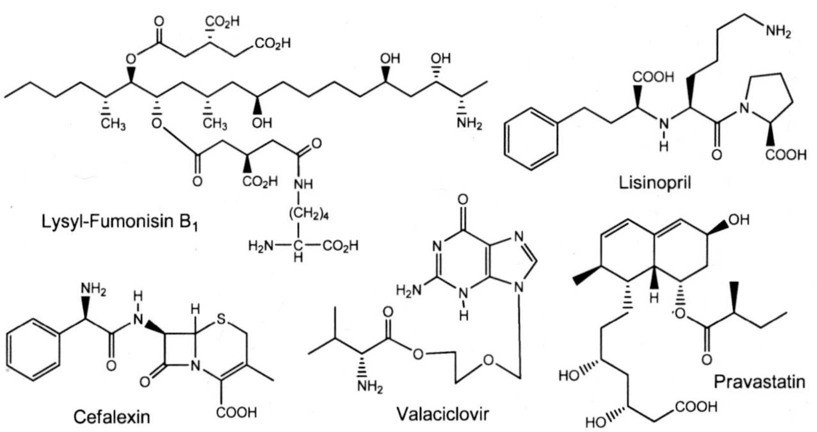

Figure 6. Comparison of the structure of lysyl-fumonisin B_1 with drugs known to be transported by the human intestinal oligopeptide transporter, hPepT1, and related transporters

b) Food Processing-induced Changes to Fumonisins Which May Increase Toxicity

As indicated above, removing the side chains from FB_1 to give HFB_1 reduces, but does not completely destroy biological activity. Side chain hydrolysis has also served to reduce concern about fumonisin toxicity in processed foods derived from corn, because the majority of corn consumed directly by humans is subjected to lime (calcium hydroxide) water treatment in a process known as nixtamalization. Nixtamalization is used to remove the pericarp from corn kernels and soften the rest of the kernel, making it easy to grind into masa, the form of corn used to make tortillas, the major starchy staple in the Mexican diet. Nixtamalization has the additional advantage that some of the fumonisins and hydrolyzed fumonisins are extracted into the lime water, which is discarded. Masa is also used to make tortilla chips and corn chips in the US.

In 1998 Humpf et al. (*74*) discovered that N-palmitoyl-HFB_1 not only retains *in vitro* toxicity, but exhibits about ten-fold greater toxicity than the original intact FB_1. This observation was confirmed (*70*) in an investigation of the structure-activity relationships for *in vitro* toxicity of N-fatty-acyl derivatives of intact and hydrolyzed fumonisins. These studies showed that optimal cytotoxicity occurred with N-fatty acyl-HFB_1 derivatives with carbon lengths of 8 to 12 (see Figure 7). Longer chain lengths (C-14 and C-16) of N-fatty acyl-HFB_1 derivatives retained most of the toxicity observed at the optimum, whereas shorter chain lengths (down to acetate, C-2) were much less active. Polyunsaturation in the fatty acid moieties of N-fatty acyl-HFB_1 derivatives resulted in slight reductions in cytotoxic activity. In contrast, all N-fatty acyl derivatives of intact fumonisins exhibited no cytotoxic activity. The observations of *in vitro* toxicity of N-fatty acyl-HFB_1 derivatives combined with the results of experiments carried out in our laboratories on abiogenic conversions of radiolabelled FB_1 and radiolabelled HFB_1 in laboratory models of tortilla chip manufacture raise concerns about the safety of nixtamalized, fried corn products in the American diet, including taco shells, corn chips and tortilla chips.

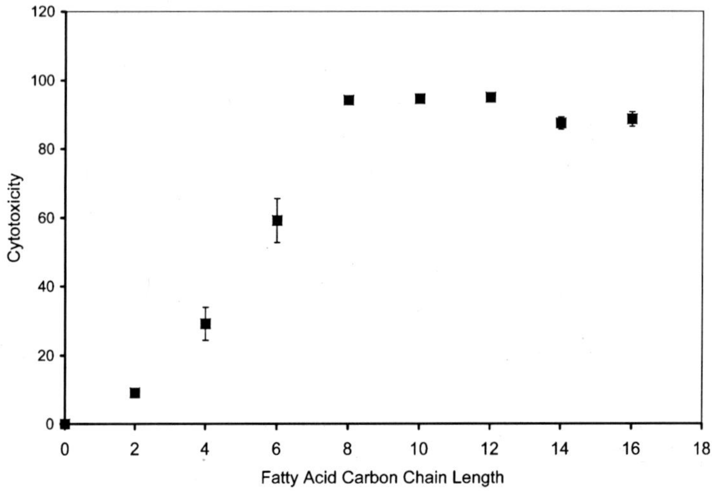

Figure 7. The effect of fatty acid chain length on cytotoxicity of N-fatty acyl hydrolyzed fumonisin B_1 derivatives. Cytotoxicity is expressed as the percent reduction in the concentration of N-fatty acyl hydrolyzed fumonisin B_1 derivative required to cause a 50 percent reductionin cell proliferation in four cultured mammalian cell lines.

It has been well established (*75*, *76*) that N-fatty acylation occurs during food processing, particularly frying. The reaction has been most extensively studied for acylation of the ε-amino groups of lysine in proteins, which has been investigated to determine if there is a significant effect on the nutritional value of fried foods. In principle, N-fatty acylation of HFB_1 in nixtamalized corn can occur by at least two mechanisms. In the simplest mechanism, acylation can occur by direct acyl transfer as a result of nucleophilic attack of free amino groups on the ester moieties of triglycerides. However, several lines of evidence suggest that the majority of N-acylation occurs with intermediate formation of fatty acid anhydrides or other reactive species by thermal decomposition of triglycerides (Figure 8). Any fatty acid anhydride formed will react rapidly with free amines, including HFB_1 to form N-fatty acyl-HFB_1 derivatives. Lime-treated corn is distinctly basic, so most amino groups are in the reactive unprotonated form. The studies in this laboratory examined the possibility that N-fatty acyl-HFB_1 derivatives would form from added radiolabled HFB_1 in a laboratory model for the frying of corn chips in pre-heated vegetable oil. Tortilla chips and corn chips are made from lime-treated corn and standard manufacturing processes always fry them in a mixture of new and pre-heated oil. Tortilla chips contain about 30% oil absorbed from the frying bath. The oil removed in the chips is continuously replaced with new oil, so that no used cooking oil is ever discarded, and all mycotoxins extracted out of the masa during frying end up in other chips, if they are not destroyed by the heat. Corn chips differ from tortilla chips in that they are fried at higher temperatures (185-210°C for corn chips *vs* 170-190°C for tortilla chips) and they are thicker, so that they retain more oil (*77, 78*). We observed in the laboratory model that essentially all tritium-labled HFB_1 underwent a chemical reaction during frying. About 90% of the radiolabelled HFB_1 was converted to N-fatty acyl-HFB_1 derivatives, which were efficiently extracted out of the chips into the frying oil. The remaining 10% of radiolabel was protein-bound in the chip or converted to polar substances extracted into the oil. Radiolabled FB_1 is also converted to N-fatty acyl derivatives (about 85% conversion) during frying in this laboratory model, but the extraction efficiency into the cooking oil was lower (about 80%).

Efficient conversion of HFB_1 to N-fatty acyl-HFB_1 derivatives with ten times the *in vitro* toxicity of FB_1 during food processing is a food safety concern that is only partially compensated for by reduction in toxicity by N-fatty acylation of intact FB_1. The observation that the conversion products are efficiently extracted into the cooking oil is of no practical importance with current manufacturing practices, because they will be absorbed into other chips along with the cooking oil. However, the phenomenon does present an opportunity to remove a potential carcinogen from food during processing, if efficient scrubbers could be developed to continuously remove the N-fatty acyl-HFB_1 derivatives from the oil during frying. However, it will first be necessary to demonstrate that N-fatty acyl-HFB_1 derivatives are both present in foods and toxic *in vivo*.

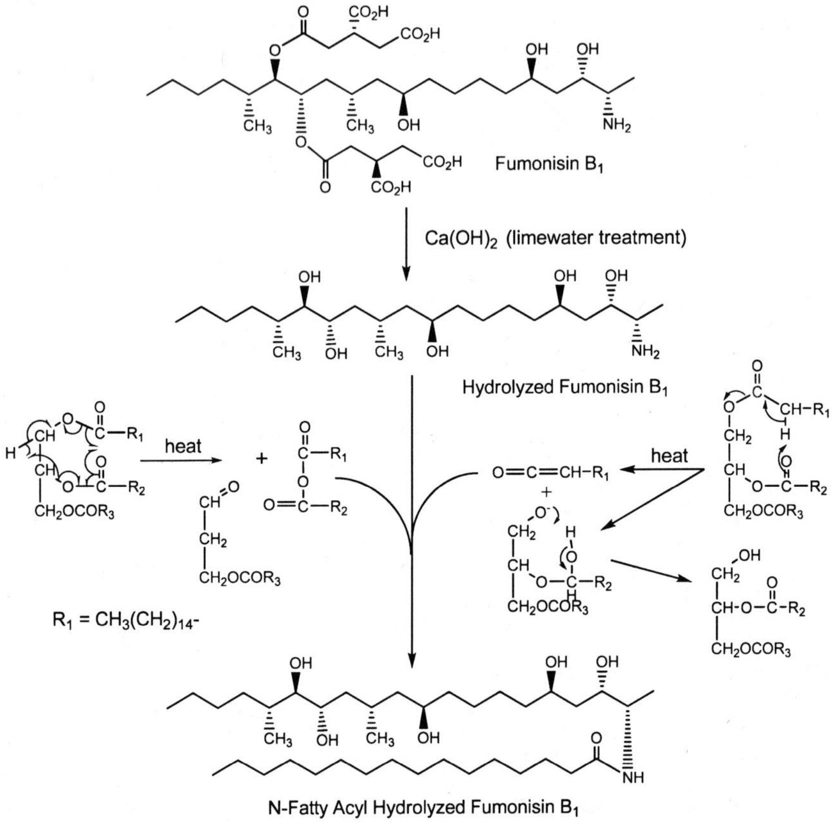

Figure 8. Proposed mechanisms for the formation of N-fatty acyl hydrolyzed fumonisin B_1 derivatives from tritium labeled hydrolyzed fumonisin B_1 in a laboratory model of corn chip manufacturing.

Conclusion

While substantial progress has been made in developing methods to reduce mycotoxin contamination in foods and feeds, more research is needed to ensure that completely safe foods and feeds will continue to be available in the future.

References

1. Binder, E. M.; Tan, L. M.; Chin, L. J.; Handl, J.; Richard, J. *Animal Feed Sci. Technol.* **2007**, *137*, 265-282.
2. Erber, E.; Binder, E.M. In *The 5th Korea Feed Ingredient Association International Symposium*, Korea Feed Ingredient Association, Seoul, Korea, July 16, 2004; pp 21-45.

3. CAST (Council for Agriculture Science and Technology). *Mycotoxins risks in plant, animal, and human systems.* CAST Task Force Report, Ames, IA, 2003; p 139.
4. Bennett, J. W.; Klich, M. *Clin. Microbiol. Rev.* **2003**, *16*, 497-516.
5. Nicholas, T. E.. In Aflatoxin and *Aspergillus flavus* in corn. Vol. 279, Dienes, U. L.; Asquith, R. L.; Dickens, J. W., Eds. So. Coop. Sel. Bull., Auburn University, Auburn, AL: 1983; pp 67–71.
6. Robens, J.; Cardwell, K. F. In *Aflatoxin and Food Safety*, Abbas, H. K., Ed.; Taylor & Francis, Boca Raton, FL, 2005; pp 1–12.
7. Wu, F., In *The Mycotoxin Factbook. Food & Feed Topics*, Barug, D.; Bhatnager, D.; van Egmond, H. P.; van der Kamp, J. W., van Osenbruggen, W. A., Visconti, A., Eds.; Wageningen Academic Publishers, The Netherlands, 2006; pp 83-94.
8. Wu, F. *Anim. Feed Sci. Technol.* **2007,** *137*, 363-374.
9. Windels, C. E. *Phytopathol.* **2000**, *90*, 17-21.
10. NTP (National Toxicology Program). 1999. *Technical Report Series, No. 496. NIH Publication No. 99-3955.* U.S. Department of Health and Human Services, National Institutes of Health, Research Triangle Park, NC.
11. Wu, F.; Miller, J. D.; Casman, E. A. In *Aflatoxin and Food Safety*, Abbas, H. K., Ed.; Taylor & Francis, Boca Raton, FL, 2005; pp 459-482.
12. Abbas, H. K.; Ed. *Aflatoxin and Food Safety*. Taylor & Francis Group, Boca Raton, FL, 2005, 585 pp.
13. Wu, F. *Environ. Sci. Technol.* **2004,** *38*, 4049-4055.
14. van Egmond, H. P.; Schothorst, R. C.; Jonker, M. A.. *Anal. Bioanal. Chem.* **2007,** *389*, 147-157.
15. Anon. (2006) *Grain Fungal Diseases & Mycotoxin Reference*, United States Department of Agriculture Grain Inspectors, Packers and Stockyards Administration: Washington, DC, 2006, p. 4.
16. Abbas, H. K.; Mirocha, C. J.; Kommedahl, T; Burnes, P. M.; Meronuck, R. A.; Gunther, R. *Phytopathology* **1988,** *78*, 1258-1260.
17. Abbas, H. K.; C. J. Mirocha, R. A. Meronuk, J. D. Pokorny, S. L Gould, and T. Kommedahl. *Appl. Environ. Microbiol.* **1988,** *54*, 1930-1933.
18. Abbas, H. K.; Cartwright, R. D.; Xie, W.; Mirocha, C. J.; Richard, J. L.; Dvorak, T. J.; Sciumbato, G. L.; Shier, W. T. *Mycopathologia* **1999,** *147*, 97-104.
19. Abbas, H. K.; Cartwright, R. D.; Windham, G. L.; Xie, W.; Shier, W. T.; Mirocha, C. J. *Bull. Inst. Compr. Agr. Sci. Kinki Univ.* **2000,** *8*, 21-34.
20. Abbas, H. K.; Williams, W. P.; Windham, G. L.; Pringle, J. C., Jr.; Xie, W.; Shier, W. T. *J. Agric. Food Chem.* **2002,** *50*, 5246-5254.
21. Abbas, H. K.; Cartwright, R. D.; Xie, W.; Shier, W. T. *Crop Protection* **2006,** *25*, 1-9.
22. Abbas, H. K.; Shier, W. T.; Cartwright, R. D. *Phytoprotection* **2008,** *88*, 41 – 50.
23. Cardwell, K. F.; King, J. G.; Maziya-Dixon B.; Bosque-Perez, N. A. *Phytopathology* **2000,** *90*, 276-284.
24. Chulze, S. N.; Ramirez, M. L.; Farnochi, M.; Pascale, M.; Visconti, A.; March, G. *J. Agric. Food Chem.* **1996,** *44,* 2797-2801.

25. Marin, S.; Sanchis, V.; Saenz, R.; Ramos, A. J.; Vinas, I.; Magan, N. *J. Appl. Microbiol.* **1998,** *84*, 25-36.
26. Payne, G. A. *Critical Rev. in Plant Sci.* **1992,** *10*, 423-440.
27. Zummo, N.; G. E. Scott. *Plant Disease* **1992,** *76*, 771-773.
28. Reddy, K. R. N.; Abbas, H. K.; Abel, C.A.; Muralidharan, K. *Toxin Rev.* **2008,** *27* in press.
29. Bruns, H. A. *J. Toxicol. - Toxin Rev.* **2003,** *22*, 153-173.
30. Bruns, H. A.; Abbas, H. K. *Maydica* **2004,** *49*, 1-7.
31. Bruns, H.A.; Abbas, H. K. *Agron. J.* **2005,** *97*, 446-451
32. Shier, W. T.; Abbas, H. K. *J. Toxicol. - Toxin - Rev.* **1999,** *18*, 323-335.
33. Buser, M. D. Abbas, H. K. *J. Toxicol – Toxin Rev.* **2001,** *20*, 179-208.
34. Buser, M. D.; Abbas, H. K. *J. Agric. Food Chem.* **2002,** *50*, 2556-2559.
35. Abbas, H. K.; Mirocha, C. J.; Rosiles, M. R.; Carvajal, M. *Cereal Chem.* **1988,** *65*, 15-19.
36. Abbas, H. K.; Mirocha, C. J.; Rosiles, M. R.; Carvajal, M. *Mycotoxin Research* **1988,** *4*, 33-36.
37. Abbas, H. K.; Cartwright, R. D.; Shier, W. T.; Abouzied, M. M.; Bird, C. B.; Rice, L. G.; Ross, P. F.; Sciumbato, G. L.; Meredith, F. I. *Plant Disease.* **1998,** *82*, 22-25.
38. Abbas, H.K.; Mirocha, C. J.; Pawlosky, R. J.; Pusch, D. J. *Appl. Environ. Microbiol.* **1985,** *50*, 482-486.
39. Huwig, A.; Freimund, S.; Kappeli, O.; Dutler, H. *Toxicol. Lett.* **2001,** *122*, 179-188.
40. Leibetseder, J. In *Biology of Nutrition in Growing Animals. Biology of Growing Animals Series, 4*, Mosenthin, R.; Zentek, J.; Zebrowska, T., Eds.; Elsevier, Amsterdam, 2005; pp 439-466.
41. Voss, K. A.; Smith, G. W.; Haschek, W. M. *Anim. Feed Sci. Technol.* **2007,** *137*, 299-325.
42. Abbas, H. K.; Accinelli, C.; Zablotowicz, R. M.; Abel, C. A.; Bruns, H. A.; Dong, Y.; Shier, W. T. *J. Agric. Food Chem.* **2008**, *56*, 7578-7585.
43. Abbas, H. K.; Zablotowicz, R. M.; Abel, C. A.; Bruns, H. *Phytopathology* **2008,** *98*, S9.
44. Abbas, H. K.; Cartwright, R. D.; Sciumbato, G. L.; Keeling, B.; Vesonder, R. F.; Shier, W. T. *Phytopathology* **2000,** *90*, S1.
45. Abbas, H. K.; Zablotowicz, R. M.; Cartwright, R. D.; Xie, W.; and Shier, W. T. *Phytopathology* **2001,** *91*, S1.
46. Bothast, R. J.; Schlicher, M. A. *Appl. Microbiol. Biotechnol* **2005,**. *67*, 19–25.
47. Mercado, C. J.; Real, M. P. N.; Del Rosario, R. R. *J. Food Sci.* **1991,**. *56*, 733–735.
48. Tabata, S.; Kamimura, H.; Ibe, A.; Hashimoto, H.; Tamura, Y. *J. Food Protection* **1994,** *57*, 42-47.
49. Philips, T. D.; Afriye-Gyawu, E.; Williams, J.; Huebner, H.; Ankrah, N. A.; Ofori-Adjei, D.; Jolly, P.; Johnson, N.; Taylor, J.; Marroquin-Cardona, A.; Xu, L.; Tang, L.,Wang, J. S. *Food Additives and Contaminants* **2008,** *25,* 134-145.
50. Shantha, T. *Natural Toxins* **1999,** *7*, 175-178.

51. Shantha, T.; Sreenivasa, M.; Rati, E. R.; Prema, V. *J. Food Safety* **1986,** *7*, 225–231.
52. Kenkel, P.; Anderson, K. 1999. *Grain handlers guide to aflatoxin. Extension Facts WF-233.* Oklahoma State University, Stillwater, OK.
53. Palencia, E.; Torres, O.; Hagler, W.; Meredith, F. I.; Williams, L. D.; Riley, R. T. *J. Nutr.* **2003,** *133*, 3200-3203.
54. Torres, O. A.; Palencia, E.; Lopez de Pratdesaba, L.; Grajeda, R.; Fuentes, M.; Speer, M.C.; Merrill, A. H.; O'Donnell, K.; Bacon, C.W.; Glenn, A. E.; Riley, R. T. *J. Nutr.* **2007,** *137*, 2723-2729.
55. ***Voss, K. A.; Bullerman, L. B.; Bianchini, A.; Hanna, M. A.; Ryu, D.*** *J. Food Protection 2008*, *71*, 2036-2041.
56. Hopmans, E.C.; Murphy, P.A. *J. Agric. Food Chem.*, **1993**, *41*, 1655-1658.
57. Cramer, B.; Humpf, H-U. 2008. *The CORNUCOPIA*, 235th ACS National Meeting, April 6-10, 2008, New Orleans, LA., Abst. #AGFD 118.
58. Bandoh, S.; Higashihara, K.; Takeuchi, M.; Ohsawa, K.; Goto, T. 2008. *The CORNUCOPIA,* 235th ACS National Meeting, April 6-10, 2008, New Orleans, LA. (Abst. #AGFD 70).
59. Abbas, H. K.; Shier, W. T. 2008. *The CORNUCOPIA*, 235th ACS National Meeting, April 6-10, 2008, New Orleans, LA., Abst. #AGFD 66.
60. Shier, W. T.; Abbas, H. K.; Mirocha, C. J. *Mycopathol.* **1991,** *116,* 97-104.
61. Hendrick, S.; Miller, K. A.; Wilson, T. M.; Murphy, P. A. *J. Agric. Food Chem.* **1993,** *41*, 1649-54.
62. Voss, K. A.; Bacon, C. W.; Meredih, F. I.; Norred, W. P. *Food Chem. Toxicol.* **1996**, *34*, 623-32.
63. Scott, P. M.; Lawrence, G. A. *Food Additives and Contaminants* **1996,** *13,* 823-832.
64. Stack, M. E. *J. AOAC International* **1998,** *81,* 737-740.
65. Hartl, M.; Humpf, H-U. *J. Agric. Food Chem.* **1999,** *47*, 5078-5083.
66. Shier, W. T. *J. Toxicol.-Toxin Rev.* **2000,** *19*, 161-187.
67. Bolger, M.; Coker, R. D.; DiNovi, M.; Gaylor, D.; Gelderblom, W.; Olsen, M.; Paster, N.; Riley, R. T.; Shephard, G.; Speijers, G. J. A. *Safety Evaluation of Certain Mycotoxins in Food*; WHO Food Additive Series, JECFA 47; World Health Organization: Geneva, Switzerland, 2001; Paper 47; pp 103–279.
68. Seiferlein, M.; Humpf, H.-U.; Voss, K. A.; Sullards, M. C.; Allegood, J. C.; Wang, E.; Merrill, A. H., Jr. *Mol. Nutr. Food Res.* **2007**, *51*, 1120-1130.
69. Shier, W. T.; Abbas, H. K.; Abou-Karam, M.; Badria, F. A.; Resch, P. A. *J. Toxicol.-Toxin Rev.* **2003,** *22*, 591-616.
70. Abou-Karam, M.; Abbas, H. K.; Shier, W. T. *J. Toxicol. – Toxin Rev.,* **2004,** *23*, 123-151.
71. Kim, E.-K.; Scott, P. M.; Lau, B. P.-Y. *Food Addit. Contam.* **2003**, *20*, 161–169.
72. Park, J. W.; Scott, P. M.; Lau, B. P.-Y.; Lewis, D. A. *Food Addit. Contam.* **2004**, *21*, 1168–1178.
73. Sadee, W.; Drubbisch, V.; Amidon, G. L. *Pharmaceut. Res.* **1995**, *12*, 1823-1837.

74. Humpf, H. U.; Schmelz, E. M.; Meredith, F. I.; Vesper, H.; Vales, T. R.; Wang E.; Menaldino, D. S.; Liotta, D. C.; Merrill, A. H. Jr. *J. Biol. Chem.* **1998,** *273*, 19060-19064.
75. Nawar,W. W. In *Chemical Changes in Food during Processing*, Richardson, T.; Finley, J.W., Eds; Avi Publishing, Westport, CN, 1985; pp. 79-105.
76. Narayan, K. A; Kummerow, F. A. *J. Am. Oil Chem. Soc.* **1963,** *49*, 339-42.
77. Gomez, M. H.; Rooney, L. W. Waniska, R. D. and Pflugfelder, R. L. *Cereal Foods World* **1987,** 32, 372-377.
78. Moreira, R. G.; Palau, J. E. and Sun, X. *Food Technology* **1995,** *49*, 146-150.

Chapter 4

The fate of ochratoxin A in soy milk and bean curd (tofu) productions

Maria H. Iha[1,*], Mary W. Trucksess[2], Jeanne I. Rader[2]

[1]Seção de Bromatologia e Química, Instituto Adolfo Lutz – Laboratório I de Ribeirão Preto, Ribeirão Preto, São Paulo, Brazil
[2]Center for Food Safety and Applied Nutrition, U.S. Food and Drug Administration, College Park, Maryland, United States

Soybeans are a good source of protein. Soy products are widely consumed, particularly in vegetarian diets. The aim of the study was to determine the distibution of OTA in bean curd (tofu), soy milk and by- products during tofu and soy milk production. The effects of washing, soaking and cooking on OTA in soybean were also determined. By-products collected from each step before and after cooking were analyzed for OTA. Soy milk and tofu were prepared and all by-products were collected and analyzed for OTA. OTA was quantified using a method consisting of methanol-sodium bicarbonate extraction, dilution with buffer, immunoaffinity column cleanup, liquid chromatographic separation and fluorescence detection. The results show that washing, soaking and cooking can reduce the OTA in final products if the water used is discarded. Approximately 50 - 80% of incurred OTA in soybeans could be reduced during the production of soy milk and tofu.

Ochratoxins are a group of chemically similar (isocoumarin derivatives) mycotoxins, produced by *Aspergillus ochraceus* and related species as well as by *Penicillium* species. The major mycotoxin in this group is ochratoxin A (OTA) (*1*). OTA is nephrotoxic in animals, and exerts immunotoxic, neurotoxic and teratogenic effects at higher dose levels. (2) OTA has been found in a wide variety of food products such as coffee, grapes, paprika (*3*), sultanas (*4*), rice (*5*), dried fruits (*6*), wine (*6, 7*), beer (*8*), and cereal grains (*9, 10*).

Soybeans are an excellent source of protein. They contain little saturated fat and significant amounts of dietary fibre (*11*). Dietary soy protein has been shown to have beneficial effects on obesity (*12*). However, soybeans can become moldy in the field or during storage, and thereby become contaminated with mycotoxins. Schollenberger et al., (*13*) studied the natural occurrence of mycotoxins in soy food marketed in Germany. The results demonstrated the possibility of a multi-mycotoxin contamination in these products. Numerous reports have shown that OTA is commonly found in corn and cereal grains worldwide. OTA is not the predominant mycotoxin in soybean products. Climate changes can influence fungi and mycotoxin contamination, therefore, it is important to study this mycotoxin. In the Republic of Croatia, Domijan et al. (*14*) investigated the presence of seed-borne fungi and OTA in dried beans and found OTA at low concentrations. In our laboratory, we found in two kinds of dried beans samples: one pink bean and one green bean *(Phaseolus vulgaris L.)* were contaminated with OTA at levels of 60 and 110 μg/kg (personal communication). The pink beans showed no apparent damage but the green beans showed seed coat damaged and had high level of split. We concluded that high initial levels in beans from field, the lack of suitable extended storage facilities and optimum storage conditions could cause molds to grow and increase the OTA production in the dried beans.

In our previous work (*15*), we investigated the effects of washing, soaking and cooking on OTA content in dried beans. The study provided evidence that discarding the washing, soaking and cooking water led to a significant reduction of OTA contamination in residual dried beans.

Soybeans can be processed in various manners and these result in productions of soy milk, okara, and tofu. Soy milk, a beverage made from soybeans, is a stable emulsion of oil, water and protein. The dried soybeans are soaked in water, ground, and cooked. The beans are then separated from the soy milk. The residual beans that are separated from soy milk are called "okara" or soy pulp. This white by-product resembles wet sawdust. Okara, which is high in protein and fiber (16), is used in Japanese cooking for soups, vegetable dishes and even salads. Tofu is prepared from soy milk. Calcium chloride or magnesium chloride is added to the soy milk and is heated to produce tofu. Tofu is high in protein and is one of the most popular oriental food ingredients (*16*).

The objective of this study was to determine the distribution of OTA in soy milk, tofu and by products during their production from soybeans.

Materials and Methods

Reagents and Materials

The following chemicals and supplies were obtained from various suppliers: LC grade methanol, acetonitrile (EM Science, Gibbstown, NJ, USA); $NaHCO_3$ (J. T. Baker, Phillipsburg, NJ); ACS grade acetic acid (Merck, Darmstadt,

Germany); phosphate buffered saline, pH 7.4 (PBS, Sigma P-3813, Sigma-Aldrich, St. Louis, MO); Tween 20 (Sigma, P-5927); Clorox (6.25% sodium hypochlorite, Clorox, Oakland, CA); magnesium chloride (Merck, Darmstadt, Germany), glass microfiber filter paper (Whatman 934AH, Whatman, Inc., Clifton, NJ); OchraTest column (G1017, Vicam, Watertown, MA). *A. ochraceus* ATCC 22947 was purchased from the American Type Culture Collection (Manasas, VA, USA).The ochratoxin immunoaffinity columns (IAC) contain monoclonal antibodies cross reactive towards OTA and should have a minimum binding capacity of not less than 100 ng OTA, and should give a recovery of not less than 80%. OTA standard – O1877 (Sigma-Aldrich Chemie Gmbh, Steinheim, Germany). An OTA solution at approximately 100 μg/mL in methanol was prepared and then 1 mL of the solution was pipetted into a 3 mL volumetric flask and diluted to volume with methanol. Absorbance was determined at 333 nm. A molar absorptivity of 6330 was used to calculate concentration of the second stock standard solution which should be approximately 30 μg/mL. The stock solutions were stored at -18° C. Working OTA Calibrant Solution were prepared daily in methanol at concentrations of 4, 2, 1, 0.5 and 0.25 ng/mL. The extraction solution was methanol - 1% sodium bicarbonate (7 + 3 v/v). The diluting solution was prepared by dissolving 1 package PBS in 1 L of water and adding 1 mL Tween 20.

Apparatus

The apparatus used in our experiments included the following: liquid chromatography (LC) system, Waters Model 2690 Alliance separate system (Waters, Milford, MA), Waters Model 2475 fluorescence detector; or , LC system from Shimadzu Instruments (Kyoto, Japan) with a fluorescence detector, a Rheodyne L.P. injector with a 50 μL loop (Rheodyne, Cotati, CA, USA); LC column (Beckman, catalogue #235332, Ultrasphere, 4.6 x 250 mm, 5 μm (Beckman Instruments, Inc., Fullerton, CA), or equilvalent; grinder (Blixer 3, Robot Coupe USA Inc., Ridgeland, MS); centrifuge (Allegra X – 22R Centrifuge, VWR International, Bridgeport, NJ); blender (Waring Laboratory, Torrrington, CT); vortex (Vortex 2 Genie, Scientific Industries, Bohemia, NJ); and orbital shaker (DS-500E, VWR, West Chester, PA).

Materials

OTA incurred soybeans (*Glicine* max L.) and soybeans were purchased from local markets and were analyzed for OTA. The soybeans were found to be free of OTA contamination (Limit of detectiton 0.1 μg/kg). Twenty- five g soybeans and 100 mL of 10% Clorox solution were placed in a sterile wide mouth flask. After 1 minute, the flask was emptied and the soybeans were rinsed 3 times with 200 mL sterile water. An appropriate amount of sterile water was added to result a total water content of 25 g (the sum of water absorbed by the beans from washing and water added). After storing at 5° C for 14 hours, the beans were inoculated with *A. ochraceus* suspension and kept at ambient temperature (about 25° C) for 3 days. Beans used to process soybean test sample were disinfected by soaking in 50 mL methanol overnight and rinsing 3 times with water. Beans were examined for mold growth and no *A. ochraceus* was found. The beans were analyzed for OTA production and were dried at 70° C for 5-6 h until the weight of the beans was reduced to 25 g.

Methods

Processing of soybean test samples

A) Washing, soaking and cooking soybeans

Liquid was collected from soybeans after 3 different processing procedures: washing, soaking and cooking (Figure 1). Six replicate test samples were carried through each processing procedure.

(a) Washing: 5 g soybeans were washed with 25 mL water for 60 min. The washing water and the beans were separated and OTA analyses were performed for both test samples;
(b) Soaking: 5g soybeans were soaked with 25 mL water for 60 min. The soaking water and the beans were separated and OTA analyses were performed for both test samples;
(c) Cooking: 5 g soybeans were cooked with 25 mL water for 60 min. After cooking, the liquid was separated from the beans and OTA analyses were performed for both test samples.

B) Soy milk and tofu production

Soy milk and tofu were prepared using a mixture of OTA-free and OTA incurred soybeans. Figure 2 shows the procedure used for production of soy milk and tofu. In our first trial (n=4), approximately 34 g of OTA-free and 34 g OTA-incurred soybeans were used, while in the 2nd and third trials (n=2), approximately 4 g OTA-incurred soybeans and 50 g of OTA-free soybeans were used. The beans were soaked in 150 mL water overnight (16 hours). The soaking water was separated from the beans and OTA was analyzed. The soaked beans were ground with 230 mL water, an aliquot was collected and OTA was analyzed. The ground beans were cooked for 10 min and filtered. The residual ground beans (okara) and an aliquot soy milk were analyzed for OTA. After warming to 70°C, 2 g magnesium chloride were added to the approximately 200 mL warm soy milk. The mixture was stirred gently. After tofu was formed, the whey was separated from the tofu. Both tofu and whey were analyzed for OTA.

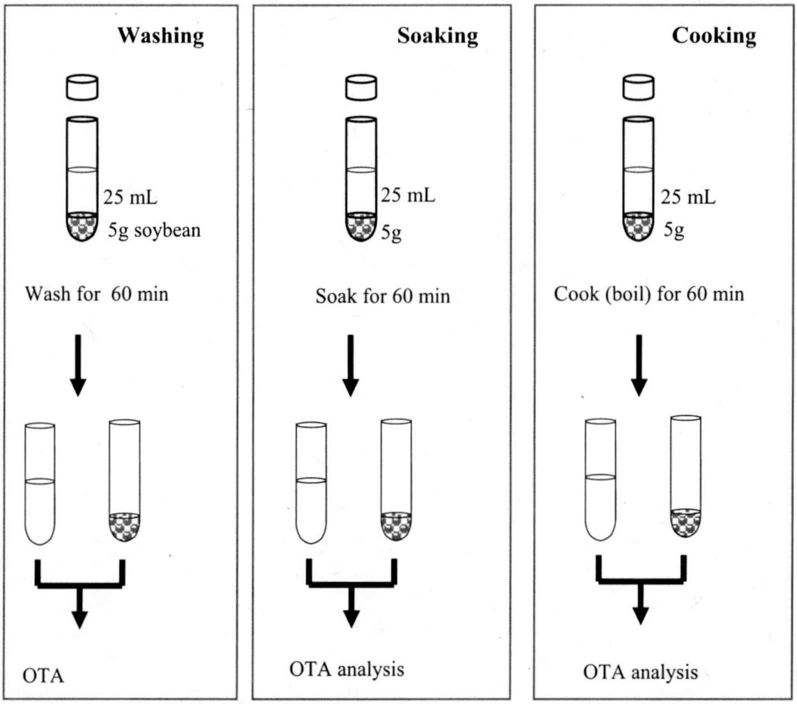

Figure 1. Processing procedures (washing, soaking, and cooking) for soybean.

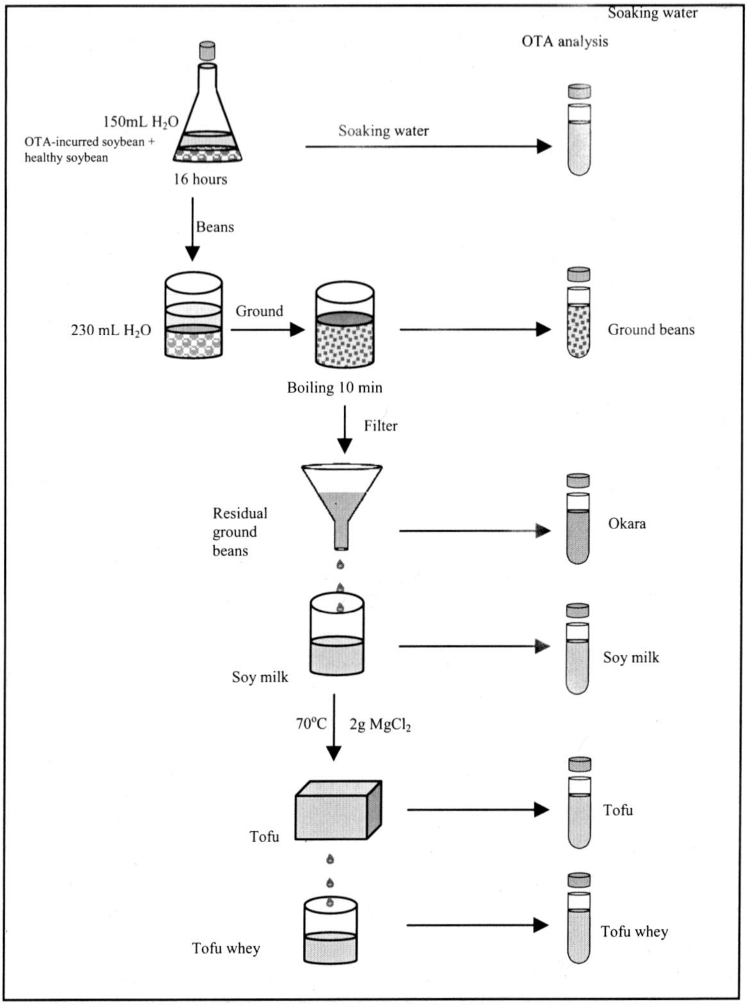

Figure 2. Procedure to produce soy milk and tofu.

Method of analysis

OTA was determined using a published method (*17*). The method consists of methanol-sodium bicarbonate extraction, dilution in buffer, immunoaffinity column cleanup, and reversed phase liquid chromatographic/fluorescence detection, separation and determination. The mobile phase was acetonitrile:water:acetic acid (47:53:1, v/v) at a flow rate of 1 mL/min. The fluorescence detector was set at an excition wavelength of 333 nm and an emission wavelength of 460 nm. The limit of detection was 0.1 μg/Kg and recovery 86%.

Liquid separated from beans after washing, soaking and cooking, from tofu processing after each step, soy milk and residual beans and tofu were all analyzed for OTA. Details of analysis are described below:

(a) Analysis of liquid separated from beans, soy milk, and tofu whey: Five mL liquid, 1 mL 5% $NaHCO_3$ and 14 mL methanol were placed in 50 mL tubes, mixed on a vortex and centrifuged for 10 min (7000 rpm, g = 5323 mm s^{-2}). The supernatant (5 mL) and 40 mL diluting solution were mixed and then filtered through a glass microfiber filter. Five milliliters of filtratrate were passed through an IAC. The column was washed with 5 mL diluting solution and 5 mL water. OTA was eluted 2 times with 1 mL MeOH. The eluate was collected into a 5-mL volumetric flask and diluted to volume with water. After mixing, the test extract was subjected to LC separation and quantitation.

(b) Analysis of beans, ground bean, okara and tofu: Appropriate amounts of methanol and 3% $NaHCO_3$ were added to 5 g test samples to result in a total volume of 25 mL extraction solvent containing a mixture of methanol:1%$NaHCO_3$ (7+3). Test samples were shaken for 30 min on a shaker. Beans were homogenized with a polytron for 2 min. The extract was centrifuged for 10 min (7000 rpm, g = 5323 mm s^{-2}. The remaining procedure was the same as above (a).

Results and Discussion

LC chromatograms of OTA standard and soybean by products: soaking water, ground beans, soy milk, okara, tofu, and whey from tofu all showed a similar pattern. Figure 3 shows typical chromatograms of OTA standard solution, test extracts of soy milk and tofu. There were no interfering peaks at the area near or at the OTA peak. The mean recovery of OTA added in pink beans at 1 µg/kg was 86% and the standard deviation (SD) was +/-3%. Our data indicate that the method was applicable to dried beans.

The soybeans used in this experiment were mixture of OTA-free and OTA-incurred soybean. The incurred beans contained OTA at levels ranging from 2000 and 4000 µg/kg (on the basis of 5 separate analyses). It was not possible to prepare a homogeneous mixture of whole beans. Bean test samples at the beginning of each procedure could contain OTA at different levels. Therefore, the unit of expression for all data is presented as percentage of OTA distribution (i.e., % OTA in liquid = [total OTA in liquid/(total OTA in liquid + total OTA in beans)] x 100).

Table I gives the distribution of OTA in soybeans and in liquid from washing, soaking and cooking. Approximately 10% of OTA partitioned from the bean into the liquid after washing for 60 min. We also obtained similar partition data for beans washed for only 2 min. Soaking for 60 min or washing for 60 min also gave similar yields of % OTA in water. Cooking was a more effective way to reduce % OTA in beans. These results are in agreement with our previous study with OTA distribution in dried beans during food processing (*15*). The partition of OTA from the beans into washing, soaking, and cooking water could reduce the amount of OTA contamination in soybean before consumption if the liquid is discarded.

There is some uncertainty to the absolute OTA distribution in liquid and beans. The nutrient compostion of OTA-naturally contaminated soybeans might be different from that of OTA incurred beans. The disinfection step in the treatment of the *A. ochraceus* infected beans probably denatured some of the protein on the surface of the beans. However, from our previous study on OTA distribution using OTA naturally contaminated bean flour, we obtained similar results to those obtained in this experiment during preparation of soy milk.

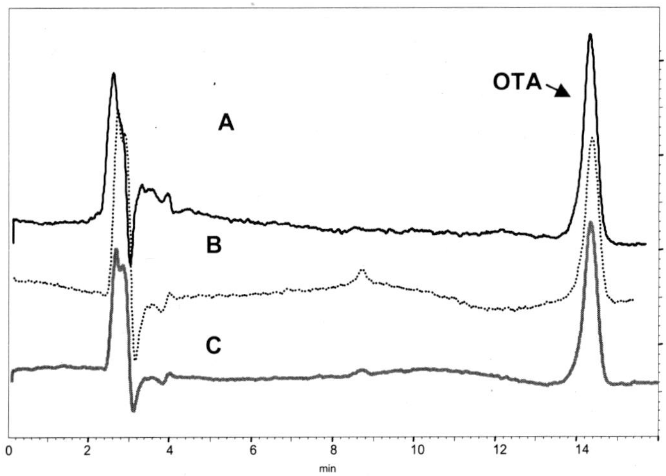

Figure 3. LC Chromatograms of (A) ochratoxin A standard solution (1 ng/mL), (B) soy milk test extract and (C) tofu test extract.

Table I. Ochratoxin A distribution in liquid and residual soybeans after processing (5 g soybeans, 25 mL water).

Procedure	time (min)	liquid separated from beans		residual beans
		liquid (mL)	OTA±SD* (%)	OTA±SD* (%)
Wash	60	17 ± 1	13±2	87±2
Soak	60	17 ± 1	9±1	91±1
Cook	60	15 ± 3	32±4	68±4

* SD, standard deviation, n=6

Results of soy milk and tofu of trial 1 are somewhat different from trial 2 and trial 3 in Table II for two reasons. 1. The same procedure was used for the three trials except larger amounts of OTA-incurred soybeans were used for the first trial. And 2. Soy milk and tofu were prepared simulating home production. With use of filtration with cheese cloth rather than filter paper, it was not possible to measure the separated liquid accurately.

Table II. Ochratoxin A distribution during tofu processing

By product	OTA total amount (μg) ± SD[1]		
	Trial 1	Trial 2	Trial 3
Soybean	197.5	26.4±0.4	20.0±0.5
Soaking water	96.0	12.9±0.3	9.7±0.4
Ground beans	101.5±6.8	13.5±0.4	10.3±0.6
Okara	22.7±0.4	3.0±0.2	1.5±0.1
Soy milk	63.9±0.9	9.1±0.9	6.9±0.3
Tofu whey	15.7±0.1	1.3±0.1	1.2±0.1
Tofu	42.8±1.9	4.5±0.3	5.0±0.1

[1]SD, standard deviation, trial 1 , n=4, trials 2 and 3, n=2.

In summary, if, during the preparation of soy milk and tofu, the soaking water and tofu whey are discarded, then the OTA level in the final products such as soy milk, okara, or tofu, could be reduced. According to this study only 20% of OTA in contaminated soybeans remained in tofu. Approximately 50% OTA that remained inside the beans after soaking and cooking were distributed in soy milk and okara, 34 and 10%, respectively. Further work could be required to support the distribution of OTA in these products. There are many ways of cooking soybeans and soybean products. Subsequently, the amount of OTA that remains in the final consumable portion could be reduced depending on the methods of food preparation.

Acknowledgements

This project was supported by an appointment to the Research Participation Program at the Center for Food Safety and Applied Nutrition administered by the Oak Ridge Institute for Science and Education through an interagency agreement between the U.S. Department of Energy and U.S. Food and Drug Administration. The equipment used in part of this project was provided by Fundação de Amparo à Pesquisa do Estado de São Paulo. This work was partly supported by the Office of Dietary Supplements, National Institutes of Health, Bethesda, MD, USA.

References

1. *Mycotoxins: Risks in plant, animal, and human systems.* Gilbert, J.; Pohland, A. E., Eds.; Council for Agricutural Science Technology: Ames, IA, 2003.
2. The European Food Safety Authority Journal, **2006**, *365*, 1 - 56
3. Almela, L.; Rabe, V.; Sanchez, B.; Torrella, F.; Lopez-Perez, J. P.; Gabaldon, J. A.; Guardiola, L. *Food Microbiol.* **2007**, *24*, 319-327.

4. Aksoy, U.; Eltem, R.; Meyvaci, K. B.; Altindisli, A.; Karabat, S. *Food Addit. Contam.* **2007**, *24*, 292-296.
5. Zinedine, A.; Soriano, J. M.; Juan, C.; Mojemmi, B.; Molto, J. C.; Bouklouze, A.; Cherrah, Y.; Idrissi, L.; El Aouad, R.; Manes, J. *Food Addit. Contam.* **2007,** *24*, 285-291.
6. Chulze S. N. ; Magnoli, C. E. ; Dalcero, A. M. *Int. J. Food Microbiol.* **2006,** *111* (Suppl 1), S5-S9.
7. Grazioli, B.; Fumi, M. D.; Silva, A. *Int. J. Food Microbiol.* **2006,** *111* (Suppl 1), S93-S96.
8. Anselme, M.; Tangni, E. K.; Pussemier, L.; Motte, J. C.; Van Hove, F.; Schneider, Y. J.; Van Peteghem, C.; Larondelle, Y. *Food Addit. Contam.* **2006,** *23*, 910-918.
9. Domijan A. M.; Peraica. M.; Cvjetković, B.; Turcin, S.; Jurjević, Z.; Ivić, D. *Acta Pharm.* **2005,** *55*, 349-356.
10. Riba, A.; Mokrane, S.; Mathieu, F.; Lebrihi, A.; Sabaou, N. *Int. J. Food Microbiol.* **2008,** *122*, 85-92.
11. Aparicio, M.; Cuenca, A. R.; Suárez, V.; Revilla, Z. *Nutr. Hosp.* **2008,** *23*, 305-312.
12. Velasquez, M. T.; Bhathena, S. J. *Int. J. Med. Sci.* **2007**, *4*(2), 72-82.
13. Schollenberger, M; Müller, H. M.; Rüfle, M.; Terry-Jara, H.; Suchy, S.; Plank, S.; Drochner, W. *Int. J. Food Microbiol.* **2007**, *113*, 142–146.
14. Domijan, A. M.; Peraica, M.; Zlender, V.; Cvjetkovic, B.; Jurjevic, Z.; Topolovec-Pintaric, S.; Ivic, D. *Food Chem. Toxicol.* **2005,** *43*, 427-432.
15. Iha, M. H.; Trucksess, M. W.; Tournas, V. H. *Food Addit. Contam.* **2009**, in press..
16. van der Riet, W. B.; Wight, W. A.; Cilleirs, J. J. L.; Datel, J. M. *Food Chem.* **1989**, *34*, 193-202.
17. Trucksess, M.W.; Giler, J. Young,; K. White,; K.D, Page SW. *J. AOAC Int.* **1999**, *82*, 85-89.

Chapter 5

Physical protection of apple products from patulin contamination

Seiji Bandoh[1], Keisuke Higashihara[1], Masahiko Takeuchi[2], Katsumi Ohsawa[3] and Tetsuhisa Goto[1]

[1] Faculty of Agriculture, Shinshu University, Nagano, Japan. [2] Agricultural Technology Institute of Nagano Farmers' Federation, Nagano, Japan, [3]Food Technology Department of Nagano Prefecture General Industrial Technology Center, Nagano, Japan.

Patulin is a mycotoxin produced by various *Aspergillus* and *Penicillium* fungi. Among these fungi, *P. expansum* often invades and damages apples and can cause patulin contamination in apple products. To prevent patulin contamination of apple products proper storage of apples before processing and physical removal of damaged portions of apples are very important. Spores of *P. expansum* were inoculated onto sound apples (mainly Fuji apples) and cultured. Decay caused by the growth of *P. expansum* and patulin content were measured. The movement of patulin within damaged apples was relatively restricted. Therefore patulin was effectively reduced in apples by removal of the physically damaged portion of the apples through either trimming or washing. However, *P. expansum* can grow and produce patulin on apples at relatively low temperatures, even at +1 °C. So for storage for longer than 3 months, apples need to be stored below 0 °C in order to prevent further production of patulin.

Introduction

Patulin is a mycotoxin produced by various *Aspergillus*, *Penicillium* and *Byssochlamys* fungi (*1*, *2*). Among these fungi, *P. expansum* often invades and damages apples and can cause patulin contamination in apple products (*1*, *3*, *4*, *5*). Patulin has antibiotic effects on both Gram negative and Gram positive bacteria (*6*). Patulin also has various toxic effects, including DNA damage (*7*), mutagenicity (*8*) and neurotoxicity (*9*). Therefore, Codex Alimentarius has recommended a patulin content of less than 50 μg/kg in apple products intended for human consumption and in many countries patulin regulations exist (*10*). Patulin can be partially removed from clear type apple juice during processing (*11*). However, patulin is relatively heat stable and tolerant to acidic conditions (*12*). Once *P. expansum* causes decay and produces patulin on apples, it is not so easy to reduce patulin contamination in product, such as straight (cloudy) type apple juice and puree. Therefore, to prevent patulin contamination of apple products both the proper storage of apples before processing and the physical removal of damaged portions of apples are very important.

Materials and Methods

Chemicals and samples

Patulin, 4-Hydroxy-4H-furo-[3,2-c]pyran-2(6H)-one, standard was purchased from Sigma-Aldrich (MO, USA) and Biopure (Tulln, Austria). Patulin standard stock solution (2 mg/mL) was made in ethyl acetate and stored at -20 °C. For the patulin working solution, an aliquot of patulin stock solution was taken into a small amber glass vial and dried under a gentle stream of N_2 gas then dissolved with water which was ajusted its pH as 4 by acetic acid. 5-hydroxymethylfurfural (5-HMF), was purchased from Acros (Geel, Belgium) and prepared same as patulin. All other chemicals were either HPLC grade or GR grade and purchased from Kanto Chemical Co. Ltd (Tokyo, Japan) and used without further purification. *P. expansum* used for this study was cultured on PDA (Potato Dextrose Agar) slant medium at 20 °C in the dark for 2 to 3 weeks.

Preparation of mold damaged apple sample

Spores of *P. expansum* were suspended in 0.1% Tween 20 solution (ca 10^6 C.F.U./mL). The surface of a sound apple was wiped with 80 % (v/v) ethanol and then spores were inoculated into the apple using a needle to about 1-2 mm below the surface. Apples were individually wrapped in aluminum foil then incubated at designated temperature in the dark. Temperature of incubators and apple samples during cultivation was monitored using a Center 309 data logger thermometer (MK Scientific, Tokyo, Japan). After cultivation, the diameter of the decayed area of apple was measured and then the apple was pressed to prepare juice for patulin analysis. Amounts of patulin in apple juices were measured using the method dicussed below. For each study 5 to 6 apples were used.

To study the relation between size of decay and patulin accumulation on apple, four strains of *P. expansum* were inoculated on four different cultivars of apple (Fuji, Ohrin, Yoko and Shinano gold) individually and cultured from +1 to 20 °C for 5 to 83 days then measued the size of visible decay and amount of patulin in the apple were measured.

Preparation of sample and extraction of patulin from apple juice

For the study of the distribution of patulin in damaged apples, each apple was marked out into 36 sections. First the apple was divided into 9 sections using a commercially available apple cutter (vertically) then each piece of apple was divided into 4 sections (horizontally) (Figure 2). Ten Fuji apples were treated in the same way. The same sections from each apple were combined and then pressed to prepare the juice for analysis.

Levels of patulin in apple juices were measured using AOAC-OMA method 995.10 (*13*, *14*) with a minor modification. Also for some samples, a method (*15*) with a more extensive clean up was used for sample preparation. Prepared samples were dissolved in water ajusted its pH as 4 with acetic acid and analyzed by LC and/or LC-MS.

Analysis of patulin.

Patulin was analyzed by a single quadropole type LC-MS (Shimadzu LCMS-2010EV) with a UV detector. For quantitative purpose, UV absorption was mainly used and MS detection was mainly used for confirmation. The LC portion of the LCMS-2010EV system was a model LC-2010CHT liquid chromatograph system (Shimadzu). Patulin was separated on an ODS column (Synergi Hydro-RP 80À, 2.0 mm i.d. x 250 mm, Phenomenex) at 40 °C. The mobile phase was a mixture of water and acetonitrile (95:5, v/v) with a flow rate of 0.2 mL/min. Patulin was monitored by a UV absorbance detector set at 276 nm. Electrospray ionization (ESI) was used to introduce the sample to the MS and the patulin was detected at m/z 153, 125 and 109 by selected ion monitoring mode. ESI was carried out at a spray voltage of -3.5 kV, temperatures of 200 °C and 250 °C were applied to the Curved Desolvation Line (CDL) and heat block with nitrogen as a nebulizer gas at a flow rate of 1.5 L/min. Voltages of CDL and the detector were at 15.0 V and 1.5 kV. Samples were analyzed in negative ion mode.

Results and Discussion

LC and LC-MS analysis

As shown in Figure 1, patulin was detected at a retention time of about 13.5 min by both UV and MS and clearly separated from HMF and other materials. The lower detection limits of patulin by UV and LC-MS (m/z=153) were 0.01 and 0.1 ng, repectively. The calibration curves by UV and MS were linear from 0.05 to 40 ng and 0.5 to 40 ng, respectively. Patulin was successively analyzed in the range from 2 to 40,000 µg/kg in apple juice.

Effects of heat pasteurization for patulin in apple juice

Apple juice was usually pasteurized at 96 °C for 1 min. When the juice was pasteurized at 105 °C for 30 min or at 125 °C for 150 seconds, about 30 % of patulin was degradated (Table 1). However after these threatments the taste of the apple juice was severely damaged. So this type of physical treatment is not useful for apple juice.

Table 1. Effect of heat treatments on patulin in apple juice

Treatment	96 °C 1 min	105 °C 30 min	105 °C 60 min	120 °C 150 sec	125 °C 150 sec	130 °C 150 sec
Patulin (%)	100	88	69	73	70	68

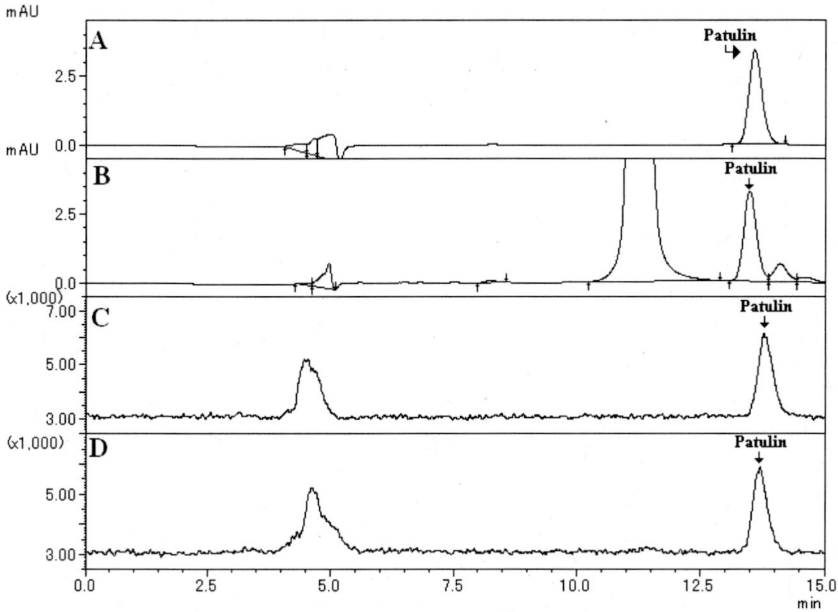

Figure 1. Chromatograms of patulin
A: patulin standard (2 ng, 276 nm), B: apple extract (80 µg/kg, 276 nm)
C: patulin standard (2ng, m/z=153), D: apple extract (80 µg/kg, m/z=153)

Distribution of patulin in damaged apple and effect of trimming for its reduction

A *P. expansum* was inoculted and cultured in ten Fuji apples. Each apple was divided into 36 pieces then correspond pieces (ten pieces for each position) were mixed and pressed. As shown in Figure 2, a large amount of patulin (1500 µg/kg) was detected from the portion that included the point at which the fungus was inoculated. Also, 930 µg/kg and 150 µg/kg of patulin were detected from the upper and lower side portion of the inoculation point and 8 to 20 µg/kg of patulin were detected from the other surrounding areas, including one section in the core area. However, no patulin was detected in the other 26 portions (16). This study was repeated three times using other apple cultivars and similar results were obtained each time.

Because it is relatively easy to manually separate the area damaged by *P. expansum* from the visibly unaffected parts, the decayed parts of apples were carefully removed from the surrounding sound area. Two layers (ca. 5 mm thickness) of visibly unaffected parts of the damaged area were also taken. High concentrations of patulin (40 mg/kg, average of three samples) were detected in the juice extracted from the decayed portions. In the first layer extending up to 5 mm from the decayed section the concentration of patulin detected was 0.14 mg/kg (1:290 to compare to decayed area). In the layer taken 5-10 mm from the decayed area, the concentration of patulin detected was only 0.003 mg/kg (1:13000).

These results clearly show that the distribution of patulin in apples that have decayed is mostly confined to the decayed areas. Therefore it should be possible to prevent patulin contamination of apple products made from decayed apples by the removal of the visibly decayed parts before processing. To confirm these results, the effects of trimming for patulin reduction were studied (*16*) and after trimming the decayed area with a knife, patulin was not detected from the remaining portion of apple.

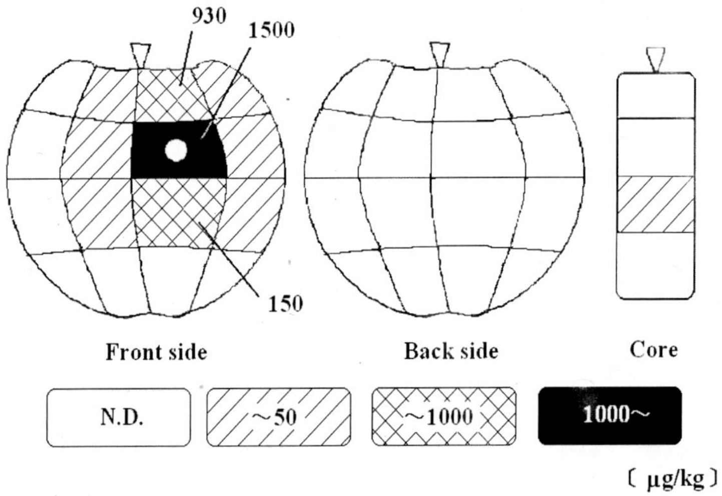

Figure 2. Distribution of patulin in P. expansum inoculated apple.

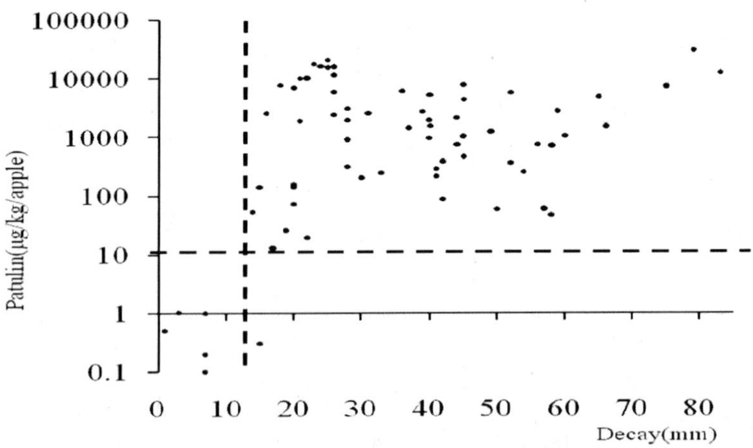

Figure 3. Relation of decay size and concentration of patulin
Each dot represents one apple.
Four apple cultivars of apple and four strains of P. expansum were used.

Relation between size of decay and patulin accumulation on apple

As shown in figure 3, there was a correlation between the decay size and the level of patulin. This result clearly showed that a small decay of less than 10 mm in diameter may not result in patulin contamination problem even if apples with such decay were to be used for processing without trimming.

Effect of storage temperature on *P. expansum* growth and patulin contamination on apple

Four strains of *P. expansum* were inoculated onto apples and stored at 20, 5, 2.5, 1.0, -0.5 and -1 °C. Apples stored at less than -2.5 °C were frozen. As shown in Table 2, *P. expansum* can grow on apples at very low temperatures, even at -0.5 °C, but no patulin was detected in this sample even 83 days after inoculation. When apples were kept at 20 °C, decay started to be observed within two days after inoculation and the decay spread about 6 mm per day. The spread of decay on apples at 1 to 5 °C was 1 to 3 mm per day after decay was first observed. Interestingly, when apples were stored at low temperature, there was some lag time before decay started that depended on the storage temperature(17). For example at 5 °C, the lag time was 10 days to two weeks and at 1 °C, 2 to 4 weeks, depending on the strains of fungi and cultivar of apple. These results show that it is important to store apples in cool conditions as soon as possible after harvest to prevent patulin contamination.

Table 2. Patulin production and decay caused by *P. expansum*

Temp. (°C)	20	5	2.5	1	-0.5	-1	-2.5
The day decay was start to observed	2	14	21	35	60	60	-
Stored period (days)	8	60	60	63	83	63	*b
Decay (mm) after storage*a	61	70	55	23	4	2	*c
Patulin (mg/kg)*a	900	1300	350	300	N.D.	N.D.	*d

*a: Average of four *P. expansum*

*b: Storage period is not applicable because frozen

*c: Apple was frozen and no mold growth was observed

*d: No analysis was done because apple was frozen

N.D.: Not Detected (less than 0.5 µg/kg in the portion analyzed)

References

1. Frank, H.K. *Ann. Nutr. Alim.* **1977**, *31*, 459-465.
2. Handbook of Secondary Fungal Metabolites, II; Cole, R.J, Schweikert, M.A. Eds.; Academic Press: San Deigo, 2003; p.758-759.
3. Scott, P.M.; Miles, W.F.; Toft, P.; Dubé, J.G. *J. Agr. Food Chem.* **1972**, *20*, 450-451.
4. Spadaro, D.; Ciavorella, A.; Frati, S.; Garibaldi, A.; Gullino, M.L. *Food Control* **2007**, *18*, 1098-1102.
5. Leggott, N.L.; Shephard, G.S. *Food Control*, **2001**, *12*, 73-76.
6. *WHO Food Additives Series No 35*; Toxicological Evaluation of Certain Food Additives and Contaminants; World Health Organization Geneva, 1996; p.377-402.
7. Liu, B.H.; Yu, F.Y.; Wu, T.S.; Li, S.Y.; Su, M.C.; Wang, M.C.; Shih, S.M. *Toxicol. Appl. Pharmacol.* **2003**, *191*, 255-263.
8. Schumacher, D.M.; Metzler, M.; Lehmann, L. *Arch Toxicol.* **2005**, *79*, 110-121.
9. Sabater-Vilar, M.; Maas, R.F.M.; Bosschere, H.D.; Ducatelle, R.; Fink-Gremmels, J. *Mycopathologia* **2004**, *158*, 419-426.
10. FAO Food and Nurition Paper 81; Worldwide regulations for mycotoxins in food and feed in 2003; FAO, Rome, 2004; p.23.
11. Gökmen, V.; Artik, N.; Acar, J.; Kahraman, N.; Poyrazoglu, E. *Eur. Food Res. Technol.*. 2001, 213, 194-199.
12. Brackett, R.E.; Marth, E.H.; *Zeitsch. Lebensm. Unters. Forsch.* **1979**, *169*, 92-94.
13. Official Methods of Analysis of AOAC International; 995.10 Patulin in apple juice, AOAC Internatinal, Gaithersburg, MD.
14. Brause, A.R.; Trucksess, M.W.; Thomas, F.S.; Page, S.W. *J. AOAC Int.* **1996**, *79*, 451-455.
15. Kawamoto, Y.; Bandoh, S.; Higashihara, K.; Miyagawa, H.; Goto, T. *World Mycotoxin J.* **2008**, *1*, 59-65.
16. Bandoh, S.; Takeuchi, M.; Ohsawa, K.; Higashihara, K.; Kawamoto, Y.; Goto, T. *Int. Biodet. Biodegr.* **2009**, *63*, 379-382.
17. Higashihara, K.; Takeuchi, M.; Bandoh, S.; Saegusa, Y.; Miyagawa, H.: Goto, T. *Mycotoxins* **2009**, *59*, 7-13.

Chapter 6

Determination of *Aspergillus* section *Flavi* and their aflatoxin and cyclopiazonic acid production patterns in naturally dried figs

Dilek Heperkan and Funda Karbancıoglu-Güler

Istanbul Technical University, Faculty of Chemical and Metallurgical Engineering, Department of Food Engineering 34469, Maslak, Istanbul/Turkey
E-mail: heperkan@itu.edu.tr

Aspergillus section *Flavi* contamination in naturally dried figs were investigated in samples collected from orchards. Isolated fungi were identified using morphological and phenotypic characteristics. Isolates in *Aspergillus* section *Flavi* from fig were tested for aflatoxin (AF) and cyclopiazonic acid (CPA) production using thin-layer chromatography (TLC)-high performance liquid chromatography (HPLC) and TLC respectively. Of fig samples collected from 115 different orchards before any treatment, forty three (37.4%) were contaminated with *Aspergillus flavus,* whereas *Aspergillus parasiticus* contamination was rare - only in 4 samples (3.5%) of dried figs for two consecutive years. *A. flavus* is the predominant species among *Aspergillus* section *Flavi* members for figs (85.7%). In some sample groups, both nontoxic and aflatoxins B_1 and B_2 producing *A. flavus* strains were found together on the same fig sample; also both *A. flavus* (producing aflatoxins B_1 and B_2) and *A. parasiticus* (producing aflatoxins B_1, B_2, G_1 and G_2) were found together on the same fig sample. All aflatoxin producing *A. flavus* isolates also produced CPA. Most of the *A. flavus* strains produced aflatoxins B_1 and B_2 and CPA at the same time during the first and second year, with ratios of 71.9% and 62.5% respectively, averaging to 68.7%. *A. flavus* strains producing only AFB_1 and CPA were rare, with a ratio of 7.9% and 0% with an average of 6.2% for the first and second year, respectively. All *A. parasiticus* strains produced only aflatoxins B_1, B_2, G_1 and G_2 but no CPA. *A. tamarii*, another member of *Aspergillus* section *Flavi*, is rarely present in Turkish figs.

Introduction

Aspergillus flavus, first described by Link in 1809, together with *A. parasiticus* are well known members of *Aspergillus* section *Flavi* (*1*). In fact *A. flavus* is used as a general name for a group of closely related species.

During the last decades, methods to discriminate *A. flavus* from *A. parasiticus* and other related species have changed and as a result, taxonomic studies have become interdisciplinary, with new species added, increasing the number of aflatoxigenic species. Raper and Fennell (*2*) considered the "*A. flavus* group" to contain nine species and two varieties based on morphology and colony characteristics. Later, Gams *et al.* (*3*) placed subgenera and sections instead of groups and put *A. flavus* and related species under the *Aspergillus* section *Flavi*. In addition to taxonomic work, Klich and Pitt (*4*) and Samson *et al.* (*1*) have introduced a simplification in the identification by considering the mycotoxigenic properties of the molds.

Recent studies (*5*) have attained significant development specifically in genetic approaches. Internal transcribed spacer (ITS) sequence analysis (*6*), rDNA sequence analysis (*7*); partial sequences of the mitochondrial cytochrome *b* gene (*8*) and restriction fragment length polymorphism (RFLP) (*9*) analysis are some selected genetic methods used in identification of *A. flavus* and related species. A new concept representing *A. flavus* and related species is being used as a result of identification based on genetic characteristics. "*Aspergillus flavus* complex" named and used by Hedayati and coworkers (*10*) currently includes 23 species or varieties, including two sexual species, *Petromyces alliaceus* and *P. albertensis*. In spite of all these developments, difficulties in discriminating *A. flavus* from *A. parasiticus* and other related species still exist.

Since *A. flavus* is not only important in the field of food and agriculture but also from the medical point of view (*10*), it continues to be a focus of interest for several disciplines. *A. flavus* causes diseases in important crops such as cereals, peanut, and cotton (*11*) as well as animals and human beings. One of the important secondary metabolites of *A. flavus*, AFB_1 was classified as Group 1 human and animal carcinogen by international agencies (*12*).

Cyclopiazonic acid (CPA), known indole tetramic acid (*13*), is another important toxic metabolite of *A. flavus*, produced by some aflatoxigenic as well as some atoxigenic isolates of *A. flavus* (*14*). CPA, a mutagenic toxin (*15*) exists in several agricultural products naturally (*16, 17*). It has also co-occurred with aflatoxins in peanuts, corn (*18,19*) and sour lime (*20*).

Fig is one of the dried fruits sensitive to mycotoxin production (*21*). Several surveys exist on mycotoxin, especially on aflatoxin presence (*22, 23*). However studies related to mold flora are rare and include general information regarding the mold types (*24*).

Aflatoxin contaminated samples having blue flourescence under UV can be separated on the production line in factories. Therefore samples collected from marketing stages do not always reflect the real contamination level. In addition, products sold in a specific region usually belong to a certain brand or company and do not represent all of the orchards in the region. The aim of this study is to determine *Aspergillus* section *Flavi* contamination in naturally dried figs and their aflatoxin and cyclopiazonic acid production potentials right after the harvesting stage. Although the presence of molds does not necessarily show the presence of mycotoxins; presence of toxigenic molds, especially during the harvest and drying stages, is thought to be an indicator for the possible presence of mycotoxins and could contribute to the solution of the problem.

Materials and Methods

Samples

A total of 115 samples of dried figs were collected from orchards during the drying stage in the Aegean Region after the 2003 and 2004 harvests. Fig samples were collected according to the EC sampling procedures (*25*). Samples were stored in polyethylene bags and stored at -18°C until analysis.

Figs were divided into 4 pieces; one part was used for direct plating, one part was used for the dilution plating, one part was kept for aflatoxin analysis and the fourth part was kept in the freezer. This article does not contain results related to mycotoxin in figs.

Enumeration, Isolation and Identification of Molds

Twenty-five grams of figs were mixed with 225 mL sterile distilled water with 0.5% agar and 0.1% peptone in a stomacher Lab-blender 400 (Seward Medical, London, UK) for 2 min. The diluted fig samples were held for 30 min at room temperature before stomaching to soften the figs. Further decimal dilutions were also made in 0.1% peptone, 0.5% agar solutions. From each dilution, 0.1 mL was pipetted in duplicate onto Dichloran Rose Bengal Chloramphenicol Agar (DRBC, Merck) for total mold count and *Aspergillus flavus* and *parasiticus* agar (AFPA, Merck) for *Aspergillus* section *Flavi* count. DRBC was incubated at 25°C for 7 days and AFPA at 30°C for 2 days. Plates with 10-100 mold colonies were counted. Actual counts converted to $\log_{10}$ counts are given in Table I and II. Fig samples for direct plating were dispensed with sterile forceps onto DRBC plates. Six pieces of figs were plated for each sample and incubated at 25°C for 5 days. After incubation, both plates were examined under a stereo microscope every fungus with a green and yellow colony color were transferred to Malt Extract Agar (MEA) and incubated at 25°C for 7 days. Cultures were stored at +4°C until analysis. MEA, Czapek Dox Agar, Czapek Yeast Extract Agar (CYA) prepared as described by Pitt and Hocking (*26*) were used for identification of isolates. The plates were incubated at 25°C and 37°C for 7 days. Molds were identified according to Pitt and Hocking (*26*) as well as Samson *et al.* (*1*), based on morphological and phenotypical characteristics.

Determination of Aflatoxigenic *Aspergillus* section *Flavi* members

Aflatoxin standard was purchased from Supelco (Pennsylvania, USA). Aflatoxin analysis was performed by thin-layer chromatography (TLC) and by high performance liquid chromatography (HPLC). The procedure was as follows. Identified *Aspergillus* section *Flavi* isolates were inoculated on Yeast Extract Agar (YES) and CYA and incubated at 25°C for 7 days. After incubation the content of the petri dishes was extracted with 50 mL 70% methanol and water in a stomacher for 5 minutes. Samples were filtered with Whatman No:1 filter paper. The extracts were then filtered through Minisart RC filters (0.45 µm pore size, 4 mm diameter, Sartorius AG, Goettingen, Germany) for HPLC analysis.

Aflatoxins were determined by HPLC with post-column derivatization according to Stroka *et al.* (*27*). The liquid chromatograph was an Agilent Technologies 1100 system equipped with a fluorescence detector (λ_{ex}: 360 nm, λ_{em}: 420 nm), quaternary pump, vacuum degasser, the Rheodyne injector with a 100 µL loop and post column derivatization system (CoBrA – cell, 100 µA current). The separation was achieved at room temperature on ODS-Hypersil C18 reversed phase column (Supelco 250 mm×4.6 mm, 5 µm particle size) with 1 mL/min flow rate of 110 µL 65% nitric acid and 132 mg potasium bromide (KBr) dissolved in 1100 mL of water-acetonitrile-methanol (6:2:3, v/v/v) as a mobile phase. The retention times for aflatoxins B_1, B_2, G_1 and G_2 were approximately 11, 9, 8 and 7 minutes, respectively. Limit of detection for aflatoxins B_1, B_2, G_1 and G_2 were 0.1, 0.09, 0.1 and 0.09 µg/kg, respectively (*28*).

For TLC analysis, three agar plugs were cut out of the colony center and placed into an eppendorf tube. Aflatoxins were extracted by adding 1 mL of chloroform to each tube by mixing for 30 s. The plates (Silicagel G60, Merck, Darmstadt, Germany) were spotted with 30 µL of the extract and developed in the solvent mixture toluene, ethyl acetate, 98% formic acid (5:4:1). Aflatoxin was evaluated under longwave UV light (366 nm) by visual comparison with standards. Limit of detection for aflatoxins was 1 µg/kg.

Determination of Cyclopiazonic acid Producing *Aspergillus* section *Flavi* members

CPA standard was purchased from Sigma-Aldrich (Steinheim, Germany). CPA analysis was carried out through a modification of the methods suggested by Horn *et al.* (*29*) and Samson *et al.* (*1*). Pure isolates were inoculated at three points on CYA at 25°C for 7 days. Three agar plugs were cut out of the colony center placed into an eppendorf tube and extracted with 1 mL of chloroform for 30 s in vortex. Thin layer plates (Silicagel G60, Merck, Darmstadt, Germany) were dipped in a 8% solution of oxalic acid and dried, the plates were spotted with 50 μL of the extract and developed in the solvent mixture toluene, ethyl acetate, 98% formic acid (5:4:1) was used as the developing solvent. To visualize CPA the plate was sprayed by Ehrlich's reagent, dried, sprayed with 50% sulfuric acid and dried again. CPA was detected as blue spots in daylight. Limit of detection for CPA was 50 ng per spot (*30*).

Results and Discussion

Total mold count in fig samples, *Aspergillus* section *Flavi* counts, isolation methods, and mycotoxigenic properties of species are given in Table I for year 2003 and Table II for year 2004. The total mold counts in fig samples were determined using DRBC. The log cfu/g counts were from 2.6 to 6.7 with a mean of 4.5. *Aspergillus* section *Flavi* count was determined using AFPA. The log cfu/g counts were from 2.0 to 5.5 with a mean of 2.5.

Thirty-two of the fig samples collected from 71 orchards (45%) were contaminated with *Aspergillus* section *Flavi.* Since more than one member of the *Aspergillus* section *Flavi* were isolated from the same fig samples, 38 isolates were obtained from the 71 orchards. Two different methods were used for isolation. During direct plating, 6 figs were randomly selected from each sample and 1/4 of the figs were placed onto DRBC plates. The number of *Aspergillus* section *Flavi* isolates determined by the direct plating method was 42. Twenty-six out of the 42 isolates could not be determined by the other method. The numbers of *Aspergillus* section *Flavi* isolates from the dilution plating method-AFPA and dilution plating method-DRBC were 12 and 16, respectively. Four isolates from AFPA and 3 from DRBC could not be determined by the other method. The total number of isolates from both methods combined (direct+dilution-AFPA, direct+dilution-DRBC) was 5 and 7, respectively.

Fourteen of the 38 isolates of *Aspergillus* section *Flavi* (36.8%) produced sclerotia. Horn *et al.* (*29*) found that a large proportion of the *A. flavus* (98%) and *A. parasiticus* (78%) obtained from soil and peanut seed, produced sclerotia. In this study sclerotia production of the strains were checked only for the first year and it was found that the number of sclerotia producing *A. flavus* and *A. parasiticus* were low (13/32 and 1/3, respectively). However, as mentioned by Horn *et al.* (*29*), sclerotium production depends on culture condition, geography, climate and crop composition and isolates identified as being nonsclerotial may produce sclerotia on other media.

Identification based on morphological and phenotypical properties of the isolates revealed that 32 out of the 38 isolates were *A. flavus*, 3 *A. parasiticus* and 3 unidentified CPA producer *Aspergillus* section *Flavi* (Table I). It is worth noting here that *A. tamarii* was isolated from one sample, since it is a CPA producer. Similarly, CPA producer *Penicillium* species were also low in Turkish figs. No data exists in studies of other researchers regarding the presence of *A. tamarii* and *Penicillium* spp. in figs originating from Turkey (*24,31*).

Table I. Mold counts, isolation methods, sclerotia producing and mycotoxigenic properties of species in fig samples for year 2003

	Mold Counts		*Method of Isolation*			*Properties of Molds*			
		Asp.[1] sect.	*Direct Plate*	*Dilution Plate*		*Sclerotia*	*Mycotoxin Production*		
No	*Total*	*Flavi*	*DRBC*	*AFPA*	*DRBC*		*AFs*	*CPA*	*Species*
1	6.00	-[2]	+	ND[3]	ND	+	+	+	*A. flavus*
2	5.48	-	+	ND	ND	+	+	+	*A. flavus*
3.1	5.90	-	+	ND	ND	+	+	+	*A. flavus*
3.2	5.90	-	+	ND	ND	ND	ND	ND	*A. flavus*
4	6.00	2.60	ND	+	ND	+	+	+	*A. flavus*
5	5.60	4.48	+	+	+	ND	+	+	*A. flavus*
6	3.30	2.00	ND	+	+	+	ND	ND	*A. flavus*
7.1	4.30	2.00	ND	+	ND	ND	+	+	*A. flavus*
7.2	4.30	2.00	+	ND	ND	+	+	ND	*A. parasiticus*
8	6.78	3.00	+	+	ND	ND	+	+	*A. flavus*
9	5.85	2.00	+	ND	+	+	+	+	*A. flavus*
10	5.30	2.00	+	ND	+	ND	+	+	*A. flavus*
11	5.60	2.30	ND	+	+	ND	ND	+	*A.* sect. *Flavi*[4]
12	2.90	2.85	+	+	+	ND	+	+	*A. flavus*
13	3.00	-	+	ND	ND	+	+	+	*A. flavus*
14	4.30	-	+	ND	ND	ND	+	+	*A. flavus*
15	5.60	-	+	ND	ND	+	+	+	*A. flavus*
16	4.48	3.00	+	+	+	ND	+	+	*A. flavus*
17	5.00	2.00	ND	+		ND	ND	ND	*A. flavus*
18	3.00	-	+	ND	ND	ND	ND	ND	*A. flavus*
19.1	3.00	2.00	+	ND	+	+	+	+	*A. flavus*
19.2	3.00	2.00	ND	ND	+	ND	ND	ND	*A. flavus*
20	3.30	2.70	+	ND	+	+	+	+	*A. flavus*
21	5.30	-	+	ND	+	ND	ND	+	*A.* sect. *Flavi*
22.1	6.30	2.30	ND	+		ND	+	+	*A. flavus*
22.2	6.30	2.30	+	ND	ND	ND	+	ND	*A. parasiticus*
23	2.60	-	+	ND	ND	ND	+	+	*A. flavus*
24	4.78	-	+	ND	ND	ND	ND	+	*A.* sect. *Flavi*
25.1	6.00	5.48	+	ND	ND	ND	+	+	*A. flavus*
25.2	6.00	5.48	ND	ND	+	ND	+	ND	*A. parasiticus*
26	4.85	2.00	ND	ND	ND	ND	+	+	*A. flavus*
27	4.30	-	+	ND	ND	+	+	+	A. flavus
28	4.30	2.00	+	+	+	ND	+	+	A. flavus
29	3.30	2.00	ND	ND	+	ND	+	+	*A. flavus*
30	6.48	-	+	ND	ND	ND	+	+	*A. flavus*
31	3.00	2.00	ND	+	+	+	+	+	*A. flavus*
32.1	3.48	2.60	+	ND	+	+	ND	ND	*A. flavus*
32.2	3.48	2.60	+	ND	ND	ND	+	+	*A. flavus*
Total			*27*	*12*	*16*	*14*	*29*	*29*	*38*

1: *Aspergillus* section *Flavi* members

2 : Determined by direct plating method

3 : Not detected

4 : Unidentified CPA producer *Aspergillus* section *Flavi* member

Aspergillus section *Flavi* of different species and mycotoxigenic profile showing different morphological properties were isolated from the same fig samples, collected from 6 orchards. Among them, *A. flavus* producing aflatoxins B_1 and B_2 and atoxigenic *A. flavus* were isolated together from two samples (sample 3 and 19). *A. parasiticus* was found positive in only three samples. Aflatoxins B_1, B_2, G_1 and G_2 producing *A. parasiticus* and aflatoxins B_1 and B_2 producing *A. flavus* were found together in these three samples (sample 7, 22 and 25).

Of the 71 fig samples, 32 were found *Aspergillus* section *Flavi* positive in 2003, corresponding to a ratio of 45%. Twenty-six fig samples (36.6%) were contaminated with aflatoxin producing *A. flavus* and *A. parasiticus*. Among them, the ratio of *A. parasiticus* was very low, it was only found in 3 of the 71 samples (4.2%). All aflatoxin producing *A. flavus* isolates also produced CPA.

The average mold count in fig samples for the second year was 4.4 log cfu/g (2.3-6.5) (Table II). The average *Aspergillus* section *Flavi* count was 2.2 log cfu/g (2.0-3.5) and was less than the first year's. Fifteen fig samples from the 44 orchards (34%) contained *Aspergillus* section *Flavi*, where the total number of isolates was 18.

The number of isolates determined by the direct plating method was 15. Ten out of the 15 isolates could not be determined by the other method. The numbers of *Aspergillus* section *Flavi* isolates determined by the dilution plating method-AFPA and dilution plating method-DRBC were 4 and 5, respectively.

The total numbers of isolates from both methods combined (direct+dilution-AFPA, direct+dilution-DRBC) were 2 and 3, respectively. *A. parasiticus* could be isolated only by DRBC for both years. As our results indicate, the direct plating method is more effective compared to the dilution plating method to determine *A. flavus*.

It was found that the total *A. flavus-A. parasiticus* count ranged from $1.4x10^3$ to $3.2x10^6$ cfu/g in 60 shelled peanut samples (*32*). According to the results of this study, the count of *A. flavus-A. parasiticus* in figs is lower compared to peanuts. It was suggested that the addition of water to produce slurry instead of dry-grinding peanuts for homogenization produced high temperatures that killed most of the *A. flavus-A. parasiticus* propagules (*31*). In this study, direct plating and dilution plating methods have been used together, as already indicated. During the dilution process, dried figs made up from 1/4 pieces were held for 30 min at room temperature before stomaching to soften the figs.

Table II. Mold counts, isolation methods and mycotoxigenic properties of species in fig samples for year 2004

	Mold counts		*Method of isolation*					
		Asp.[1] *sect.*	*Direct Plate*	*Dilution Plate*		*Mycotoxin Production*		
No	*Total*	*Flavi*	*DRBC*	*AFPA*	*DRBC*	*AFs*	*CPA*	*Species*
33	3.00	2.00	+	ND[3]	+	+	+	*A. flavus*
34	3.00	-[2]	+	ND	ND	+	+	*A. flavus*
35.1	4.30	3.48	+	ND	+	+	+	*A. flavus*
35.2	4.30	3.48	+	ND	ND	+	ND	*A. parasiticus*
36	2.30	-	+	ND	ND	+	+	*A. flavus*
37	3.00	-	+	ND	ND	+	+	*A. flavus*
38	3.00	2.30	+	ND	ND	+	+	*A. flavus*
39	2.30	-	+	ND	ND	+	+	*A. flavus*
40	4.30	2.00	ND	+	ND	ND	ND	*A. flavus*
41.1	4.00	2.00	ND	ND	+	ND	ND	*A. flavus*
41.2	4.00	2.00	+	ND	ND	+	+	*A. flavus*
42	5.48	-	+	ND	ND	+	+	*A .flavus*
43	4.30	2.00	ND	+	+	ND	ND	*A. flavus*
44	5.30	2.00	+	+	ND	ND	ND	*A. flavus*
45.1	3.70	2.00	+	ND	ND	ND	ND	*A. flavus*
45.2	3.70	2.00	+	ND	+	+	+	*A. flavus*
46	6.48	2.00	+	ND	ND	ND	+	*A.* sect. *Flavi*[4]
47	4.48	2.00	+	+	ND	ND	ND	*A. flavus*
Total			*15*	*4*	*5*	*11*	*11*	*18*

1: *Aspergillus* section *Flavi* members

2 : Determined by direct plating method

3 : Not detected

4: Unidentified CPA producer *Aspergillus* section *Flavi* member

Sixteen of the 18 isolates were *A. flavus*, one was *A. parasiticus*, and one was CPA producing unidentified *Aspergillus* section *Flavi* (Table II). *Aspergillus* section *Flavi* with different morphological properties, mycotoxigenic profiles and species were isolated from the same fig samples collected from 3 orchards. Among them, in two samples (sample 41 and 45), both aflatoxins B_1 and B_2 producing and non toxin producing *A. flavus* were isolated together. *A. parasiticus* was found only in one sample (sample 35) together with aflatoxins B_1 and B_2 producing *A. flavus*. Similarly, for the previous year, all the samples containing *A. parasiticus* also contained *A. flavus*. However it is difficult to make a general conclusion since the number of samples containing *A. parasiticus* is very few.

The ratio of *Aspergillus* section *Flavi* positive fig samples was lower in the second year compared to the first year. Of the 44 samples, 15 were *Aspergillus* section *Flavi* positive (34%). Similar to the first year, *A. parasiticus* occurrence was very low for the second year, it was only observed in 1 out of the 44 fig samples (2.3%).

All aflatoxin producing *A. flavus* produced CPA, in 10 of the 44 samples (22.7%). The distribution of the aflatoxin and cyclopiazonic acid producing *Aspergillus* section *Flavi* isolates among dried fig samples are given in Table III. The mycotoxigenic profiles of *Aspergillus* section *Flavi* (excluding *A. parasiticus*) found in fig samples were as follows for the first and second year, respectively: aflatoxin producers, 36.6% and 22.7% with an average of 31.3% (all aflatoxin producing *A. flavus* isolates also produced CPA); strains producing only CPA, 4.2% and 2.3% with an average of 3.5%; total CPA producers [(AF+CPA)+(CPA)] 40.8% and 25% with an average of 34.8%. Lee and Hagler (*33*) found that aflatoxin/CPA-producing isolates of *A. flavus* were common in maize, and only one of the 12 aflatoxin producers did not produce CPA. Trucksess *et al.* (*34*) found that 6 out of 31 isolates (19.3%) of *A. flavus* obtained from different foods were aflatoxin producers whereas 19 out of 31 (61.3%) were CPA producers.

Table III. Mycotoxigenic properties of Aspergillus section Flavi isolated from dried figs for two consecutive years

	Numbers of Aspergillus section Flavi					
Year	*Isolates*	*Aflatoxins[1]* *B_1,B_2,G_1,G_2*	*CPA[2]*	*Aflatoxin* *B_1* *+CPA*	*Aflatoxins* *B_1, B_2* *+CPA*	*Atoxigenic*
1st year	38	3	3	3	23	6
2nd year	18	1	1	0	10	6
Total	*56*	*4*	*4*	*3*	*33*	*12*

1: *A.parasiticus* (did not produce CPA)

2: Unidentified CPA producer *Aspergillus* section *Flavi* members

The number of *Aspergillus* section *Flavi* isolates isolated from 71 and 44 dry fig samples during the first and second year, respectively was 38 and 18 with a total of 56. The aflatoxigenic profiles of the isolates for the first and second year, respectively were AFB_1 at 7.9% and 0%; aflatoxins B_1 and B_2 at 60.5% and 55.5%; aflatoxins B_1, B_2, G_1 and G_2 at 7.9% and 5.5% (Table III).

According to the outcome of the two-year study, *A. flavus* was predominant among the *Aspergillus* section *Flavi* members in figs: 84.2% and 88.9% for the first and second year, respectively with an average of 85.7%. *A. flavus* was previously also found to be predominant in other products, such as wheat and soybean (*35*), and maize and cottonseed (*36*). Only in peanuts *A. flavus* and *A. parasiticus* were equally common (*35*). *A. parasiticus* is quite rare in figs. Pitt (*37*) also reported that *A. parasiticus* is less widely distributed.

In another survey on Turkish figs, fig samples from 11 different orchards in the Aegean region have been collected after harvesting till storage to establish *A. flavus-A. parasiticus* contamination in 1989. *A. flavus-A. parasiticus* contamination has been found to be 41% for ripe fig samples collected directly from the trees, 42% for sun dried fig samples, 33% for fig samples obtained from various store houses, 25% for fig samples from different processing plants and 25% for fig paste samples (*31*). Our results in agreement with this study.

The ratio of aflatoxigenic *Aspergillus flavus* isolates was 81.3% for the first year and 62.5% for the second year, with an average of 75%. Figs are comparable to peanuts in terms of the high aflatoxigenic *A. flavus* ratio. Incidence of aflatoxigenic *A. flavus* strains was 69% in peanuts (*35*). It was also observed in other studies that the aflatoxigenic *A. flavus* ratio was high for figs. Toxin production of *A. flavus* isolated from figs were stated to be 67% (*24*) and 72% (*31*).

CPA production frequency of *Aspergillus* section *Flavi* members were 29 of the 38 isolates (76.3%) and 11 of the 18 isolates (61.1%) for the first and second year respectively, with an average of 71.4%. Eleven of the 19 isolates (57.9%) produced both aflatoxin and CPA, 12 isolates (63%) produced aflatoxin and 14 isolates (74%) produced CPA. The ratio of CPA producing *A. flavus* isolated from different products was higher compared to our results, such as 94% in peanuts, 93% in wheat and 73% in soybeans (*35*). Horn *et al.* (*29*) stated that CPA-producing isolates ranged from 23% (*36*) to the high percentage (93%). Lee and Hagler (*33*) found that aflatoxin/CPA-producing isolates of *A. flavus* were common in maize as well.

It was also determined that the ratio of the isolates concurrently producing aflatoxins B_1 and B_2 and CPA was the highest for both the first and second year, with an average of 68.8%. A few of the CPA producing *Aspergillus* section *Flavi* isolates produced only CPA, their ratios were 7.9% and 5.6% for the first and second year, respectively.

A. parasiticus ratio in dry figs was very low for the first (4.2%) and second year (2.3%), respectively. All of the *A. parasiticus* isolates produced only aflatoxins B_1, B_2, G_1 and G_2 but not CPA. In other studies, none of the *A. parasiticus* isolates produced CPA (*38*). Unidentified *Aspergillus* section *Flavi* members which were CPA producers in figs were also low, with ratios 7.9% for the first year and 5.6% for the second year, averaging to 7.1%.

A. tamarii produces CPA, but not aflatoxins (*39*). *A. tamarii* was divided into types A (CPA producer) and B (non CPA producer) based on morphological differences and the lack of vegetative compatibility between the two types (*29*). In this two-year study, we isolated only one strain of *A. tamarii* from 115 figs. The strain was a CPA producer and we did not check for other characteristics. Doster and Michailides (*40*) found that bright greenish yellow fluorescence under UV was produced in figs infected by *A. tamarii* and *A. alliaceus* as well as *A. flavus* and *A. parasiticus*. They also explained that the reason why *A. tamarii* was not found in BGYF figs from Turkey could be that *A. tamarii* may be less common in Turkey than in USA. They also stated that the levels of *A. tamarii* in Turkish fig orchards were unknown, so they compared their results with Turkish pistachio orchards where *A. tamarii* does occur at very low levels and is much rarer than *A. flavus*. This study also indicates that the predictions related to *A. tamarii* by Doster and Michailides (*40*) are realistic. *A. tamarii* ratio in Turkish originated figs is 1/115. In the USA however, *A. tamarii* affects approximately the same number of figs as *A. flavus* (*40*).

Conclusion

Figs, one of the earliest fruits cultivated on earth, are considered as one of the dry fruits sensitive to mycotoxin production. Since the fig fruit has a soft outer skin, it is vulnerable to physical injury and could be easily damaged. Shelf life of fresh fig is very short, even when the fruits are carefully picked and placed in cases by hand, the fruits lose their freshness within a few days. Therefore figs are usually utilized by drying. Drying takes place naturally under the sun.

Aflatoxin in dried figs has been determined in previous surveys. However, it is advised to take the samples before any sorting process to see the real amount of aflatoxin contamination. Based on samples collected during two consecutive years in this study, our results indicate that 41% of the fig samples on the average are contaminated by *Aspergillus* section *Flavi* members. Among them, *A. flavus* is predominant in figs and constitutes 85.7% of the isolates on the average. *A. parasiticus* is very rare (7%). More than one members of *A.* section *Flavi* were isolated in some samples. Among them, both aflatoxins B_1 and B_2 producing and non toxin producing *A. flavus* were isolated together (four samples). All the samples containing *A. parasiticus* also contained *A. flavus* producing aflatoxins B_1 and B_2. Due to the low occurrence of *A. parasiticus*, it is expected to find B aflatoxins more frequently than G aflatoxins among Turkish figs from the Aegean region.

Cyclopiazonic acid (CPA) is another mycotoxin that can be found together with aflatoxin in figs, and its contamination level should also be determined. According to our study, all aflatoxin producing *A. flavus* produce CPA in figs with an average occurrence of 75%. Another important result of this study is that the origin of CPA in Turkish figs is *A. flavus*, not *A. tamarii* or *Penicillium* spp.

Identification of molds have been carried out with morphological and phenotypic methods. If available, molecular methods can also be very helpful. Although mold presence does not indicate mycotoxin presence all the time, toxigenic molds found during harvest and drying is a strong indication of mycotoxin presence in the product and could help its control.

References

1. Samson, R. A.; Hoekstra, E. S.; Lund, F.; Filtenborg, O.; Frisvad, J. C. In *Introduction to food- and airborne fungi*, Samson, R. A.; Hoekstra, E. S.; Frisvad, J. C.; Filtenborg, O., Eds.; Centraalbureau Voor Schimmelcultures, Utrecht, Netherlands, 2002; pp 283-297.
2. Raper, K. B.; Fennell, D. I. *The genus Aspergillus*; Williams & Wilkins Co.: Baltimore, MD, 1965; 686 pp.
3. Gams, W.; Christensen, M.; Onions, A. H. S.; Pitt, J. I.; Samson, R. A. In *Advances of Penicillium and Aspergillus systematics*; Samson, R. A.; Pitt, J.I., Eds.; Plenum Publishers: New York, 1985; pp 55-62.
4. Klich, M. A.; Pitt, J. I. Differentiation of *Aspergillus flavus* from *A. parasiticus* and other closely related species. *Trans. Brit. Mycol. Soc.* **1988,** *91,* 99-108.

5. Paterson, R. R. M. Identification and quantification of mycotoxigenic fungi by PCR. *Process Biochem.* **2006,** *41*, 1467-1474.
6. Kumeda, Y.; Asao, T. Single-strand conformation polymorphism analysis of PCR-amplified ribosomal DNA internal transcribed spacers to differentiate species of *Aspergillus* section *Flavi. Appl. Environ. Microbiol.* **1996,** *62*, 2947-2952.
7. Peterson, S.W. In *Integration of modern taxonomic methods for Penicillium and Aspergillus classification*, Samson, R. A.; Pitt, J.I., Eds.; Harwood Academic Publishers: Amsterdam, The Netherlands, 2000; pp 323-355.
8. Wang, L.; Yokoyama, K.; Takahasi, H.; Kase, N.; Hanya, Y.; Yashiro, K.; Miyaji, M.; Nishimura, K. Identification of species in *Aspergillus* section *Flavi* based on sequencing of the mitochondrial cytochrome *b* gene. *Int. J. Food Microbiol.* **2001,** 71,75-86.
9. Montiel,D.; Dickinson, M. J.; Lee, H. A.; Dyer, P. S.; Jeenes, D. J.; Roberts, I. N.; James, S.; Fuller, L. J.; Matsuchima K.; Archer, D.B. Genetic differentiation of the *Aspergillus* section *Flavi* complex using AFLP fingerprints. *Mycol. Res.* **2003,** *107*,1427–1434.
10. Hedayati, M. T.; Pasqualotto, A. C.; Warn, P. A.; Bowyer, P.; Denning, D. W. *Aspergillus flavus*:human pathogen,allergen and mycotoxin producer. *Microbiology*. **2007,** *153,* 1677-1692.
11. CAST. Council for Agricultural Science and Technology. *Mycotoxins: Risks in Plant, Animal and Human Systems.* Task Force Report No. 139. Ames, Iowa, USA, 2003; 199 pp.
12. IARC. International Agency for Research on Cancer. *Some naturally occuring substances: food items and constituents, heterocyclic aromatic amines and mycotoxins.* Monograph 56. Lyon International Agency for Research on Cancer. 1993. 571 pp.
13. Dorner, J. W., In *Mycotoxins and Food Safety*; DeVries, J. W.; Trucksess, M. W.; Jackson, L. S., Eds.; Kluwer Academic/Plenum Publishers: New York, 2002; pp 107-116.
14. Horn, B.W.; Dorner, J.W. Regional differences in production of aflatoxin B_1 and cyclopiazonic acid by soil isolates of *Aspergillus flavus* along a transect within the United States. *Appl. Environ. Microbiol.* **1999,** *65,* 1444–1449.
15. Antony, M.; Shukla, Y.; Janardhanan, K. K. Potential risk of acute hepatotoxicity of kodo poisoning due to exposure to cyclopiazonic acid *J. Ethnopharmacol.* **2003,** *87,* 211-214.
16. Lansden, J. A.; Davidson, J. I. Occurence of cyclopiazonic acid in peanuts. *Appl. Environ. Microbiol.* **1983,** *45,* 766–769.
17. Dorner, J.W.; Cole, R. J.; Erlington, D. J.; Suksupath, S.; McDowell, G.H.; Bryden, W.L. Cyclopiazonic acid residues in milk and eggs. *J. Agri. Food Chem.* **1994,** *42,* 1516-1518.
18. Urano, T.; Trucksess, M. W.; Beaver, R. W.; Wilson, D. M.; Dorner, J. W.; Dowell, F. E. Co-occurence of cyclopiazonic acid and aflatoxins in corn and peanuts. *J. AOAC Int.* **1992,** *75,* 838-841.

19. Pinto, V. F.; Patriarca, A.; Locani, O.; Vaamonde, G. Natural co-occurence of aflatoxin and cyclopiazonic acid in peanuts grown in Argentina. *Food Addit. Contam.* **2001,** *18*, 1017-1020.

20. Bamba, R.; Sumbali, G. Co-occurence of aflatoxin B1 and cyclopiazonic acid in sour lime (Citrus aurantifolia Swingle) during post-harvest pathogenesis by *Aspergillus flavus*. *Mycopathologia.* **2005,** *159,* 407-411.

21. Morton, S. G.; Eadie, T.; Llewellyn, G. C. Aflatoxigenic potential of dried figs, apricots, pineapples, and raisins. *J. AOAC.* **1979,** *62,* 958-962.

22. Sharman, M.; Patey, A. L.; Bloomfield, D. A.; Gilbert, J. Surveillance and control of aflatoxin contamination of dried figs and fig paste imported into the United Kingdom. *Food Addit. Contam.* **1991,** *8,* 299–304.

23. Iamanaka, B.T.; Castle de Menezes, H.; Vicente, E.; Leite, R. S. F.; Taniwaki, M. H. Aflatoxigenic fungi and aflatoxins occurence in sultanas and dried figs commercialized in Brazil. *Food Control.* **2007,** *18,* 454-457.

24. Karapınar, M.; Gönül, D. *Symposium on Moulds and Mycotoxins in Foodstuffs,* Istanbul, Turkey,1989. pp 22-29.

25. EC. European Commission, *Commission Directive 98/53/EC of 16 July 1998 laying down the sampling methods and the methods of analysis for the official control of the levels for certain contaminants in foodstuffs, Off. J. Eur. Commun.* **1998**, *L 201,* 93.

26. Pitt, J. I.; Hocking, A. D. *Fungi and food spoilage*; Blackie Academic&Professional: London, UK, 1997; 593 pp.

27. Stroka, J.; Anklam E.; Jörissen, U.; Gilbert, J. Immunoaffinity column cleanup with liquid chromatography using post-column bromination for determination of aflatoxins in peanut butter, pistachio paste, fig paste and paprika powder: Collaborative study. *J. AOAC Int.* **2000**, *83*, 320-340.

28. Mecik, N. MSc. thesis, Istanbul Technical University, Institute of Science and Technology, Istanbul, Turkey, 2007.

29. Horn, B. W.; Grene, R. L.; Sobolev, V. S.; Dorner, J. W.; Powell, J. H.; Layton, R. C. Association of morphology and mycotoxin production with vegetative compatibility groups in *Aspergillus flavus*, *A. parasiticus*, and *A. tamarii*. *Mycologia.* **1996,** *88,* 574-587.

30. Somuncuoglu, S. MSc. thesis, Istanbul Technical University, Institute of Science and Technology, Istanbul, Turkey, 2007.

31. Heperkan, D. The importance of mycotoxins and a brief history of mycotoxin studies in Turkey. *ARI The Bulletin of The Istanbul Technical University*. **2006,** *54,* 18-27.

32. Dorner, J. W. Simultaneous quantitation of *Aspergillus flavus /A. parasiticus* and aflatoxins in peanuts. *J. AOAC Int.* **2002,** *85,* 911-916.

33. Lee, Y.J.; Hagler, W. M. Aflatoxin and cyclopiazonic acid production by *Aspergillus flavus* isolated from contaminated maize. *J. Food Sci.* **1991,** *56,* 871-872.
34. Trucksess, M.W.; Mislivec, P. B.; Young, K.; Bruce, V. R.; Page, S. W. Cyclopiazonic acid production by cultures of *Aspergillus* and *Penicillium* species isolated from dried beans, corn meal, macaroni, and pecans. *J. AOAC.* **1987,** *70,* 123-126.
35. Vaamonde, G.; Patriarca, A.; Pinto, V. F.; Comerio, R.; Degrossi, C. Variability of aflatoxin and cyclopiazonic acid production by *Aspergillus* section *Flavi* from different substrates in Argentina. *Int. J. Food Microbiol.* **2003,** *88,79-84.*
36. Pitt, J.I. In *Mycotoxins and Phycotoxins*, Njapau, H.; Trujillo, S.; van Egmond, H.P.; Park, D.L., Wageningen Academic Publishers, Wageningen, The Netherlands, 2006; pp 33-42.
37. Pitt, J. I. Toxigenic fungi, which are important? *Med. Mycol.* **2000,** *38,* 17-22.
38. Sobolev, V. S.; Horn, B. W.; Dorner, J. W.; Cole, R. J. Liquid chromatographic determination of major secondary metabolites produced by *Aspergillus* species from section *Flavi. J. of AOAC Int.* **1998,** *81,* 57-60.
39. Dorner, J. W. Production of cyclopiazonic acid by *Aspergillus tamarii* Kita. *Appl. Environ. Microbiol.* **1983,** *46,*1435-1437.
40. Doster, M. A.; Michailides, T. J. Production of bright greenish yellow fluorescence in figs infected by *Aspergillus* species in California orchards. *Plant Dis.* **1998,** *82,*669-673.

Molecular Biology and Biosynthesis

Chapter 7

Elimination and control of aflatoxin contamination in agricultural crops through *Aspergillus flavus* genomics

Thomas E. Cleveland, Deepak Bhatnagar, Jiujiang Yu

USDA/ARS, Southern Regional Research Center, Food and Feed Safety Research Unit, New Orleans, Louisiana 70124, USA

Aflatoxins are the most carcinogenic mycotoxins produced by *Aspergillus flavus* and *A. parasiticus*. The primary objectives of our *A. flavus* genomics program are to reduce and eliminate aflatoxin contamination in food and feed and control fungal infection in preharvest crops such as corn, cotton, peanut and tree nuts. *A. flavus* EST and whole genome sequencing projects for this fungus project were completed. Three different types of *A. flavus* microarrays have been constructed and used in gene profiling and functional genomics studies. In order to control aflatoxin contamination of agricultural crops, genes that are potentially involved in aflatoxin formation and fungal infection have been identified through genomics tools.

Within the genus *Aspergillus*, *Aspergillus flavus* is the most infamous species because its ability to produce the most carcinogenic secondary metabolites aflatoxins (*1*). After *A. fumigatus*, *A. flavus* is the second leading cause of invasive and non-invasive aspergillosis in humans and animals (*2*, *3*). *A. flavus* is also a weak pathogen of agricultural crops such as corn, cotton, peanut and treenuts (*4*). It can contaminate these crops with aflatoxins in the field before harvest and cause spoilage of post harvest grains during storage. Under weather conditions favorable for its growth, *A. flavus* can cause a significant ear rot on maize, resulting in significant economic losses to farmers (*5-7*). *A. flavus* produces aflatoxins B_1 and B_2, while *Aspergillus parasiticus*, produces aflatoxins B_1, B_2, G_1, and G_2. When moderate to high levels of aflatoxins are consumed, acute liver damage, acute necrosis, cirrhosis, or in severe cases, acute liver failure and death may occur (*8*, *9*). Low doses with long term dietary exposure to aflatoxins is a major risk of hepatocellular carcinoma (*10*, *11*). Aflatoxin contamination of agricultural commodities poses a potential risk to livestock and human health (*1*, *12*) and also has significant economic implications for the agricultural industry worldwide because of restrictions limiting the trade of contaminated crops. For these reasons, extensive studies on the occurrence, biosynthesis, and toxicity of aflatoxins have been made (*12-14*). In this chapter, we summarize the current progress on genetics and genomics of *A. flavus* for solving aflatoxin contamination in food and feed.

Aflatoxin biosynthesis in *Aspergillus flavus*

Aflatoxin biosynthetic pathway

The aflatoxin biosynthetic pathway is the focus of many scientists worldwide. The establishment of aflatoxin biosynthetic pathway was attributed by the hallmark discovery of a mutant that accumulates brick-red pigment in *A. parasiticus* (*15-17*). This pigment, norsolorinic acid, is the earliest, stable aflatoxin precursor (*18*, *19*). Since then the major aflatoxin pathway intermediates have been identified and elucidated through molecular genetics (*14*, *20*).

Aflatoxins are formed after a series of highly organized oxidation-reduction reactions (*21-24*). Aflatoxin biosynthesis starts from the formation of hexanoyl CoA from malonyl CoA, followed by formation of a decaketide anthraquinone (*20*, *22*). In the synthesis of the polyketide from acetyl CoA, two fatty acid synthases (FAS) and a polyketide synthase (PKS) are involved (*25*). Norsolorinic acid (NOR) is the first stable aflatoxin intermediate identified in the pathway (*19*, *26*). The general accepted aflatoxin biosynthetic pathway scheme is: a hexanoyl CoA precursor → norsolorinic acid, NOR → averantin, AVN → hydroxyaverantin, HAVN → Oxoaverantin, OAVN → averufin, AVF → hydroxyversicolorone, HVN → versiconal hemiacetal acetate, VHA → versiconal, VAL → versicolorin B, VERB → versicolorin A, VERA → demethyl-sterigmatocystin, DMST → sterigmatocystin, ST → O-methylsterigmatocystin, OMST → aflatoxin B_1 and aflatoxin G_1. After the VHA step, there is a branch point in the pathway that leads to aflatoxins B_2 and G_2 formation (*21*, *24*, *27*).

Aflatoxin biosynthetic pathway gene and gene cluster

The genes involved in aflatoxin formation have been cloned and most of them have been characterized (*14*, *21*). The first aflatoxin pathway gene identified was *aflD* (*nor-1*) by complementation studies that selected for the characterisitic red color of norsolorinic acid. Most recently, the *nadA* encoding a cytosol enzyme was reported to be involved in later step of aflatoxin formation for the conversion of NADA (an intermediate between *O*-methyl sterigmatocystin (OMST) and aflatoxin G_1) to aflatoxin G_1 together with two other microsome enzymes (*28*).

The discovery that *aflD* (*nor-1*) and *aflM* (*ver-1*) genes were linked with the regulatory gene *aflR* in a common cosmid clone provided the initial evidence of the clustering of the aflatoxin pathway genes (*29*, *30*). The aflatoxin pathway gene cluster was initially established when 9 cloned genes, including *aflD* (*nor-1*), *aflR*, *aflM* (*ver-1*), and *aflP* (*omtA*), were mapped to within 75 kb DNA region by overlapping cosmid clones in *A. parasiticus* and *A. flavus* (*31*). The completed aflatoxin pathway gene cluster was established when an 82 kb DNA sequence harboring a total of 29 aflatoxin biosynthetic pathway genes (or ORFs) and 4 sugar utilization genes was reported (*21*). Interestingly a partially duplicated aflatoxin gene cluster was identified in *A. parasiticus*, consisting of seven duplicated genes, named *aflR2*, *aflJ2*, *adhA2, estA2*, *norA2, ver1B,* and *omtB2* respectively (*32*, *33*).

Regulation of aflatoxin biosynthesis

A positive regulatory gene named *aflR* is identified in both the aflatoxin and sterigmatocystin pathways in *A. parasiticus, A. flavus* and *A. nidulans* (*34*, *35*). It activates toxin pathway gene transcription (*36*). The *aflR* gene, coding for a sequence specific zinc binuclear DNA-binding protein, is required for transcriptional activation of most, if not all, of the aflatoxin pathway genes (*36-39*). The AflR protein has major domains typical of fungal and yeast Gal4-type transcription factors (*36*): a N-terminal cysteine-rich stretch, (Cys_6-Zn_2) DNA-binding domain (*36*, *39*); an arginine-rich (RRARK) nuclear localization domain; and a transcription activation domain in the C-terminus (*37*, *38*). Aflatoxin pathway gene transcription is activated when the AflR protein binds to the palindromic sequence 5'-TCGN5CGA-3' (also called AflR binding motif) in the promoter region of structural genes (*40-42*) in *A. parasiticus, A. flavus* and *A. nidulans*. *A. sojae* was found to contain a defective *aflR* transcription activation domain due to early termination of 62 amino acids from its C-terminus (*43*, *44*). Thus, with the absence of the functional regulatory protein, no induction of aflatoxin can occur in this non-toxigenic species used for fermentation in food industry. Adjacent to the *aflR* gene, a divergently transcribed gene, *aflS* (*aflJ*), was also found to be involved in the regulation of transcription (*45*). The AflS protein binds to the carboxy terminal region of AflR and may affect AflR activity (*46*).

The *laeA*, (for loss of *aflR* expression), a newly identified regulatory gene located outside of the aflatoxin pathway gene cluster, was identified to regulate *aflR* gene expression (*47*, *48*). Disruption of *laeA* resulted in loss not only of *aflR* gene expression for ST synthesis, but also expression of genes involved in penicillin biosynthesis in *A. nidulans*, as well as genes involved in gliotoxin biosynthesis in *A. fumigatus* (*47*). Thus the *laeA* appears to be involved in the global regulation of some secondary metabolite production in several fungal species.

The *veA*, is reported to regulate fungal development and affect aflatoxin/sterigmatocystin production (*49*). Unlike the *aflR* and *aflS*, the *laeA* and *veA* are located outside the known aflatoxin/sterigmatocystin pathway gene cluster. In the three sequenced Aspergilli, *A. flavus* NRRL 3357, *A. oryzae* RIB40, and *A. fumigatus* CE10, these regulatory genes are orderly organized. The *laeA* gene is located in an 80-100 kb gene cluster and the genes in the cluster are well conserved in the three species. The level of conservation and homology among the three species indicates that this gene is important in fungal development and/or secondary metabolism.

Most recent investigation in genetic regulation of fungal development and secondary metabolite production, Dr. Keller and collaborators identified the heterotrimeric velvet complex VelB/VeA/LaeA connecting light-responding developmental regulation and control of secondary metabolism. In *A. nidulans*, light inhibits sexual reproduction as well as secondary metabolism. Deletion of either *velB* or *veA* results in defects in both sexual fruiting-body formation and the production of secondary metabolites. The *velB* is expressed during sexual development, while *veA* is primarily expressed in the dark. The gene product VeA bridges VelB to LaeA, the nuclear master regulator of secondary metabolism (*50*).

Identification of genes that are involved in aflatoxin formation through *A. flavus* genomics

Aspergillus flavus genomics

Genomics is the process of determining the entire genetic contents of an organism by high throughput sequencing of the DNA in a living cell and identification of all of the genes through bioinformatics. Genomics tools allow scientists to study genetics of an organism at the genome scale to determine the location, structure and biological function of genes. Significant progress has been made in *Aspergillus flavus* genomics to understand the mechanisms of aflatoxin formation and fungal infection.

Aspergillus flavus Expressed Sequenced Tags is the first genomic effort completed at The Food and Feed Safety Research Unit of Southern Regional Research Center, USDA/ARS. A normalized cDNA expression library using wild type strain NRRL 3357 (ATCC# 20026) was constructed commercially. Over 26,110 cDNA clones from such library were sequenced at The Institute for Genomic Research (TIGR), now named J. Craig Venter Institute (JCVI). A total of 7,218 unique *A. flavus* EST sequences were identified from 19,618 ESTs generated (*51*). These EST sequences have been released to the public at the NCBI GenBank Database (http://www.ncbi.nlm.nih.gov/). The *A. flavus* Gene Index was constructed at TIGR (http://www.tigr.org) which is currently maintained and curated by The Dana Farber Cancer Institute (http://compbio.dfci.harvard.edu/tgi).

The *A. flavus* whole genome sequencing (using wild type strain NRRL 3357) was funded by The USDA/NRI Microbial Genome Sequencing Project and The Food and Feed Safety Research Unit of Southern Regional Research Center, USDA/ARS (*52*). The sequencing has been completed to 5 X coverage at JCVI under the supervision of Dr. William C. Nierman by a shotgun approach and Sanger sequencing protocol. Automated annotation of the genome was completed using tools trained on the available genomic sequences of *A. oryzae* and *A. fumigatus* as well as *A. flavus* and *A. oryzae* Expressed Sequence Tags (EST). The scaffold sizes of the DNA sequence range from 4.5 Mb to 1.0 Kb, and over 99.6 % of the genome is represented in the 16 largest scaffolds, which correspond to the 16 predicted arms of the 8 *A. flavus* chromosomes based on optical map constructed for *A. oryzae* genome. Due to high degree of sequence similarity between *A. flavus* and *A. oryzae*, the *A. flavus* genome matches well to *A. oryzae* and 99.6% of the predicted genes in *A. flavus* were assigned. The estimated genome size is 36.8 Mb, which is similar to that for *A. oryzae* (36.7 Mb) (*53*), but larger than that for *A.* nidulans (30.1 Mb) (*54*) or *A. fumigatus* (29.4 Mb) (*55*). The availability of the *A. oryzae* whole genome sequence (*53*) provided not only the sequence data but the chromosomal structure for comparison with *A. flavus*. The sequence data have been deposited with the NCBI GenBank database (http://www.ncbi.nlm.nih.gov) and are also available through the *Aspergillus flavus* website (http://www.aspergillusflavus.org) (*52*).

Aspergillus flavus genomics for identifying genes involved in aflatoxin formation

Studies on aflatoxin biosynthesis through molecular cloning in *A. flavus* and *A. parasiticus* led to the identification of 29 clustered genes within a 82 kb DNA region on the chromosome (*21*). Transcript mapping of the ESTs to the cluster region of the genome, additional five new transcripts (*hypA*, *hypB*, *hypC*, *hypD*, and *hypE*) were identified within the gene cluster, which could be involved in aflatoxin biosynthesis.

With the completion of the *A. flavus* genome sequencing, it is estimated that there are over 12,000 functional genes, which is similar to those of other *Aspergillus* species (*53-56*). Annotation of the *A. flavus* genome sequence and EST data revealed numerous categories of genes encoding for enzymes possibly involved, directly or indirectly, in aflatoxin production or other secondary metabolites, such as in global regulation, signal transduction, pathogenicity, virulence, and fungal development (*51*). *A. flavus* genome is predicted to contain 25 polyketide synthases (PKS), 18 non-ribosomal peptide synthases (NRPS), 77 ABC transporters, more than 125 cytochrome P450 monooxygenases, and enzymes such as fatty acid synthases (FAS), carboxylases, dehydrogenases, reductases, oxidases, oxidoreductases, epoxide hydrolases, oxygenases, and methyltransferases (*21*). Within the aflatoxin biosynthetic pathway gene cluster there is a single gene encoding the PKS and at least 5 genes encoding cytochrome P450 monooxygenases. No other PKS is known to be involved in aflatoxin biosynthesis. Other categories of genes potentially involved in aflatoxin production are genes for global regulation, signal transduction, pathogenicity, virulence, oxidative stress, and fungal development. The genes for mitogen-activated protein kinase (MAPK), MAPK kinase (MAPKK) and MAPKK kinase (MAPKKK) in stress responses could be good candidates involved in global regulation (*57*, *58*).

Functional genomics for studying the regulatory mechanism of aflatoxin formation

Acquiring the large number of gene sequences facilitates the construction of genechip or microarray. Microarrays are robust tools used for functional genomics studies. In recent years, several formats of *A. flavus* microarrays have been constructed. The first spot gene array consisting of 753 gene features, including known aflatoxin pathway genes *aflD* (*nor-1*) and *aflP* (*omtA*) and regulatory gene *aflR*, was constructed by the laboratory of Gary Payne, North Carolina State University (*59*). A 5,002 gene-elements *A. flavus* amplicon microarray was constructed at TIGR by the Food and Feed Safety Research Unit, USDA/ARS, Southern Regional Research Center. This microarray has been updated to a 5,031 gene-element array including newly identified genes of interest. A comprehensive whole genome *A. flavus* oligo microarray has also been constructed at TIGR by the Food and Feed Safety Research Unit, USDA/ARS, Southern Regional Research Center. All of the 11,820 *A. flavus* unique genes, 278 unique genes present in *A. oryzae* but absent in *A. flavus*, and 10 genes cloned from corn that show resistance against *A. flavus* infection, have been spotted on this whole genome microarray. An additional Affymetrix GeneChip microarray funded by a grant from USDA/NRI awarded to a Consortium was designed and constructed by Affymetrix Company. This Affymetrix array contains all of the *A. flavus* genes, *A. oryzae* unique genes, plus additional genes of interest from corn, *Fusarium* species, mouse and human genomes. These *A. flavus* microarray resources provide a platform for functional genomic studies in the fungus and promise a bright future for the elimination of mycotoxins in the food chain. Gene profiling using these microarrays, performed at the labs of USDA, North Carolina State University, and JCVI, has thus far, identified hundreds of genes that are significantly up or down regulated under various growth conditions in the fungus (*49*, *58-66*). These genes could be important in aflatoxin formation.

Gene expression studies on an *aflR* disruptant mutant compared with wild type using microarrays in *A. parasiticus* SRRC 143 (SU-1) identified 23 highly expressed genes in the wild type (*61*). Eighteen of the genes are known aflatoxin biosynthetic genes, three of these genes (*hypB*, *aflY*, and *nadA*) are in or adjacent to the established aflatoxin gene cluster (*14*, *21*), and the last two genes (*hlyC* and *niiA*) are located outside the aflatoxin gene cluster. All of the aflatoxin biosynthetic genes have a putative consensus AflR binding site (5 - TCGSWNNSCGR-3) from a typical 100-300 bp (*41*, *42*) up to 2.3 kb upstream in their promoter regions, except for *aflR*. The *niiA* gene is a member of the nitrogen assimilation gene cluster whch is divergently transcribed with *niaD*. While the *hlyC* gene is a homolog of alpha-hemolysin gene from *Aeromonas hydrophila*, located approximately 1.5 Mb from the aflatoxin gene cluster and has a putative AflR binding site approximately 1.8 kb upstream of the putative coding region. *hlyC* has no apparent role in aflatoxin production but may play a role in animal pathogenesis by Aspergilli. This is the first evidence that AflR regulates genes outside the cluster.

Aflatoxin production is affected by temperature through altering the transcriptional profile in *A. flavus* (*63*). Temperature shift from 28 C to 37 C of an aflatoxin producing culture quickly stops aflatoxin biosynthesis. This implies the regulation or inactivation of one or more of the pathway enzymes at post-translational level at 37 C. Majority of genes are more highly expressed at 28 C relative to 37 C. Transcript profiling at 28 C vs. 37 C using microarray identified a total of 144 genes differentially expressed. Of these, 103 were more highly expressed at 28 C, approximately 25% of the 103 genes were found likely to be involved in secondary metabolism. Most of the aflatoxin genes were more highly expressed at 28 C relative to 37 C. However the regulatory genes *aflS*, *aflR* and *aflR* antisense were relatively constant at both temperature conditions. Data suggest that the failure to produce aflatoxin at 37 C is not due to the effect of temperature on the transcription of the pathway regulatory genes, *aflR* and *aflS*. One explanation is that less AFLR may be produced at 37 C. Another possibility is that AFLR is non-functional at higher temperatures or AFLS and AFLR are unable to interact to function normally (*46*). There is also a possibility that other factors in addition to the non-functionality of AFLR affect aflatoxin production at elevated temperatures. One or more of the pathway enzymes could become unstable under higher temperature.

Aflatoxin formation is also modulated by carbon source. In *A. parasiticus* when the culture was shifted from a low sugar medium, Yeast Extract (YE), to a high sugar medium, Yeast Extract Sucrose (YES) caused temporary reduction in aflatoxin production and then increased to a 10-fold at 24 hrs time point. Microarray experiment identified a total of 2,120 highly expressed genes including most of the aflatoxin pathway genes, the sugar cluster genes, and the putative transport gene. Of the 2120 genes, 56 of them were found to be significant. Reduction in aflatoxin production is accompanied by a decrease in transcript levels of *aflD*, *aflE*, and *aflF* (*nor-1*, *norA*, and *norB*). The aflatoxin pathway genes *aflD* (*nor1*), *aflO* (*omtB*), and *aflP* (*omtA*) were consistently correlated with the aflatoxin production, but not the *aflS* (*aflJ*), *aflR*, and *aflR* antisense sequence (*65*).

The regulation of aflatoxin biosynthesis is different in *A. flavus* NRRL 3357 vs. *A. parasticus* SRRC 143, in response to individual amino acid, such as tryptophan and tyrosine (*64*). YES media supplemented with 50 mM tryptophan was found to significantly reduce aflatoxin B_1 and B_2 biosynthesis in *A. flavus,* but significantly increased B_1 and G_1 biosynthesis in *A. parasiticus*. Tyrosine has the opposite effect of tryptophan resulting in increased aflatoxin production. YES medium supplemented with 50 mM tyrosine doubled or tripled aflatoxin B_1 and B_2 biosynthesis in *A. flavus* and aflatoxins B_1, B_2 and G_2 biosynthesis in *A. parasiticus*, but reduced aflatoxin G_1 production. Further studies using these microarray resources for a genome-wide gene profiling and functional analysis in relation to aflatoxin formation will surely reveal the secret of aflatoxin production and the regulation. The knowledge will help researchers to find effective strategies in controling aflatoxin contamination of food and feed.

Comparative genomics in studying fungal evolution

Several *Aspergillus* genomes have been sequenced. These include *A. fumigatus* (*55*), *Neosartorya ficheri* (anamorph *A. fisheri*), *A. oryzae* (*53*), *A. nidulans* (*54*), and *A. niger* (*67*). This will facilitate research on basic biology, infection mechanism, host-fungus interaction, mycotoxin synthesis, genetic regulation, and evolution of these *Aspergillus* species through comparative genomic studies of these closely related *Aspergillus* species. Due to the high degree of similarity between *A. flavus* and *A. oryzae*, comparative studies on the two genomes becomes the major focus of scientists not only on aflatoxin elimination but food fermentation. There is a high degree of DNA correspondence and synteny between these two species. Each species has an entire aflatoxin biosynthetic gene cluster located near the telomere on chromosome III. The cluster in *A. flavus* resides within 82 kb region at the end of the chromosome III. The major physical difference observed between the two species is a translocation event in *A. flavus* between chromosomes II and VI. The break site is associated with a family of uncharacterized repeat elements. Other major differences between the two genomes are the unique genes of about 300 in each species and a large deletion in *A. flavus* that spans 250 kb.

References

1. Jelinek, C. F.; Pohland, A. E.; Wood, G. E. Worldwide occurrence of mycotoxins in food and feeds--an update. *J. Assoc. Off. Anal. Chem.* **1989,** *72*, 223-230.
2. Denning, D. W. Invasive aspergillosis. *Clin. Infect. Dis.* **1998,** *26*, 781-805.
3. Denning, D. W.; Riniotis, K.; Dobrashian, R.; Sambatakou, H. Chronic cavitary and fibrosing pulmonary and pleural aspergillosis: Case series, proposed nomenclature and review. *Clin. Infect. Dis.* **2003,** *37(Suppl 3)*, S265-280.
4. Klich, M. A. Soil fungi of some low-altitude desert cotton fields and ability of their extracts to inhibit aspergillus flavus. *Mycopathologia* **1998,** *142*, 97-100.
5. Richard, J. L.; Payne, G. A. In *Mycotoxins: Risks in Plant, Animal and Human Systems*, Council for Agricultural Science and Technology (CAST), Ames, IA, Ames, IA, 2003.
6. Robens, J. F. The costs of mycotoxin management to the USA: Management of aflatoxins in the United States. *APSnet Feature* *http://www.apsnet.org/online/feature/mycotoxin/top.html* **2001**, 2-8.
7. Robens, J. F.; Cardwell, K. The cost of mycotoxin management in the United States. In *Aflatoxin and Food Safety*, Abbas, H. K., Ed. CRC Press: Boca Raton, FL, 2005; pp 1-12.
8. Fung, F.; Clark, R. F. Health effects of mycotoxins: a toxicological overview. *J. Toxicol. Clin. Toxicol.* **2004,** *42*, 217-234.

9. Lewis, L.; Onsongo, M.; Njapau, H.; Schurz-Rogers, H.; Luber, G.; Kieszak, S.; Nyamongo, J.; Backer, L.; Dahiye, A. M.; Misore, A.; DeCock, K.; Rubin, C. Aflatoxin contamination of commercial maize products during an outbreak of acute aflatoxicosis in eastern and central Kenya. *Environ. Health Perspect.* **2005,** *113*, 1763-1767.
10. Bressac, B.; Kew, M.; Wands, J.; Ozturk, M. Selective G to T mutations of p53 gene in hepatocellular carcinoma from southern Africa. *Nature* **1991,** *350*, 429-431.
11. Wogan, G. N. Aflatoxins as risk factors for hepatocellular carcinoma in humans. *Cancer Res.* **1992,** *52*, 2114-2118.
12. Bennett, J. W.; Klich, M. Mycotoxins. *Clin. Microbiol. Rev.* **2003,** *16*, 497-516.
13. Van Egmond, H. P. Current situation on regulations for mycotoxins. Overview of tolerances and status of standard methods of sampling and analysis. *Food Addit. Contam.* **1989,** *6*, 139-188.
14. Yu, J.; Bhatnagar, D.; Cleveland, T. E. Completed sequence of aflatoxin pathway gene cluster in *Aspergillus parasiticus*. *FEBS Lett.* **2004,** *564*, 126-130.
15. Lee, L. S.; Bennett, J. W.; Goldblatt, L. A.; Lundin, R. E. Norsolorinic acid from a mutant strain of *Aspergillus parasiticus*. *J. American Oil Chemists Society* **1971,** *48*, 93-94.
16. Bennett, J. W.; Lee, L. S.; Vinnett, C. H. The correlation of aflatoxin and norsolorinic acid production. *J. Am.Oil Chem. Soc.* **1971,** *48*, 368-370.
17. Bennett, J. W.; Kronberg, F.; Gougis, G. Pigmented isolates from anthraquinone-producing mutants of *Aspergillus parasiticus*. *Am. Soc. Microbiol.* **1976,** *76*, 6.
18. Barnes, S. E.; Dola, T. P.; Bennett, J. W.; Bhatnagar, D. Synthesis of sterigmatocystin on a chemically defined medium by species of *Aspergillus* and *Chaetomium*. *Mycopathologia* **1994,** *125*, 173-178.
19. Bennett, J. W.; Chang, P. K.; Bhatnagar, D. One gene to whole pathway: the role of norsolorinic acid in aflatoxin research. *Adv. Appl. Microbiol.* **1997,** *45*, 1-15.
20. Minto, R. E.; Townsend, C. A. Enzymology and molecular biology of aflatoxin biosynthesis. *Chem. Rev.* **1997,** *97*, 2537-2556.
21. Yu, J.; Chang, P. K.; Ehrlich, K. C.; Cary, J. W.; Bhatnagar, D.; Cleveland, T. E.; Payne, G. A.; Linz, J. E.; Woloshuk, C. P.; Bennett, J. W. Clustered pathway genes in aflatoxin biosynthesis. *Appl. Environ. Microbiol.* **2004,** *70*, 1253-1262.
22. Bhatnagar, D.; Ehrlich, K. C.; Cleveland, T. E. Oxidation-reduction reactions in biosynthesis of secondary metabolites. In *Mycotoxins in Ecological Systems*, Bhatnagar, D.; Lillehoj, E. B.; Arora, D. K., Eds. Marcel Dekker: New York, 1992; Vol. 10, pp 255-285.
23. Townsend, C. A. Progress towards a biosynthetic rationale of the aflatoxin pathway. *Pure Appl. Chem.* **1997,** *58,* 227-238.
24. Yabe, K.; Chihaya, N.; Hamamatsu, S.; Sakuno, E.; Hamasaki, T.; Nakajima, H.; Bennett, J. W. Enzymatic conversion of averufin to

hydroxyversicolorone and elucidation of a novel metabolic grid involved in aflatoxin biosynthesis. *Appl. Environ. Microbiol.* **2003,** *69*, 66-73.

25. Watanabe, C. M.; Townsend, C. A. Initial characterization of a type I fatty acid synthase and polyketide synthase multienzyme complex NorS in the biosynthesis of aflatoxin B_1. *Chem. Biol.* **2002,** *9*, 981-988.
26. Bennett, J. W.; Silverstein, R. B.; Kruger, S. J. Isolation and characterization of two nonaflatoxigenic classes of morphological variants of *Aspergillus parasiticus*. *J. Am. Oil Chem. Soc.* **1981**, *58*, 952-955.
27. Yu, J.; Chang, P. K.; Ehrlich, K. C.; Cary, J. W.; Montalbano, B.; Dyer, J. M.; Bhatnagar, D.; Cleveland, T. E. Characterization of the critical amino acids of an *Aspergillus parasiticus* cytochrome P-450 monooxygenase encoded by *ordA* that is involved in the biosynthesis of aflatoxins B_1, G_1, B_2, and G_2. *Appl. Environ. Microbiol.* **1998,** *64*, 4834-4841.
28. Cai, J.; Zeng, H.; Shima, Y.; Hatabayashi, H.; Nakagawa, H.; Ito, Y.; Adachi, Y.; Nakajima, H.; Yabe, K. Involvement of the nadA gene in formation of G-group aflatoxins in Aspergillus parasiticus. *Fungal Genet. Biol.* **2008,** *45*, 1081-1093.
29. Trail, F.; Mahanti, N.; Rarick, M.; Mehigh, R.; Liang, S. H.; Zhou, R.; Linz, J. E. Physical and transcriptional map of an aflatoxin gene cluster in *Aspergillus parasiticus* and functional disruption of a gene involved early in the aflatoxin pathway. *Appl. Environ. Microbiol.* **1995,** *61*, 2665-2673.
30. Skory, C. D.; Chang, P. K.; Linz, J. E. Regulated expression of the *nor-1* and *ver-1* genes associated with aflatoxin biosynthesis. *Appl. Environ. Microbiol.* **1993,** *59*, 1642-1646.
31. Yu, J.; Chang, P. K.; Cary, J. W.; Wright, M.; Bhatnagar, D.; Cleveland, T. E.; Payne, G. A.; Linz, J. E. Comparative mapping of aflatoxin pathway gene clusters in *Aspergillus parasiticus* and *Aspergillus flavus*. *Appl. Environ. Microbiol.* **1995,** *61*, 2365-2371.
32. Chang, P. K.; Yu, J. Characterization of a partial duplication of the aflatoxin gene cluster in *Aspergillus parasiticus* ATCC 56775. *Appl. Microbiol. Biotechnol.* **2002,** *58*, 632-636.
33. Liang, S. H.; Skory, C. D.; Linz, J. E. Characterization of the function of the ver-1A and ver-1B genes, involved in aflatoxin biosynthesis in *Aspergillus parasiticus*. *Appl. Environ. Microbiol.* **1996,** *62*, 4568-4575.
34. Chang, P. K.; Cary, J. W.; Bhatnagar, D.; Cleveland, T. E.; Bennett, J. W.; Linz, J. E.; Woloshuk, C. P.; Payne, G. A. Cloning of the *Aspergillus parasiticus apa-2* gene associated with the regulation of aflatoxin biosynthesis. *Appl. Environ. Microbiol.* **1993,** *59*, 3273-3279.
35. Payne, G. A.; Nystrom, G. J.; Bhatnagar, D.; Cleveland, T. E.; Woloshuk, C. P. Cloning of the *afl-2* gene involved in aflatoxin biosynthesis from *Aspergillus flavus*. *Appl. Environ. Microbiol.* **1993,** *59*, 156-162.
36. Chang, P. K.; Ehrlich, K. C.; Yu, J.; Bhatnagar, D.; Cleveland, T. E. Increased expression of *Aspergillus parasiticus aflR*, encoding a sequence-specific DNA-binding protein, relieves nitrate inhibition of aflatoxin biosynthesis. *Appl. Environ. Microbiol.* **1995,** *61*, 2372-2377.
37. Chang, P. K.; Yu, J.; Bhatnagar, D.; Cleveland, T. E. Repressor-AFLR interaction modulates aflatoxin biosynthesis in *Aspergillus parasiticus*. *Mycopathologia* **1999,** *147*, 105-112.

38. Chang, P. K.; Yu, J.; Bhatnagar, D.; Cleveland, T. E. The carboxy-terminal portion of the aflatoxin pathway regulatory protein AFLR of *Aspergillus parasiticus* activates *GAL1::lacZ* gene expression in *Saccharomyces cerevisiae*. *Appl. Environ. Microbiol.* **1999,** *65*, 2508-2512.
39. Woloshuk, C. P.; Foutz, K. R.; Brewer, J. F.; Bhatnagar, D.; Cleveland, T. E.; Payne, G. A. Molecular characterization of *aflR*, a regulatory locus for aflatoxin biosynthesis. *Appl. Environ. Microbiol.* **1994,** *60*, 2408-2414.
40. Ehrlich, K. C.; Montalbano, B. G.; Cary, J. W. Binding of the C6-zinc cluster protein, AFLR, to the promoters of aflatoxin pathway biosynthesis genes in *Aspergillus parasiticus*. *Gene* **1999,** *230*, 249-257.
41. Ehrlich, K. C.; Cary, J. W.; Montalbano, B. G. Characterization of the promoter for the gene encoding the aflatoxin biosynthetic pathway regulatory protein AFLR. *Biochim. Biophys. Acta.* **1999,** *1444*, 412-417.
42. Fernandes, M.; Keller, N. P.; Adams, T. H. Sequence-specific binding by *Aspergillus nidulans* AflR, a C6 zinc cluster protein regulating mycotoxin biosynthesis. *Mol. Microbiol.* **1998,** *28*, 1355-1365.
43. Matsushima, K.; Chang, P. K.; Yu, J.; Abe, K.; Bhatnagar, D.; Cleveland, T. E. Pre-termination in *aflR* of *Aspergillus sojae* inhibits aflatoxin biosynthesis. *Appl. Microbiol. Biotechnol.* **2001,** *55*, 585-589.
44. Takahashi, T.; Chang, P. K.; Matsushima, K.; Yu, J.; Abe, K.; Bhatnagar, D.; Cleveland, T. E.; Koyama, Y. Nonfunctionality of *Aspergillus sojae aflR* in a strain of *Aspergillus parasiticus* with a disrupted *aflR* gene. *Appl. Environ. Microbiol.* **2002,** *68*, 3737-3743.
45. Meyers, D. M.; Obrian, G.; Du, W. L.; Bhatnagar, D.; Payne, G. A. Characterization of *aflJ*, a gene required for conversion of pathway intermediates to aflatoxin. *Appl. Environ. Microbiol.* **1998,** *64*, 3713-3717.
46. Chang, P. K. The *Aspergillus parasiticus* protein AFLJ interacts with the aflatoxin pathway-specific regulator AFLR. *Mol. Genet. Genomics* **2003,** *268*, 711-719.
47. Bok, J.-W.; Keller, N. P. LaeA, a regulator of secondary metabolism in *Aspergillus* spp. *Eukaryot. Cell* **2004,** *3*, 527-535.
48. Butchko, R. A.; Adams, T. H.; Keller, N. P. *Aspergillus nidulans* mutants defective in *stc* gene cluster regulation. *Genetics* **1999,** *153*, 715-720.
49. Cary, J. W.; OBrian, G. R.; Nielsen, D. M.; Nierman, W. C.; Harris-Coward, P.; Yu, J.; Bhatnagar, D.; Cleveland, T. E.; Payne, G. A.; Calvo, A. M. Elucidation of *veA*-dependent genes associated with aflatoxin and sclerotial production in *Aspergillus flavus* by functional genomics. *Appl. Microbiol. Biotechnol.* **2007,** *76*, 1107-1118.
50. Bayram, O.; Krappmann, S.; Ni, M.; Bok, J. W.; Helmstaedt, K.; Valerius, O.; Braus-Stromeyer, S.; Kwon, N. J.; Keller, N. P.; Yu, J. H.; Braus, G. H. VelB/VeA/LaeA complex coordinates light signal with fungal development and secondary metabolism. *Science* **2008,** *320*, 1504-1506.
51. Yu, J.; Whitelaw, C. A.; Nierman, W. C.; Bhatnagar, D.; Cleveland, T. E. *Aspergillus flavus* expressed sequence tags for identification of genes with putative roles in aflatoxin contamination of crops. *FEMS Microbiol. Lett.* **2004,** *237*, 333-340.
52. Payne, G. A.; Yu, J.; Nierman, W. C.; Machida, M.; Bhatnagar, D.; Cleveland, T. E.; Dean, R. A. A first glance into the genome sequence of

Aspergillus flavus. In *The ASPERGILLI: Genomics, Medical Aspects, Biotechnology, and Research Methods*, Osmani, S.; Goldman, G., Eds. CRC Press: Boca Raton, FL, 2007; Vol. 26, pp 15-23.

53. Machida, M.; Asai, K.; Sano, M.; Tanaka, T.; Kumagai, T.; Terai, G.; Kusumoto, K.; Arima, T.; Akita, O.; Kashiwagi, Y.; Abe, K.; Gomi, K.; Horiuchi, H.; Kitamoto, K.; Kobayashi, T.; Takeuchi, M.; Denning, D. W.; Galagan, J. E.; Nierman, W. C.; Yu, J.; Archer, D. B.; Bennett, J. W.; Bhatnagar, D.; Cleveland, T. E.; Fedorova, N. D.; Gotoh, O.; Horikawa, H.; Hosoyama, A.; Ichinomiya, M.; Igarashi, R.; Iwashita, K.; Juvvadi, P. R.; Kato, M.; Kato, Y.; Kin, T.; Kokubun, A.; Maeda, H.; Maeyama, N.; Maruyama, J.; Nagasaki, H.; Nakajima, T.; Oda, K.; Okada, K.; Paulsen, I.; Sakamoto, K.; Sawano, T.; Takahashi, M.; Takase, K.; Terabayashi, Y.; Wortman, J. R.; Yamada, O.; Yamagata, Y.; Anazawa, H.; Hata, Y.; Koide, Y.; Komori, T.; Koyama, Y.; Minetoki, T.; Suharnan, S.; Tanaka, A.; Isono, K.; Kuhara, S.; Ogasawara, N.; Kikuchi, H. Genome sequencing and analysis of *Aspergillus oryzae*. *Nature* **2005,** *438*, 1157-1161.
54. Galagan, J. E.; Calvo, S. E.; Cuomo, C.; Ma, L. J.; Wortman, J. R.; Batzoglou, S.; Lee, S. I.; Basturkmen, M.; Spevak, C. C.; Clutterbuck, J.; Kapitonov, V.; Jurka, J.; Scazzocchio, C.; Farman, M.; Butler, J.; Purcell, S.; Harris, S.; Braus, G. H.; Draht, O.; Busch, S.; D'Enfert, C.; Bouchier, C.; Goldman, G. H.; Bell-Pedersen, D.; Griffiths-Jones, S.; Doonan, J. H.; Yu, J.; Vienken, K.; Pain, A.; Freitag, M.; Selker, E. U.; Archer, D. B.; Penalva, M. A.; Oakley, B. R.; Momany, M.; Tanaka, T.; Kumagai, T.; Asai, K.; Machida, M.; Nierman, W. C.; Denning, D. W.; Caddick, M.; Hynes, M.; Paoletti, M.; Fischer, R.; Miller, B.; Dyer, P.; Sachs, M. S.; Osmani, S. A.; Birren, B. W. Sequencing of *Aspergillus nidulans* and comparative analysis with *A. fumigatus* and *A. oryzae*. *Nature* **2005,** *438*, 1105-1115.
55. Nierman, W. C.; Pain, A.; Anderson, M. J.; Wortman, J. R.; Kim, H. S.; Arroyo, J.; Berriman, M.; Abe, K.; Archer, D. B.; Bermejo, C.; Bennett, J.; Bowyer, P.; Chen, D.; Collins, M.; Coulsen, R.; Davies, R.; Dyer, P. S.; Farman, M.; Fedorova, N.; Fedorova, N.; Feldblyum, T. V.; Fischer, R.; Fosker, N.; Fraser, A.; Garcia, J. L.; Garcia, M. J.; Goble, A.; Goldman, G. H.; Gomi, K.; Griffith-Jones, S.; Gwilliam, R.; Haas, B.; Haas, H.; Harris, D.; Horiuchi, H.; Huang, J.; Humphray, S.; Jimenez, J.; Keller, N.; Khouri, H.; Kitamoto, K.; Kobayashi, T.; Konzack, S.; Kulkarni, R.; Kumagai, T.; Lafon, A.; Latge, J. P.; Li, W.; Lord, A.; Lu, C.; Majoros, W. H.; May, G. S.; Miller, B. L.; Mohamoud, Y.; Molina, M.; Monod, M.; Mouyna, I.; Mulligan, S.; Murphy, L.; O'Neil, S.; Paulsen, I.; Penalva, M. A.; Pertea, M.; Price, C.; Pritchard, B. L.; Quail, M. A.; Rabbinowitsch, E.; Rawlins, N.; Rajandream, M. A.; Reichard, U.; Renauld, H.; Robson, G. D.; Rodriguez de Cordoba, S.; Rodriguez-Pena, J. M.; Ronning, C. M.; Rutter, S.; Salzberg, S. L.; Sanchez, M.; Sanchez-Ferrero, J. C.; Saunders, D.; Seeger, K.; Squares, R.; Squares, S.; Takeuchi, M.; Tekaia, F.; Turner, G.; Vazquez de Aldana, C. R.; Weidman, J.; White, O.; Woodward, J.; Yu, J. H.; Fraser, C.; Galagan, J. E.; Asai, K.; Machida, M.; Hall, N.; Barrell, B.; Denning, D. W. Genomic sequence of the pathogenic and allergenic filamentous fungus *Aspergillus fumigatus*. *Nature* **2005,** *438*, 1151-1156.

56. Yu, J.; Cleveland, T. E. *Aspergillus flavus* genomics for discovering genes involved in aflatoxin biosynthesis. In *Polyketides Biosynthesis, Biological Activity, and Genetic Engineering*, Rimando, A. M.; Baerson, S. R., Eds. American Chemical Society: Washington, DC, 2007; Vol. 955, pp 246-260.
57. Kim, J. H.; Yu, J.; Mahoney, N.; Chan, K. L.; Molyneux, R. J.; Varga, J.; Bhatnagar, D.; Cleveland, T. E.; Nierman, W. C.; Campbell, B. C. Elucidation of the functional genomics of antioxidant-based inhibition of aflatoxin biosynthesis. *Int. J. Food Microbiol.* **2008,** *122*, 49-60.
58. Kim, J. H.; Campbell, B. C.; Molyneux, R.; Mahoney, N.; Chan, K. L.; Yu, J.; Wilkinson, J. R.; Cary, J.; Bhatnagar, D.; Cleveland, T. E. Gene targets for fungal and mycotoxin control. *Mycotoxin Res.* **2006,** *22*, 3-8.
59. OBrian, G. R.; Fakhoury, A. M.; Payne, G. A. Identification of genes differentially expressed during aflatoxin biosynthesis in *Aspergillus flavus* and *Aspergillus parasiticus*. *Fungal Genet. Biol.* **2003,** *39*, 118-127.
60. Price, M. S.; Conners, S. B.; Tachdjian, S.; Kelly, R. M.; Payne, G. A. Aflatoxin conducive and non-conducive growth conditions reveal new gene associations with aflatoxin production. *Fungal Genet. Biol.* **2005,** *42*, 506-518.
61. Price, M. S.; Yu, J.; Nierman, W. C.; Kim, H. S.; Pritchard, B.; Jacobus, C. A.; Bhatnagar, D.; Cleveland, T. E.; Payne, G. A., The aflatoxin pathway regulator AflR induces gene transcription inside and outside of the aflatoxin biosynthetic cluster. *FEMS Microbiol. Lett.* **2006,** *255*, 275-279.
62. Chang, P. K.; Wilkinson, J. W.; Horn, B. W.; Yu, J.; Bhatnagar, D.; Cleveland, T. E. Genes differentially expressed by *Aspergillus flavus* strains after loss of aflatoxin production by serial transfers. *Appl. Microbiol. Biotechnol.* **2007,** *77*, 917-925.
63. OBrian, G. R.; Georgianna, D. R.; Wilkinson, J. R.; Yu, J.; Abbas, H. K.; Bhatnagar, D.; Cleveland, T. E.; Nierman, W. C.; Payne, G. A. The effect of elevated temperature on gene transcription and aflatoxin biosynthesis. *Mycologia* **2007,** *99* (2), 232-239.
64. Wilkinson, J. R.; Yu, J.; Bland, J. M.; Nierman, W. C.; Bhatnagar, D.; Cleveland, T. E. Amino acid supplementation reveals differential regulation of aflatoxin biosynthesis in *Aspergillus flavus* NRRL 3357 and *Aspergillus parasiticus* SRRC 143. *Appl. Microbiol. Biotechnol.* **2007,** *74*, 1308-1319.
65. Wilkinson, J. R.; Yu, J.; Abbas, H. K.; Scheffler, B. E.; Kim, H. S.; Nierman, W. C.; Bhatnagar, D.; Cleveland, T. E. Aflatoxin formation and gene expression in response to carbon source media shift in *Aspergillus parasiticus*. . *Food Addit. Contam.* **2007,** *24*, 1051-1060.
66. Yu, J.; Ronning, C. M.; Wilkinson, J. R.; Campbell, B. C.; Payne, G. A.; Bhatnagar, D.; Cleveland, T. E.; Nierman, W. C. Gene profiling for studying the mechanism of aflatoxin biosynthesis in *Aspergillus flavus* and *A. parasiticus*. *Food Addit. Contam.* **2007,** *24*, 1035-1042.
67. Pel, H. J.; de Winde, J. H.; Archer, D. B.; Dyer, P. S.; Hofmann, G.; Schaap, P. J.; Turner, G.; de Vries, R. P.; Albang, R.; Albermann, K.; Andersen, M. R.; Bendtsen, J. D.; Benen, J. A.; van den Berg, M.; Breestraat, S.; Caddick, M. X.; Contreras, R.; Cornell, M.; Coutinho, P. M.; Danchin, E. G.; Debets, A. J.; Dekker, P.; van Dijck, P. W.; van Dijk, A.; Dijkhuizen, L.; Driessen, A. J.; d'Enfert, C.; Geysens, S.; Goosen, C.;

Groot, G. S.; de Groot, P. W.; Guillemette, T.; Henrissat, B.; Herweijer, M.; van den Hombergh, J. P.; van den Hondel, C. A.; van der Heijden, R. T.; van der Kaaij, R. M.; Klis, F. M.; Kools, H. J.; Kubicek, C. P.; van Kuyk, P. A.; Lauber, J.; Lu, X.; van der Maarel, M. J.; Meulenberg, R.; Menke, H.; Mortimer, M. A.; Nielsen, J.; Oliver, S. G.; Olsthoorn, M.; Pal, K.; van Peij, N. N.; Ram, A. F.; Rinas, U.; Roubos, J. A.; Sagt, C. M.; Schmoll, M.; Sun, J.; Ussery, D.; Varga, J.; Vervecken, W.; van de Vondervoort, P. J.; Wedler, H.; Wosten, H. A.; Zeng, A. P.; van Ooyen, A. J.; Visser, J.; Stam, H. Genome sequencing and analysis of the versatile cell factory *Aspergillus niger* CBS 513.88. *Nat. Biotechnol.* **2007,** *25*, 221-31.

Chapter 8

Microorganisms and their enzymes for detoxifying mycotoxins posing a risk to livestock animals

I. Rodrigues[1], E. M. Binder[2] and, G. Schatzmayr[3*]

[1]BIOMIN Holding GmbH, Industriestrasse 21, 3130 Herzogenburg, Austria
[2] ERBER AG, Technopark 1, 3430 Tulln, Austria
[3]BIOMIN Research Center, Technopark 1, 3430 Tulln, Austria

Occurrence of mycotoxins is ubiquitous. Even with the use of prevention techniques, it is virtually impossible to avoid their presence in agricultural commodities. The toxicity of these fungal metabolites brings serious risks upon humans and animals. Mycotoxicoses are animal or human diseases caused by mycotoxin ingestion, inhalation or skin-contact. In animals, these range from immunosuppression and performance effects to hepatotoxic, nephrotoxic, neurotoxic, dermal, carcinogenic, reproductive, teratogenic and gastro-intestinal effects depending on animal-, environmental- and toxin-related factors. A suitable mycotoxin risk management should take the different chemical structures of mycotoxins into consideration as a successful strategy for one mycotoxin, may fail in the elimination of another. Biotransformation and biodegradation are mycotoxin-specific methods which rely in microorganisms and enzymes' capacity of metabolization or degradation of mycotoxins into less or non-toxic metabolites prior to their resorption in the gastro-intestinal tract. Some microorganisms have shown biotransformation capacity both *in vitro* and *in vivo*, representing effective mycotoxin risk management tools in animal feed.

Although mycotoxin-related problems had been already described in medieval times, when people affected by ergot-alkaloids exhibited swollen members with burning sensations, with subsequent necrosis and loss of the extremities, modern mycotoxicology begun quite recently with the discovery of aflatoxins in the early 1960s (1). Turkey-X disease was responsible for the death of more than 100 000 young turkeys in poultry farms in England and followed the consumption of groundnuts infected with *Aspergillus flavus*. Ever since then, the impact of mycotoxins in human and animal health has been the focus of many scientific studies and several reports are available concerning prevention, decontamination and minimization of mycotoxins risk (2, 3).

Unlike primary metabolites, which are essential for fungi growth, mycotoxins are secondary metabolites produced by filamentous fungi (moulds) in the final stages of exponential growth phase (4). These chemical compounds are toxic to vertebrates upon ingestion, inhalation and/or dermal contact. Reports by Hawksworth (5,6) describe the existence of more than 70,000 species of fungi; however, this number might represent only 5% of the world's total fungal species, estimated to be 1.5 million. Nevertheless, the majority of the known toxigenic species falls into three recognized genera: *Aspergillus*, *Penicillium* and *Fusarium (1)*. Mycotoxin contamination often begins in the field and continues throughout harvest, transportation and storage, depending on the activity and colonization levels of fungi which are in turn determined by the prevailing environmental conditions and the nutritional components of the food matrix (7). In general, fungi are divided in two main groups, field fungi and storage fungi, depending if they occur more frequently on the field or after harvest, respectively. However, even if this terminology has been commonly used, conditions for growth of a specific organism can occur in either the field or during storage (8), especially because even the same fungal genus contains species that differ greatly in their optimum temperature for growth and for their parasitic abilities. Important mycotoxins produced by *Aspergillus* fungi include aflatoxin B1, B2, G1 and G2 and ochratoxin A. The latter can also be produced by species belonging to the *Penicillium* genus. *Fusarium* mycotoxins commonly impacting the health and productivity of animals are type-A (T-2 toxin, HT-2 toxin and diacetoxyscirpenol (DAS)) and type-B (deoxynivalenol (DON) and nivalenol) trichothecenes, zearalenone (ZON) and fumonisins.

Reports on the worldwide occurrence of mycotoxins in commodities, feed and feed ingredients are available (9, 10, 11). Mycotoxins occurrence is ubiquitous, not only geographically but also in terms of commodities. European samples for example are more frequently contaminated with DON, ZON and T-2 whereas materials from Asia and the Pacific tend to be contaminated with DON, ZON, fumonisins and aflatoxins (9). Nevertheless, agricultural trade globalization and climate change might also have a role on the contamination pattern, leading to the occurrence of mycotoxins in regions where originally they would not be found at.

Approaches to prevent mycotoxicoses include pre-, harvest and post-harvest strategies. The latter comprise physical, chemical, biological methods and use of adsorbents. Pre-harvest strategies have been fastidiously reviewed by Jouany (2) and will not be object of discussion in this paper. Furthermore, pre-harvest control of mould growth is greatly compromised by the inability of man to control climate, a critical factor on mould contamination and mycotoxin production. Physical and chemical treatments of contaminated grains comprehend innumerous treatments which practical use is questionable due to the high costs associated, limited efficacy and possible formation of toxic compounds. The use of adsorptive materials has been studied and applied in animal feeds for decades as a method to bind mycotoxins thus making them unavailable for absorption by the animal. As reviewed by Huwig and team (3) the efficacy of these adsorbents against fusariotoxins such as zearalenone, fumonisins and trichothecenes is limited or near zero. This has also been confirmed by *in vivo* studies (12, 13).An efficient way of detoxification of mycotoxins that can poorly be bound is microbial or enzymatic detoxification (14).

In this chapter, the latest advances regarding these biological methods for detoxification of mycotoxins will be thoroughly discussed.

Biotransformation and biodegradation of mycotoxins

Biotransformation stands for the conversion of mycotoxins into less toxic molecules by enzymes or microorganisms. As in the case of physical adsorption, this degradation takes place in the gastro-intestinal tract of the animal consuming mycotoxin contaminated feed.

Initial research in this field remounts to 40 years ago, when Ciegler *et al* (15) isolated the first bacteria with aflatoxin-detoxification capability. However, for this mycotoxin, the application of the abovementioned adsorbents was reported to result in almost total protection against aflatoxicosis (3). After that, several research reports about the degradation of other mycotoxins, such as ochratoxins, trichothecenes, zearalenone and fumonisins, were published but for only a few of them the practical use as feed additives has been possible.

Trichothecenes biotransformation and non-toxicity of the new metabolite

Trichothecenes are a group of more than 200 sesquiterpenes characterized by the 12,13-epoxy-trichothec-9-ene ring system. The epoxy moiety possessed by all trichothecenes has been shown to play a crucial role in their toxicity (14, 16).

In comparison to monogastric species, ruminants are generally considered to be less susceptible to the adverse effects caused by contamination of feeds with mycotoxins (17). Based on this knowledge, many detoxification studies on trichothecenes were conducted in the past two decades using rumen fluid (18, 19, 20). First detoxification studies used mixed cultures of anaerobic microorganisms; however, the single strain capable of removing (de-epoxidation the 12,13-epoxy-group to form a double bond was only later on described (21). The gram-positive non-motile, non-spore forming, strictly anaerobic novel bacteria strain from *Eubacterium* sp. (*Eubacterium* BBSH 797) was able to transform DON into de-epoxy-DON (DOM-1), a metabolite first described by Yoshizawa et al (22). For the use of BBSH 797 as a feed additive, the fermentation and stabilization processes were optimized with respect to fast growth of the microbe and high biotransformation activity of the resulting product. For enhancement of stability during storage and within the gastrointestinal tract, a three-step encapsulation process was implemented. The toxicity of this new compound was tested using a chicken lymphocyte proliferation assay (LPA) (14). At a DON concentration of 0.15 μg/mL proliferation of lymphocytes was lower in comparison with the control.After adding 0.3 μg/mL to the cells, only one third of them could proliferate whereas at a concentration of 0.63 μg/mL DON the growth of lymphocytes stopped. In the case of DOM-1 only a concentration of 116 μg/mL inhibited proliferation of lymphocyte cells completely. If the concentration values at which the growth of the lymphocytes stopped are compared, it is possible to infer that DOM-1 toxicity is approximately 200 times lower than the parent compound DON. These results are in accordance with previous studies where the biotransformation of DON to de-epoxy-DON had already shown significant loss of cytotoxic activity by MTT (3-(4,5-dimethylthiazol-2-yl)-2,5-diphenyltetrazolium bromide) cell-culture test (23).

Ochratoxin A biodegradation and non-toxicity of the new metabolite

Ochratoxin A consists of a dihydroisocoumarin moiety (the pentaketide-derived ochratoxin α) linked through the carboxyl group to phenylalanine (24). After its discovery in 1965 by van der Merwe *et al* (25) several experiments followed with the aim of finding a suitable microorganism for its detoxification. The Gram-positive soil bacterium *Phenylobacterium immobile* was shown to degrade ochratoxin A into a less toxic metabolite (26). Later on, *Acinetobacter calcoaceticus* was found to have the same activity by cleaving the amide bond, thus transforming OTA into ochratoxin alpha (OTα) (27). Other reports show that filamentous fungi from *Aspergillus* and *Rhizopus* genera are able to degrade OTA. While A. *niger* could degrade more than 90% of OTA after a 6-day incubation, the *Rhizopus* isolates could degrade about 90% of OTA in about 12 days (28). Despite these scientific findings, a suitable mycotoxin-degrading microorganism with practical application in mycotoxin-contaminated feeds was only later on discovered. Following a screening of more than 20 OTA-cleaving microorganisms, the yeast strain associated with the hindgut of lower termites, *Trichosporon mycotoxinivorans MTV* was isolated and described by Schatzmayr *et al* (29, 30). OTA was incubated with several OTA-degrading isolates (31)and their degradation rate was compared with that of *Phenylobacterium immobile* (26). The fastest degraders belonged to the genera of *Stenotrophomonas* and *Trichosporon.* Anaerobic isolates were slower in OTA deactivation (complete degradation was detected after 20 hours of incubation). In this experiment *Phenylobacterium immobile* was the "slowest" strain in terms of OTA-cleavage (31). After being stabilized and applied as lyophilized powder in the feed, the mycotoxin-degrading strain has to regain activity in the gastrointestinal (GI) tract very rapidly, since time available for detoxification is very limited. Therefore, growth and mycotoxin-degradation activity of *Trichosporon mycotoxinivorans MTV* was optimized. Lyophilized cells grown in an optimized culture medium were able to degrade OTA (200 μg/L) completely into OTα within 1 hour of incubation, whereas cells grown in a standard yeast medium needed more than 5 hours. Studies were developed concerning the toxicity on the degradation metabolite, OTα (14). Growth of macrophages was depressed from 0.741 to 2.222 μg/mL of OTA. At concentrations above 6.667 μg/mL of OTA their growth was completely inhibited. Contrarily, concentrations up to 20 μg/mL of OTα did not affect macrophages growth (14). These results were in accordance with those of other scientific studies were OTα was shown to be non-toxic or at least 500 times less toxic than OTA (32, 33).

Zearalenone biodegradation and non-toxicity of the new metabolite

Toxicity of zearalenone relies on the molecule's resemblance with the sexual female hormone oestradiol, which enables it to couple with the oestrogenic receptors acting as an oestrogen agonist in the brain resulting in severe effects on the reproductive system (34). The history of zearalenone detoxification attempts began in 1988 when the fungus *Clonostachys rosea (*also known as *Gliocladium roseum)* was reported to open its lactone ring thus changing its structure and removing its ability to bind to the oestrogenic-binding sites. Two microorganisms, *Rhodococcus erythropolis* and *Norcardia globulera* were later on patented as ZON-degrading strains (35). A practical application of these strains has never been reported.

Following the studies made with OTA, the yeast strain *Trichosporon mycotoxinivorans MTV* was incubated with ZON and its activity screened.This led to the reduction of the toxic metabolite into carbon dioxide or into a non toxic metabolite (14). α- and β-zearalenol, which are more estrogenic, could not be detected at the end of the degradation study. The non-toxicity of the newly formed compound was later on tested by an E-screen assay, commonly used system for evaluating the ability of chemicals to induce a hormonal response, based on the ability of MCF-7 cells to proliferate in the presence of estrogens. The cell proliferation of the human estrogen-receptor-positive breast cancer cell line, MCF-7 was quantified by using the colorimetric MTT (3-(4,5-dimethylthiazol-2-yl)-2,5-diphenyltetrazolium bromide) assay. As this cell line is estrogen-dependent, by incubating it with ZON and separately with the product of ZON incubation with *Trichosporon mycotoxinivorans MTV* it was possible to compare the growth activity of both inoculates in relation to control. Following the incubation with the test compounds the medium was discarded and 0.6 mg/ml MTT solution was added to the wells. After 4 h incubation the MTT solution was discarded and the purple insoluble formazan solubilized by adding 100 μl/well of lysis buffer (0.5% sodium dodecyle sulphate, 36 mM HCl, and isopropanol acid). After mixing the optical density (OD) was measured at 594 nm using a micro plate reader (BIORAD Model 3550) equipped with a spectrophotometer (BioRad, Veenendaal, the Netherlands). Cell proliferation rate was expressed as (A_{595} treated cells/A_{595} of appropriate control) x 100.The standard curve for zearalenone (continuous line in Figure 1) in the E-screen assay indicates a log-normal cell proliferation over the concentration range from 10^{-6} to 10^{-11} Mol/L. At higher concentrations ZON becomes toxic to the cells; therefore this concentration range was adopted for the studies of the test compounds. The sample resulting from the ZON incubation with *Trichosporon mycotoxinivorans MTV* (dashed line in Figure 1) did lack the ability to induce any response indicating an estrogenic activity (remaining at baseline values), thus confirming the stable degradation of zearalenone into a non-estrogenic metabolite.

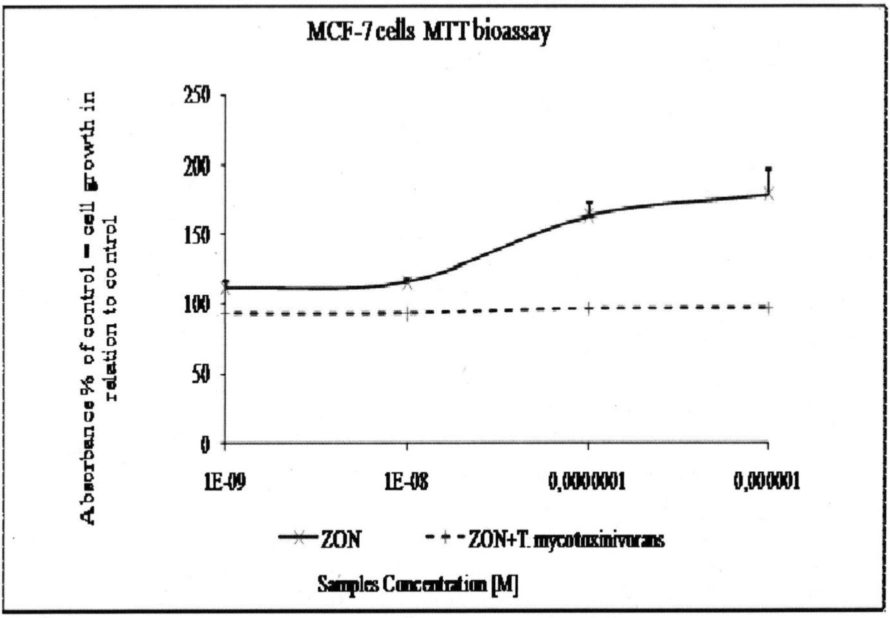

Figure 1. Results of the E-screen assay. The continuous line represents the standard proliferation curve for the incubation of zearalenone with the MCR-7 cell line. The dashed line represents the proliferation curve for the incubation of the MCR-7 cell line with the product of zearalenone and T. mycotoxinivorans incubation.

Fumonisins biodegradation

Fumonisins (FUM) are a group of *Fusarium* toxins which were isolated for the first time in 1988, from moldy maize samples originated from an area in Africa with a high incidence of esophageal cancer (36). The structural similarity of fumonisins to the sphingoid bases sphinganine and sphingosine is critical to their ability to disrupt sphingolipid metabolism (37). Sphingolipids are basically important for the membrane and lipoprotein structure and also for cell regulation and communication (second messenger for growth factors). They are found in great amounts in the brain and in nervous tissue. As in the case of other mycotoxins, also the biotransformation of fumonisins has been an object of scientific research. Two species of "black yeast" found widely in plant debris, *Exophiala spinifera* and *Rhinocladiella atrovirens* and a gram negative bacterium from stalk tissue were reported capable of metabolizing fumonisins (38). However, these microorganisms cannot be used as feed additives since they can only perform successfully if fumonisins are the only carbon source present in the media. In addition, E. *spinifera* is pathogenic and a known causal agent of mycoses. Microorganisms have been isolated from pig intestines, bovine rumen, soil and contaminated maize and screened for FUM-degrading capacity (39). Degradation of the mycotoxin was detected in soil and maize samples. The most promising strains were taxonomically characterized and further tests have been performed to screen the degradation ability in different pH-values, oxygen, temperature, toxin concentration and complex media. One strain was able to completely transform FUM at three different pH-values (5, 7 and 9). All isolates degraded FUM under aerobic conditions. When oxygen was limited, only 2 strains (#144 and 5bfII(2)b) completely degraded FUM in the liquid medium. One strain was capable of completely metabolizing the mycotoxin up to a concentration level of 100mg/L. Degradation experiments in complex media, such as artificially contaminated wheat, maize and beer, revealed two strains (# 144 and 5bfII(2)b) which were capable of completely degrading 2 mg/L FUM (Table 1).

Table 1. FUM degradation ability of FUM-degrading isolates in complex media

Strain	*FUM degradation (%)*		
	wheat	maize	beer
Negative control	0	0	22
# 135	0	0	0
# 144	100	100	100
# 151	5	0	0
# 152	0	0	2
5bfII(2)b	100	100	100
T2	100	100	0

SOURCE: Reproduced from Reference 40. Copyright 2006 Wageningen Academic Publishers. Reproduced with permission.

Confirming *in vitro* findings with *in vivo* experiments

Although *in vitro* studies represent the basis of research and product development, successful *in vivo* experiments are crucial to support the effectiveness of any creation on the field. Unfortunately, trials with mycotoxins are very difficult to run and often in controlled experiments, animals show high tolerance to contaminated feeds (17). Nonetheless, the effectiveness of the abovementioned bacteria and yeast strains has been shown in the field.Highly significant ($P<0.001$) results were obtained in piglet trials contaminated with 2.5 mg/kg DON (41). Animals supplemented with 1.24×10^6 CFU/g *Eubacterium* BBSH 797 had a feed conversion ratio (FCR) of 1.6 against the 2.0 shown by non-supplemented animals. In broilers, the mortality rate decreased from 19.4 to 5.5% when animals fed 10.5 mg/kg DON and 0.2 mg/kg acetyl-DON were supplemented with the same bacteria at an inclusion rate of 2.2×10^6 CFU/g feed. Politis *et al* (42) proved the positive impact of T. *mycotoxinivorans* on the performance and immune system of broiler chickens contaminated with 500 μg/kg of OTA. This yeast strain, at inclusion rates of 10^4, 10^5 and 10^6 CFU/g of feed blocked the detrimental effects of OTA on several immune parameters in broilers, showing to be equally effective even at different levels. In a comparative trial (43) the only feed additive capable of counteracting the adverse effects on performance caused by the dietary administration of 2 ppm T-2 toxin in broilers was Mycofix®, the one based on the enzymatic inactivation of the 12,13-epoxide ring of trichothecenes, at an inclusion rate of 2.0 g/kg of feed. More recently, the multi-organ toxicity of a combination of 1 ppm DON and 250 ppb ZON in pigs was partially or completely overcome when animals were supplemented with these microorganisms incorporated in 1.5 kg of the commercial product Mycofix® (44). Different criteria assessed included growth performance, serum biochemistry parameters, alveolar macrophage activities, antibody titers, cytokine secretion profile and histopathological observations.

Conclusions

Due to their disseminated presence and hazardous effects, mycotoxins represent a serious threat for animal and human health. Their different chemical structures however, dictate the need for different approaches for their counteraction. The specificity, irreversibility and ability of biotransformation to convert toxic molecules into non-toxic metabolites makes it an important tool in the management of mycotoxins that can be poorly bound by adsorbents. Nonetheless, even if the microorganisms show good *in vitro* results, they have to fulfill a handful of pre-requisites in order to be considered and used successfully in animal feeds. Not only they need to show a rapid degradation of the mycotoxin into less or non-toxic metabolites, but also to keep their activity at different pH values and complex environments with the presence of metabolites and nutrients. Their non-toxicity must be assured and the possibility of being applied as a lyophilisate should be granted for a practical use in animal diets.

So far few microorganisms have shown the capacity of degrading mycotoxins and from those, even less can be used safely as animal feed additives. *Eubacterium* BBSH 797, *Trichosporon mycotoxinivorans MTV* and bacterium #144 are able to biotransform trichothecenes, ochratoxin A, zearalenone and fumonisin B1 into non toxic metabolites before their absorption by the animal.

References

1. *Mycotoxins, Risks in Plan, Animal and Human Systems*; Council of Agricultural Science and Technology; Task Force Report 139; Ames, IA, 2003.
2. Jouany, J. P. Methods for preventing, decontaminating and minimizing the toxicity of mycotoxins in feeds. *Anim. Feed Sci. Tech..* **2007**, 137, 342-362
3. Huwig, A.; Freimund, S.; Käppeli, O.; Dutler, H. Mycotoxin detoxication of animal feed by different adsorbents. *Toxicol. Lett.* **2001**, 122, 179-188.

4. Jay, J. M. *Modern Food Microbiology*; 6th edition; Aspen Publication, MD, 2000.
5. Hawksworth, D. L. The fungal dimension of biodiversity: magnitude, significance, and conservation. *Mycol. Res.* **1991**, 95, 641-655.
6. Hawksworth, D. L.; Kirk, P. M.; Sutton, B. C.; Pegler, D. N. *Ainsworth and Bisby's Dictionary of the Fungi*; 8th edition. CAB International: Wallingford, United Kingdom, 1995, p 616.
7. Magan, N.; Sanchis, V.; Aldred, D. *Fungal biotechnology in Agricultural, Food and Environmental Applications*; CRC: New York, 2003; pp 311-323.
8. Santin, E. Mould Growth and Mycotoxin Production. In: *The Mycotoxin Blue Book;* Nottingham University Press, 2005; pp 225-234.
9. Binder, E. M.; Tan, L. M.; Chin, L. J.; Handl, J.; Richard, J. Worldwide occurrence of mycotoxins in commodities, feeds and feed ingredients *Anim. Feed Sci. Tech.* **2003**, 137, 265 – 282.
10. Chin, L. J.; Tan, L. M.; Wegleitner, K. Occurrence of mycotoxins in Asian feed samples. *Asian Poultry Magazine.* May, 2006, p 20.
11. Rodrigues, I. Mycotoxin contamination by region and feed type. *Feed Business Asia.* Jan/Feb, 2008, p 36.

12. Dänicke S., Goyarts T., Valenta H. On the specific and unspecific effects of a polymeric glucomannan mycotoxin adsorbent on piglets when fed with uncontaminated or with Fusarium toxins contaminated diets. *Arch. Anim. Nutr.* 2007, 61, 266 – 275.
13. Döll S., Gericke S., Dänicke S., Raila J., Ueberschär K. H., Valenta H., Schnurrbusch U., Schweigert F. J., Flachowsky G. The efficacy of a modified aluminosilicate as a detoxifying agent in Fusarium toxin contaminated maize containing diets for piglets. *J. Anim. Phys. Anim. Nutr.* 2005, 89, 342 - 358
14. Schatzmayr, G.; Zehner, F.; Schatzmayr, D.; Klimitsch, A.; Loibner, A. P.; Binder, E. M. Microbiologicals for deactivating mycotoxins. *Mol. Nutr. Food Res.* **2006**, 50, 543-551
15. Ciegler, A.; Lillehoj, E. B.; Peterson, R. E.; Hall, H. H. Microbial detoxification of aflatoxin. *Applied Microbiology*, **1966**, 14, 934 – 939.
16. He, P.; Young, L. G.; Forsberg, C. Microbial transformation of deoxynivalenol (vomitoxin). *Appl. Environ. Microb.* **1992**, 58, 3857-3863.
17. Fink-Gremmels, J. The role of mycotoxins in the health and performance of dairy cows. *Vet J*, **2008**, 176, 84-92.
18. King, R. R.; McQueen, R. E.; Levesque, D.; Greenhalgh, R. Transformation of deoxynivalenol (vomitoxin) by rumen micro- organisms. *J. Agr. Food Chem.*, **1984**, 32, 1181 – 1183
19. Westlake, K.; Mackie, R. I.; Dutton, M. F. T-2 toxin metabolism by ruminal bacteria and its effect on their growth. *Appl. Environ. Microb.* **1987**, 53, 587 – 592
20. Swanson, S. P.; Nicoletti, J.; Rood, H. D.; Buck, W. B.; Cote, L. M.; Yoshizawa, T. Metabolism of three trichothecene mycotoxins, T-2 toxin, diacetoxyscirpenol and deoxynivalenol, by bovine rumen microorganisms. *J. Chromatogr.* **1987**, 414, 335 – 342

21. Binder, J.; Schatzmayr, G.; Heidler, D.; Ellend, N.; Braun, R.; Binder, E. M. Microbial Inactivation of Trichothecenes - Development of a Feed Additive In *Proceedings of the 5th Feed Production Conference*, edited by G. Piva; Tipolitografia Benedettina: Parma, Italy, 1998; pp. 71-87.
22. Yoshizawa T., Takeda H., Ohi T.Structure of a Novel Metabolite from Deoxynivalenol, a Trichothecene Mycotoxin, in Animals. *Agric. Biol. Chem.* 1983, 47 , 2133 – 2135.
23. Kollarczik, B.; Gareis, M.; Hanelt, M. In vitro transformation of the Fusarium mycotoxins deoxynivalenol and zearalenone by the normal gut microflora of pigs. *Nat. Toxins* **1994**, 2, 105 – 110
24. Harris, J. P.; Mantle, P. G. Biosynthesis of ochratoxins by Aspergillus ochraceus. *Phytochemistry* **2001**, 58, 709 – 716
25. van der Merwe, K. J.; Steyn, P. S.; Fourie, L.; Scott, D. B.; Theron, J. Ochratoxin A, a toxic metabolite produced by Aspergillus ochraceus. *J. Nature* **1965**, 205, 1112 – 1113
26. Wegst, W.; Lingens, F. Bacterial degradation of ochratoxin A. *FEMS Microbiol. Lett.* **1983**, 17, 341 – 344
27. Hwang, C. A.; Draughon, F. A. Degradation of Ochratoxin A by Acinetobacter calcoaceticus. *J. Food Protect.* **1994**, 57, 410 – 414
28. Varga, J.; Péteri, Z.; Tábori, K.; Téren, J.; Vágvölgyi, C. Degradation of ochratoxin A and other mycotoxins by Rhizopus isolates. *Int. J. Food Microbiol.* **2005**, 99, 321 – 328
29. Schatzmayr, G.; Heidler, D.; Fuchs, E.; Mohnl, M.; Täubel, M.; Loibner, A. P.; Braun, R.; Binder, E. M. Investigation of different yeast strains for the detoxification of ochratoxin A. *Mycot. Res.* **2003**, 19, 124 - 128
30. Molnar, O.; Schatzmayr, G.; Fuchs, E.; Prillinger, H. Trichosporon mycotoxinivorans MTV sp. nov., A new yeast species useful in biological detoxification of various mycotoxins. *System. Appl. Microbiol.* **2004**, 27, 661 – 671
31. Schatzmayr, G.; Schatzmayr, D.; Pichler, E.; Täubel, M.; Loibner, A. P.; Binder, E. M.. A novel approach to deactivate ochratoxin A In: *Mycotoxins and Phycotoxins*, H. Njapau Ed.; Wageningen Academic Publishers: Wageningen, The Netherlands, 2006; pp 279-288.
32. Chu, F. S.; Noh, I.; Chang, C. C. Structural requirements for ochratoxin intoxication. *Life Sci.* **1972**, 11, 503 – 508
33. Bruinink, A.; Rasonyi, T.; Sidler, C. Differences in neurotoxic effects of ochratoxin A, ochracin and ochratoxin-α in vitro. *Nat. Toxins* **1998**, 6, 173 – 177
34. Gromadzka, K.; Waskiewicz, A.; Chelkowski, J.; Golinski, P. Zearalenone and its metabolites: occurrence, detection, toxicity and guidelines. *W. Mycot. J.* **2008**, 1, 209 – 220
35. Duvick, J.; Rood, T. Zearalenone detoxification compositions and methods. US Patent 5,846,812, 1998
36. Gelderblom,W.C.A.; Jaskiewicz, K.; Marasas,W.F.O.; Thiel, P.G.; Horak, R.M.; Vleggaar, R.; Kriek, N.P. Fumonisins--novel mycotoxins with cancer-promoting activity produced by Fusarium moniliforme.*Environ. Microbiol.* **1988**, 54, 1806–1811.

37. Voss, K. A., Smith G. W., Haschek W. M. Fumonisins: Toxicokinetics, mechanism of action and toxicity. *Anim. Feed Sci. Tech.* **2007**, 137, 299 – 325
38. Duvick, J.; Rood, T.; Maddox, J.; Gilliam, J. Detoxification of mycotoxins in planta as a strategy for improving grain quality and disease resistance: identification of fumonisin-degrading microbes from maize. *Molecular Genetics of Host-Specific Toxins in Plant Disease;* Series: Developments in Plant Pathology, 13; Proceedings of the 3rd Tottori International Symposium on Host-Specific Toxins: Daisen, Tottori, Japan, 1998, 369 – 381
39. Moll, W. D.; Täubel, M.; Kunz-Vekiru, E.; Loibner, A. P.; Braun, R.; Krska, R.; Schatzmayr, G. Biodegradation of Fumonisin B1. 30th Mycotoxin Workshop: Utrecht, Netherlands, 2008, Abstract p. 153
40. Schatzmayr, G.; Täubel, M.; Vekiru, E.; Moll, W. D.; Schatzmayr, D.; Binder, E. M.; Krska, R.; Loibner, A. P. *Detoxification of mycotoxins by biotransformation. The Mycotoxin Factbook - Food & Feed Topics*, Wageningen Academic publishers: Wageningen, Netherlands, 2006; pp. 363 – 375
41. Binder, E. M.; Heidler, D.; Schatzmayr, G.; Thimm, N.; Fuchs, E.; Schuh, M.; Krska, R.; Binder, J.; Mallmann, C. A. Microbial Detoxification of Mycotoxins in Animal Feed. In: *Proceedings of the 10th Intern. IUPAC Symposium on Mycotoxins and Phycotoxins*, de Koe, W.J.; Samson, R.A.; van Egmond, H.P.; Gilbert, J.; Sabino M., Eds. Ponden and Looijen: Wageningen, the Netherlands, pp. 271-277.
42. Politis, I.; Fegeros, K.; Nitsch, S.; Schatzmayr, G.; Kantas, D. Use of Trichosporon mycotoxinivorans MTV to suppress the effects of ochratoxicosis on the immune system of broiler chicks. *Brit. Poultry Sci.* **2005**, 46, 1 - 8
43. Diaz, G. J.; Cortés, A.; Roldán, L. Evaluation of the Efficacy of Four Feed Additives Against the Adverse Effects of T-2 Toxin in Growing Broiler Chickens. *J. Appl. Poultry Res.* **2005**, 14, 226 - 231
44. Cheng, Y. H.; Weng, C. F.; Chen, B. J.; Chang, M. H. Toxicity of different Fusarium mycotoxins on growth performance, immune responses and efficacy of a mycotoxin degrading enzyme in pigs. *Anim. Res.* **2006**, 55, 1 – 12

Chapter 9

Transgenic Expression of Lytic Peptides in Food and Feed Crops to Control Phytopathogens and Preharvest Mycotoxin Contamination

K. Rajasekaran[1], J.M. Jaynes[2], and J.W. Cary[1]

[1]Southern Regional Research Center, Agricultural Research Service, United States Department of Agriculture, New Orleans, LA 71024
[2]College of Agricultural, Environmental and Natural Sciences, Tuskegee University, Tuskegee, AL 36088.

Transgenic crops are widely cultivated in several countries to control crop losses due to insects and weeds. However, disease resistant transgenic crops that can withstand infections due to fungal and bacterial pathogens are not yet available due to several reasons. The primary reasons are 1) host plant-pathogen interaction is a very complex phenomenon and it is often crop/variety or pathogen/strain–specific.; 2) natural antimicrobial proteins and peptides are subject to digestion by proteases, lack specificity and may be toxic to non-target plant and animal species; 3) large scale production of antimicrobial proteins and peptides are very expensive and 4) microbial pathogens can develop resistance to natural proteins and peptides. Recent advances in combinatorial chemistry and automated peptide synthesis have paved the way for rational design of stable, potent, and novel synthetic peptides with target-specific biological activity. Some of these lytic synthetic peptides have been already expressed in transgenic plants with varying degrees of success towards control of phytopathogens including some fungal pathogens that produce mycotoxins. This review gives a brief account of recent developments regarding the use of lytic peptides in transgenic crops to control yield losses due to pathogens and mycotoxin contamination.

The welfare of humanity is inextricably linked with the efficient cultivation of food and feed crops. Microbial plant diseases account for more than 16% loss in agricultural production (*1*). History of agricultural civilization includes stark examples of devastating crop losses due solely to unexpected microbial phytopathogens (*1, 2*). In addition to reduction in crop yield and quality, some fungal pathogens such as *Aspergillus* spp. and *Fusarium* spp. cause food and feed safety concerns because of their ability to produce potent mycotoxins - aflatoxins and fumonisins, respectively (*3-6*). Susceptibility of cultivated crops to plant diseases is exacerbated by several factors such as monoculture of genetically uniform high yielding varieties under large acreage; lack of disease resistant genotypes in the germplasm, and development of resistance in microbial populations to commonly used chemical pesticides. In this regard, development of transgenic crops is very attractive as they maximize crop productivity and quality and is less dependent on use of toxic chemicals that can cause irreparable damage to human health and the environment in the long run. Overexpression of native or heterologous antifungal peptides to enhance host plant protection has been the subject of several reviews (*1, 5, 6, 7*). Several of these natural peptides possess nonspecific toxicity to non-target organisms and are subject to proteolytic degradation. The advent of automated peptide synthesizers and combinatorial peptide chemistry over the past decade has made it possible for rational synthesis of stable and target-specific peptides to overcome some of the problems associated with lytic peptides. More than twenty years ago, it was recognized that certain single genes, encoding for potent natural antimicrobial peptides (lytic peptides), might offer a means to improve the disease resistance of plants utilizing current molecular techniques (*8, 9*). It was first achieved in the early 90's (*10, 11*) and since that time numerous papers have been published demonstrating the efficacy of this approach in enhancing plant disease resistance (*2, 7, 12*). In spite of these efforts so far, success has not resulted in release of commercially viable disease-resistant crops although several field tests have been conducted, as listed by the USDA-APHIS Biotechnology Regulatory Services (http://www.isb.vt.edu/cfdocs/fieldtests1.cfm). The advantages of lytic peptides to control broad-spectrum microbial pathogens in agriculture is very appealing and we provide here a brief discussion on the structure of lytic peptides as related to their antimicrobial activity, advantages and disadvantages of lytic peptides, construction of vectors for both nuclear and plastid transformation and antimicrobial effects of transgenic plants expressing lytic peptides.

Lytic Peptides

Lytic peptides are small proteins that are major components of the antimicrobial defense systems of numerous species. They are a ubiquitous feature of nearly all multi-cellular and some single-cellular life forms. They generally consist of between 10-40 amino acid sequences, which have potential for forming discrete secondary structures. Often, they exhibit the property of amphipathy. An amphipathic α-helix may be depicted as a cylinder with one curved face composed primarily of nonpolar amino acids while the other face is composed of polar amino acids. Most of the lytic peptides that have been described in the literature seem to fall into one of three different classes based on the arrangement of amphipathy and high positive charge density within the molecule:

1) Cecropins (35 amino acids in length and derived from the Giant Silk Moth), N-terminal half amphipathic while the C-terminal half mostly hydrophobic (*13*);

2) Magainins (23 amino acids in length and derived from the African Clawed Frog), amphipathic the full-length of the molecule (*14*); and

3) Melittin (26 amino acids in length and derived from the Honeybee), C-terminal half amphipathic with the N-terminal half primarily hydrophobic (*15*).

The conservation of these physical properties is requisite for activity, but the requirements seem to be somewhat nonspecific in terms of amino acid sequence. For example, we have synthesized highly sequence divergent analogs for each of the peptide classes and have found some of them to be more active and less toxic than their natural counterparts (Jaynes, unpublished data).

Decades ago, the original intent in our laboratory was to utilize the gene encoding a close homolog of Cecropin B (SB-37) to augment bacterial disease resistance in plants. However, during the course of our studies, a new highly sequence divergent peptide was synthesized (Shiva-1) and was shown to possess a more potent lytic activity than SB-37 (*16*). The enhanced bioactivity of Shiva-1 was the first indication that modifications made in the primary sequence of lytic peptides would not destroy the peptide's activity provided certain physical characteristics of the peptide were conserved. Indeed, this was a paradigm-shifting moment in understanding of the structure/function relationship of these incredibly interesting natural molecules and allowed us to pursue the design of novel molecules with enhanced activities. Another example of sequence modification resulting in increased potency of natural peptides is provided in MSI-99, an analog of magainin-II that displayed more positive charge and antibacterial and antifungal activity than its predecessor (*17, 18*).

Amphipathy, Hydrophobicity, & Charge Density: Some of the Physical Properties that Unify Protein Structure and Function

To best illustrate the physical connections between proteins and peptides, it is necessary to display their sequences in ways that make it easier to visualize structural differences and similarities. There are a number of physical features that appear to be important in modulating the activity of peptides:

1. Degree of amphipathy
2. Length of amphipathy
3. Heterogeneity of amphipathic section
4. Placement of amphipathic section (N or C terminal)
5. "+" Charge density (less or more)

6. Hydrophobicity of amphipathic section
7. Presence of hydrophobic tail
8. Length of hydrophobic tail
9. Hydrophobicity of tail
10. Placement of hydrophobic tail (N or C terminal)
11. Absence, presence, & position of "+" charged center
12. Absence or presence & position of flanking sequence
13. Predominating secondary structure
14. Termini modification (N-acetylation, C-amidation)
15. Surface area of hydrophilic and hydrophobic faces
16. Steric or volume considerations.

One can distinguish these characteristics by viewing the amino acids in ways that visually accentuate the differences in their physical attributes. In this respect, it is instructive to ponder the evolution of protein structure and the fact that, generally speaking, only 20 different amino acids are found in proteins. These are: alanine (A), arginine (R), aspargine (N), aspartic acid (D), cysteine (C), glutamic acid (E), glutamine (Q), glycine (G), histidine (H), isoleucine (I), leucine (L), lysine (K), methionine (M), phenylalanine (F), proline (P), serine (S), threonine (T), tyrosine (Y), valine (V), and tryptophan (W). There are a few exceptions, but these 20 are the only ones that are represented in the genetic code (and are called the protein amino acids). So, they are the only ones that matter, at least for the sake of this discussion.

Why did life select these 20 when there are at least 500 other amino acids that have been found to occur in the natural world? Why are these so special? Living processes are never profligate; energy is never expended for little purpose. The simple answer is that these 20 amino acids provide enough unique physico-chemical information that no more are needed for protein structure *(19)*. Chemically speaking, these amino acids differ in very subtle, profound, and important ways. The remarkable diversity of life found on the Earth, in large measure, is merely a reflection of the differences in these molecular building blocks when they are assembled into proteins. With the addition of water, vitamins, minerals, lipids, carbohydrates, and nucleic acids, an individual organism is built, from bacterium to human being. The sheer majesty of the physical world that derives from this relatively simple chemical alphabet is overwhelming. Each living entity produces sets of molecules unique to its type or specie and arranges them in novel ways. During a normal life span, a human may produce as many as 100,000 unique protein transcripts that are constructed from the 20 protein amino acids *(20)*. These separate proteins come together, some, only at specific stages of life, to form a singular, functional human being.

Taking these special 20 amino acids and viewing just two of their seemingly simple properties: hydrophobicity and volume differences can give one an appreciation of the significant chemical refinements that they must represent (Table 1).

Table 1. Hydrophobicity/Hydrophilicity and volume of amino acids

AA	Volume*	D (Å)	Hydrophobicity	Hydrophobicity /Hydophilicity %	Luminosity
F	189.9	7.13	3.7	100	128
M	162.9	6.78	3.4	92	138
I	166.7	6.83	3.1	84	149
L	166.7	6.83	2.8	76	159
V	140.0	6.44	2.6	70	166
C	108.5	5.92	2.0	54	187
W	227.8	7.58	1.9	51	190
A	88.6	5.53	1.6	43	201
T	116.1	6.05	1.2	32	214
G	60.1	4.86	1.0	27	221
S	89.0	5.54	0.6	16	235
P	122.7	6.16	-0.2	2	254
Y	193.6	7.18	-0.7	6	249
H	153.2	6.64	-3.0	25	225
Q	143.9	6.50	-4.1	34	213
N	117.7	6.08	-4.8	39	206
E	138.4	6.42	-8.2	67	170
K	168.6	6.85	-8.8	72	164
D	111.1	5.96	-9.2	75	159
R	173.4	6.92	-12.2	100	128

The single letter code for the amino acids (AA) is shown on the previous page *Note: the total volume, in cubic angstroms, is derived from the van der Waals' radii occupied by the amino acid when it is in a protein. Hydrophobicity is in kcal/mol and is the amount of energy necessary to place the amino acid, when in an α-helical protein, from the membrane interior to its exterior. Luminosity helps assign the density of cyan (hydrophobic amino acids) or magenta (hydrophilic amino acids) to each glyph of the "molecular" font (Molly) that was developed and is described in the text Data derived from *(21)*.

The structural clues they provide in determining protein functionality are available, if we just look at them in the right way. For at least the last 2 billion years, life has found 20 amino acids, combined in different ways, to be adequate to meet all the challenges that it has faced on this planet. All the protein questions that will ever be asked can be answered by natural selection; and life, since the dawn of "biological" time, has been "compelled" to solve just a miniscule number of structural problems. By this it is meant that there might be, let's say, a trillion different proteins that have ever existed on the earth (there probably have been far fewer). By applying combinatorial mathematics to the 20 amino acids, we derive, in practical terms, an almost infinite number of possible combinations that becomes an even bigger number as the length of the protein is increased. For example, if we assume the maximum length for a protein is 200 amino acids, then, the total number of different proteins possible can be derived from the formula *(22)* found below (sum of a finite geometric series):

$$\sum_{i=2}^{200} 20^i = a^1 \frac{1-r^n}{1-r}$$

"a^1" is the first term, "n" is the number of terms, and "r" is the common ratio of the series increase, i.e., it goes up by a factor of 20 each time (the number of different protein amino acids). When one goes through the arithmetic, the number of possible combinations of proteins, from two amino acids in length to 200, is 8.458×10^{257}. A huge number to say the least, particularly, when one considers that the total number of atoms of matter in the universe is estimated to be less than 10^{100}! *(23)*. Also, it should be noted that there are many proteins far larger than 200 amino acids in length. The point of this exercise is that life, in 2 billion years of existence, has not significantly diminished the total number of possible assembled amino acid combinations (proteins) that can do all of the different jobs required by all living organisms. That is the power of evolution; biology will derive suitable answers to any question, given enough time. By studying the predominating 20 protein amino acids in certain ways, we can gain insight into the structural principles that govern all of protein biochemistry and then, as our awareness increases, subtle connections are discovered and seeming disparities can be replaced by recognizable physical commonalties. The unity of the protein structure/function paradigm will continue to emerge as our understanding deepens. After all, every protein that ever existed has been tempered in the "forge" of natural selection. The recognizable similarities of protein structure, even taken from widely divergent species, should not come as a surprise---all of life's processes are interconnected throughout their numerous levels of complexity.

In order to visualize differences and similarities in protein structure more easily, a molecular font was designed (Molly) that is more representative of the chemical nature of the amino acids (see below). To do this, spheres were substituted with circles with diameters equal to:

$$2\sqrt[3]{\frac{Volume}{4/3\pi}}$$

for each particular amino acid. The equation is a rearrangement of the formula for the volume of a sphere: $4/3\ \pi\ r^3$. Then, setting the largest volume to 1, the smaller ones were proportionally reduced. Thus, the size of the circle is directly related to amino acid volume and, the differences shown between the amino acids in Molly, then, are visually accurate. To increase the information of the representation, the hydrophobicity or hydrophilicity of each amino acid was converted to a color scale. The most hydrophobic amino acids are the most intense cyan color while those that are less hydrophobic are proportionally less concentrated cyan. Conversely, those amino acids that are most hydrophilic possess the deepest magenta color. Likewise, a graduated scale of less intense magenta color is used for those amino acids of lower hydrophilic character. From this scale, it can be seen that, as amino acids become less hydrophobic or less hydrophilic, they become less pigmented and, therefore, more likely to be "exchangeable" within the protein structure. Also, implicit in this scheme is that, within a particular hue, i.e., amongst hydrophobic amino acids or hydrophilic amino acids, of very similar properties, exchanges would be more likely to occur (generating the variability one observes in proteins of similar function from evolutionarily distant organisms). Of course, changes would be within the specific structural constraints imposed on each particular protein for it to retain its functionality---natural selection, at it again. Most of the amino acid glyphs possess a mnemonic symbol that further characterizes its chemical properties. For example, charged amino acids have a "+" or a "-" sign incorporated within their glyph, the thickness of which, is related to the dissociation constant of their ionizable protons, other symbols aid in identifying the rest of the amino acids (Figure 1).

With the above in mind, we will discuss in the next section lytic peptides and the basic design parameters that have guided in the construction of novel peptides.

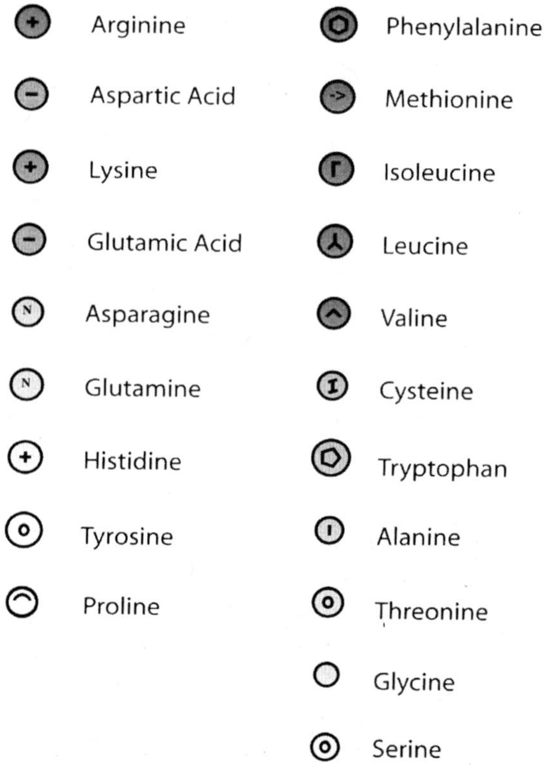

Figure 1. Amino acids in lytic peptides. The symbols represent a molecular alphabet
(see page 1 of color insert)

Lytic Peptide Design Parameters

Lytic peptides are small basic proteins that appear to be major components of the antimicrobial defense systems of a number of animal species including insects, amphibians, and mammals. They consist of around 12 to less than 50 amino acids, which have potential for forming amphipathic α-helices or partial β-pleated sheets (locked by disulfide linkages); and thus, can interact with all cell types at the membrane surface. This interaction can result in no observable cellular effect, temporary cell impairment, death, cell proliferation, or other activities (Jaynes, unpublished). That is why these molecules are more than lytic peptides. Four distinct types of lytic peptides have been discovered over the last several decades; examples of each type are melittin, cecropins, magainins, and defensins. The properties of naturally occurring lytic peptides suggest at least three distinct α-helical classes consisting of different arrangements of amphipathic and hydrophobic regions (Figure 2).

Figure 2. α-helical lytic peptide classes (see page 2 of color insert)

The green band on the cylinders indicates the amino-terminus of the peptide while the gray band represents the carboxy-terminus. The cyan color represents regions that are predominately hydrophobic and the magenta color signifies regions that are hydrophilic. Representative examples of natural peptides, which fit this classification system are: melittin-class 1, cecropins-class 2, and magainins-class 3 (note, more than 90% of all the known natural peptides fall within this classification system, data not shown). Therefore, separate synthetic peptides can be subdivided into distinct classes based on what has been observed in Nature.

Some examples of natural lytic peptides and their sequence as cast in the Molly motif are listed below, along with representative optimized analogs. These are shown in a typical linear array and are read from left to right (Figure 3).

There are several natural lytic peptides that assume a β-conformation, examples of which are the defensins and protegrins. These peptides can assume this shape because of intra-disulfide linkages that lock them into this form, an absolute requisite for activity. We have designed a novel class of peptides that form β-sheets without the necessity of disulfide linkages. An example, D4E1 is shown in Figure 4.

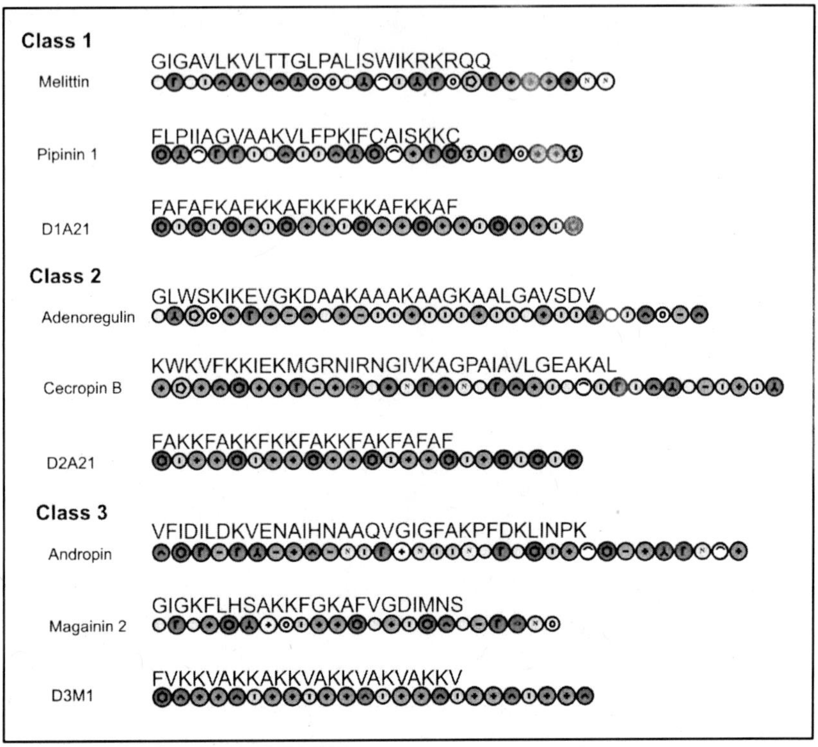

Figure 3. Natural lytic peptides
(see page3 of color insert)

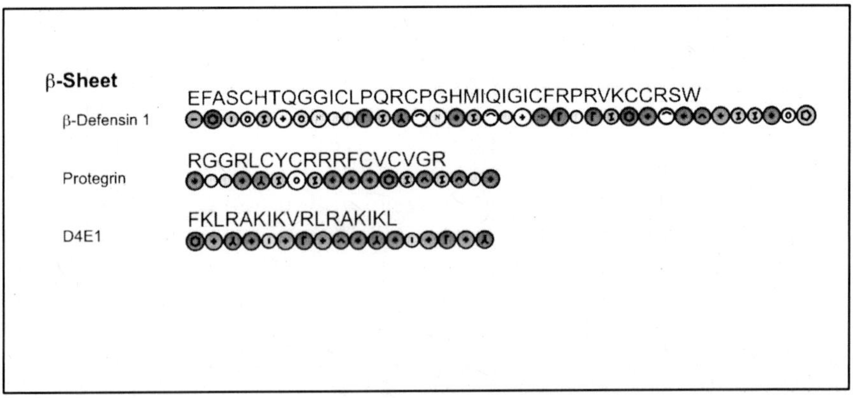

Figure 4. β-sheet peptides
(see page 3 of color insert)

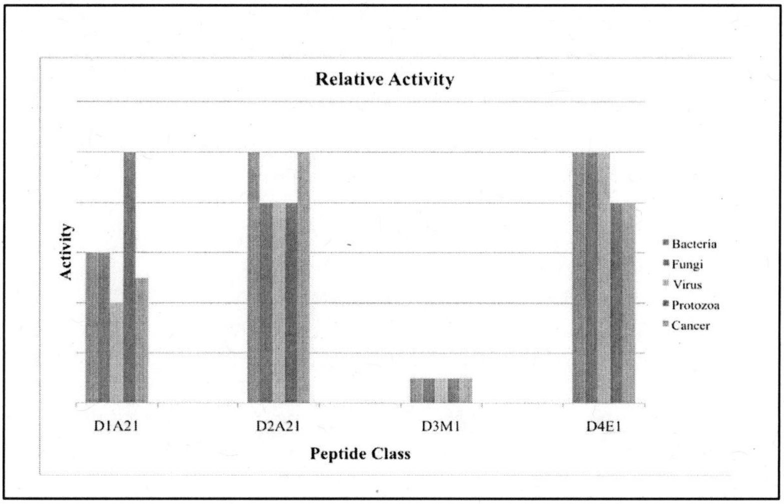

Figure 5. Relative Activity of selected synthetic lytic peptides (Jaynes, unpublished)
Note: Activity is the concentration of the peptide necessary to kill 50 % of the microorganisms. Therefore, the longer the histogram the higher the activity, the longest yielding 100% kill. The concentrations of the peptides are around 1 μM to 25 μM.
(see page 4 of color insert)

It was surmised that simply alternating hydrophobic and hydrophilic amino acids would render an amphipathic beta sheet, which indeed, is the case (*24*). The columnar array of hydrophobic and positive charged amino acids is apparent when the peptide adopts an amphipathic β-form. However, the width of the columns is narrower but overall length is greater than a peptide that adopts an amphipathic α-helix conformation of the same number of amino acids.

All classes of lytic peptides differ somewhat in activity (note that Class 3, magainin class, is far less active in all cases than are the other lytic peptide classes). Figure 5 illustrates, in general terms, their activity spectrum. For the most part, the idealized designed peptides exhibit higher levels of activity with reduced toxicity (*25-28*).

Genetic Engineering of Food crops

Gene Constructs for Plant Transformation

Well-characterized antimicrobial peptides from different sources as indicated above or synthetic peptides can be transferred to other food and feed crops to enhance host plant resistance to phytopathogens. Whenever possible, it is useful to evaluate the antimicrobial traits of unknown peptides by assaying the chemically synthesized pure peptides against as many microbial pathogens as possible (e.g., *29*) prior to actual transgenic work that is usually very time-consuming. Antifungal genes coding for promising peptides can be transferred to other crops by any one of the available transformation methods (*30, 31*). The most widely used nuclear transformation method employs a soil-borne *Agrobacterium* (*32*) which itself has been transformed to include a binary vector containing the gene of interest (antifungal genes) and selectable markers (Figure 6A). Alternatively, plasmids containing the gene of interest can be delivered into plant cell nucleus or organelles (e.g., chloroplast) by a biolistic device or "gene gun" (*33*). The latter method has become the method of choice in transforming chloroplasts (Figure 6B) and other organelles. Other methods of transformation (e.g., electroporation, microinjection, somatic hybridization and others) are rarely employed outside laboratory conditions and, as such, not discussed in detail here.

When engineering a plant transformation vector the investigator has a number of different classes of promoters to choose from depending on which tissues and at what developmental stage expression of the transgene is desired (reviewed in *34*). Often the gene is placed under the control of a constitutive promoter such as the CaMV 35S (*35*) or potato ubiquitin 3 promoter (*36*). Placing the gene under control of a constitutive promoter usually ensures that all tissues will harbor the antifungal protein/peptide though there is some difference in levels of transgene expression in different tissues and at different developmental stages. In many cases it is desirable to place expression of the antifungal gene under a tightly-controlled promoter that will allow inducible or tissue-specific expression of the transgene thus reducing the metabolic load on the host plant imparted by constitutive promoters hence decreasing the chances for reduced plant growth and yield. Examples of tissue-specific promoters include the cottonseed α-globulin B gene promoter (*37*) and the barley lemma (*lem1*) gene promoter (*38*) that demonstrate seed-specific expression. Seed is often the target of infection of mycotoxigenic fungi and therefore seed-specific expression of the antifungal gene should provide the greatest levels of protection. There are a number of classes of inducible promoters including the pathogen/wound-inducible promoters such as the maize proteinase inhibitor (*mpi*) gene promoter (*39*) and the poplar *win3.12T* gene promoter (*40*). These promoters respond to mechanical and insect damage to plant tissues and also to fungal infection. In the case of aflatoxin contamination in crops such as peanut, maize, tree nut and cottonseed, entry of the aflatoxigenic fungus *Aspergillus flavus* to the seed is usually facilitated by boring insects so these types of inducible promoters would provide activation of antifungal gene expression at a very early stage of fungal invasion and only at the site of wounding/infection thus reducing the chances of any deleterious effects on plant growth and development.

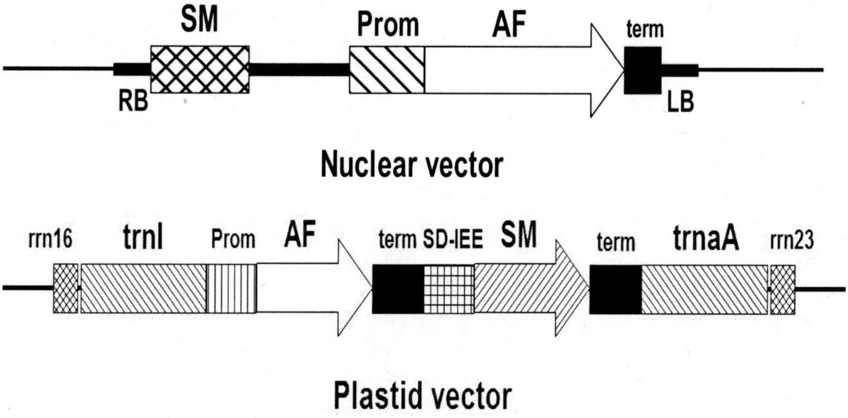

Figure 6. Generic plasmid vectors for nuclear and plastid transformation Schematic diagram of generic plasmid vectors used for nuclear and plastid transformation of plant tissues.

[A] Binary vector for nuclear transformation: the main elements for these vectors 1) Left and right T-DNA border sequences (LB and RB) required for integration of the transgenes into the plant genome; 2) SM, selectable marker gene; Examples of SMs include antibiotic resistance genes such as nptII (kanamycin resistance), hptII (hygromycin resistance), or resistance genes to various herbicides such as glyphosate, sulfonylurea, imidazolinones, and phosphonothricin; 3) Promoter: These may include constitutive promoters such as enhanced CaMV 35S or ubiquitin control regions and wound or pathogen inducible promoters such as PINII or mpi; 4) transcriptional terminator (term) such as nopaline synthase (nos) terminator sequence; AF- antifungal peptide gene of interest as listed in the text.

[B] Plastid vector: the main elements in a plastid transformation vector usually include 1) Regions of homologous plastid DNA flanking the transgenes; The plastid rrn16-trnI and trnA-rrn23 gene regions are often used as flanking DNA for integration of the transgenes into the trnI-trnA intergenic region of the plastid genome via homologous recombination; 2) plastid promoter: The choice of plastid promoter can vary but two commonly used sequences are the Prrn ribosomal RNA operon promoter which drives gene expression in both green and non-green tissues or the psbA photosystem II D1 gene promoter that drives high level gene expression under light conditions; 3) 3' UTR and transcriptional terminators (term); 4) SD-IEE region: sequence that provides a Shine-Dalgarno region and intercistronic expression element that allow for the generation of stable, translatable monocistronic mRNAs thus facilitating transgene stacking in operons (for a description of vectors for plastid transformation, see Lutz et al. (41); 5) SM, selectable marker gene: This is usually the aadA gene encoding spectinomycin resistance or the nptII gene encoding kanamycin resistance; AF- antifungal peptide gene of interest as listed in the text.

Often, despite the plethora of available promoters and transformation vectors, there is still the problem of lack of significant levels of expression of the transgene and hence lack of enough production of the antifungal protein/peptide to be efficacious in inhibiting the growth of the mycotoxigenic fungus. This can often be traced to events such as gene silencing due to multiple integrations of the transgene or position effects due to integration of the transgene within regions of the genome that are transcriptionally inactive. These events are common to nuclear transformation. An alternative to nuclear transformation is the transformation of the plastid genome in plants (*42, 43*). Plastids are present in both green and non-green plant tissues. Plastid transformation eliminates many of the drawbacks associated with nuclear transformation (Table 2). Gene silencing does not occur in plastids nor is there any position effect as integration is site-specific due to homologous recombination of flanking plastid DNA sequences that are present in the vector (*43*). Another advantage of plastid transformation is the ability to generate significant levels of the transgene product due to high numbers of plastids and hence plastid genomes present in each cell, resulting in a very high number of functional transgene copies. Additionally, multiple genes can be expressed in an operon-like fashion from one promoter thus allowing for "stacking" of genes in a single transformant. From an environmental standpoint, plastid transformation is preferable to nuclear transformation as in most angiosperm plant species plastid genes are maternally inherited and therefore transgenes present in these plastid genomes are not susceptible to dispersal to other plant species via pollen dissemination (*42, 44*). One of the major hurdles left with respect to plastid transformation technology is the development of efficient transformation systems for specific plant species. Though plastid transformation has been achieved in many plant species there are still a great number that have not been transformed up to this time (*43*). This is especially true for many crop plants that are susceptible to mycotoxin contamination such as maize, peanut, wheat, barley, grape, and tree nuts. In many cases difficulty in transforming these species arises because non-green tissues must be used as targets for biolistic transformation. Little is know about the regulation of the transcriptional and translational machinery in pro-plastids present in these tissues and pro-plastids are much smaller than that of chloroplasts found in green tissues thus making successful integration of the transgene more difficult (*43*). Given time, plastid transformation protocols will most probably be optimized for these plant species.

Table 2. Comparison of Chloroplast and Nuclear Genetic Engineering

Transgenic Traits	Chloroplast Genome	Nuclear Genome
Transgene copy number	up to 10,000 copies/cell	usually less than 10 copies/cell
Level of gene expression	high abundance w/ high accumulation of foreign protein	gene expression often too low to be efficacious
Gene transcription	genes can be arranged in operon-form allowing multiple genes to be expressed from one promoter	difficult to stack multiple genes into one construct. Requires multiple promoters
Position effect	site-specific recombination eliminates positional effects on transgene expression	random insertion into genome results in variable transgene expression levels
Gene containment	maternal inheritance results in high level of containment-transgene not carried by pollen	possible outcrossing via pollen drift
Toxicity of foreign proteins	potential for minimization of adverse effects of transgenic proteins due to compartmentation	accumulation of toxic proteins in the cytosol may be deleterious to host

Genetic Engineering for disease resistance

Genetic engineering of host plants for resistance to microbial diseases has been an ongoing area of investigation that has seen many small-scale laboratory/greenhouse successes but success with respect to wide-scale commercialization is yet to be seen (*2, 7, 12, 45-47*). Several reports are available on the efficient nuclear expression of peptides for controlling microbial plant pathogens. It is noteworthy that expression of lytic peptides did not alter the morphology or flowering of transgenic plants (*18, 48, 49*). We provide below some selected reports on antimicrobial effects of transgenic plants expressing peptides of plant or non-plant origin including synthetic peptides.

Peptides of Plant–origin

Huffaker et al. (*50*) identified peptide elicitors derived from the plant itself that activate defensive genes against pathogens. For example, they isolated a 23-aa peptide from *Arabidopsis*, called AtPep1, which activates transcription of the defensive gene defensin (PDF1.2) and activates the synthesis of H_2O_2, both being components of the innate immune response. The peptide is derived from a 92-aa precursor encoded within a small gene that is inducible by wounding, methyl jasmonate, and ethylene. Constitutive expression of the AtPep1 precursor gene PROPEP1 in transgenic *Arabidopsis* plants causes a constitutive transcription of PDF1.2. The transgenic plants exhibited increased root development compared to control plants and an enhanced resistance toward the root pathogen *Pythium irregulare*. Kanzaki et al. (*51*) overexpressed the wasabi (Japanese horseradish) defensin gene, a plant defensin effective against the rice blast fungus, in transgenic rice (*Oryza sativa* cv. Sasanishiki). Transformants exhibited resistance to rice blast caused by *Magnaporthe grisea* in T2 and T3 generations from which they were able to detect the 5-kDa peptide, corresponding to the processed form of the wasabi defensin in the total protein fraction extracted from the T3 progeny. Enhanced quantitative resistance to *Leptosphaeria maculans* (causal agent of blackleg disease) was observed by Kazan et al. (*52*) in canola (*Brassica napus* L.) by transgenic expression of a novel antimicrobial peptide MiAMP1, originally isolated from the seeds of *Macadamia integrifolia*. Ko et al. (*53*) transformed apple scion cultivar `Galaxy' and the apple rootstock M.26 with a cDNA clone of the gene encoding attacin gene (from *Cecropia* moth) to enhance resistance to *Erwinia amylovora*, the bacterium that causes fire blight. Although *in vitro* assays indicated that attacin was partially degraded in the intercellular fluid of apple leaves they observed enhanced resistance to fire blight. Constitutive over-expression of a seed specific hevein-like antimicrobial peptide from *Pharbitis nil* in transgenic tobacco plants enhanced resistance to the oomycete *Phytophthora parasitica*, the causal agent of black shank disease (*54*). Francois et al. (*55*) developed a method for expression in *Arabidopsis* of a transgene encoding a cleavable chimeric polyprotein. The polyprotein precursor consists of a leader peptide and two different antimicrobial proteins (AMPs), DmAMP1 originating from *Dahlia merckii* seeds and RsAFP2 originating from *Raphanus sativus* seeds, which are linked by an intervening sequence ("linker peptide") originating from a natural polyprotein occurring in seed of *Impatiens balsamina*. The chimeric polyprotein was found to be cleaved in transgenic *Arabidopsis* plants and the individual AMPS were secreted into the extracellular space. Both AMPS were found to exert antifungal activity *in vitro*. They also observed that the amount of AMPs produced in plants transformed with some of the poly-protein transgene constructs was significantly higher compared with the amount in plants transformed with a transgene encoding a single AMP, indicating that the polyprotein expression strategy may be a way to boost expression levels of small proteins. Recently, Swathi Anuradha et al. (*56*) expressed defensin gene from mustard in transgenic plants to obtain fungal resistance – *Fusarium moniliforme* and *Phytophthora parasitica* pv. *nicotianae* in transgenic tobacco and *Pheaoisariopsis personata* and *Cercospora arachidicoloa* which cause late leaf spot disease in peanut.

Peptides of non-plant origin

Several small MW peptides isolated from organisms other than plants have been shown to be effective antifungal agents (*57-60*). Some examples include the cecropins (*61*) and magainins (*14*) of insect and amphibian origins respectively, and their synthetic analogs (*18, 29, 48 ,62, 63*). The antimicrobial peptide, cecropin, from Cecropia moth has been introduced into several crops – tobacco (*11, 64*), potato (*65*), rice (*66*), apple (*67*) for the control of bacterial pathogens. Attacin gene from the same moth species has been incorporated into apple (*16, 68*) and potato (*65*) for bacterial pathogen control. Li et al. (*62*) introduced a magainin analog into tobacco and demonstrated both fungal and bacterial control. As a possible approach to enhance plant resistance, a DNA coding for a modified esculentin-1, a 46-residue antimicrobial peptide present in skin secretions of *Rana esculent*a, was introduced into *Nicotiana tabacum* (*69*). The antimicrobial peptide was isolated from the intercellular fluids of healthy leaves of transgenic plants, suggesting that it was properly processed, secreted outside cells and accumulated in the intercellular spaces. The morphology of transgenic plants was unaffected. Challenging these plants with bacterial or fungal phytopathogens demonstrated enhanced resistance up to the second generation. Moreover, transgenic plants also displayed insecticidal properties.

Synthetic peptides

Certain small lytic peptides, redesigned from that of natural peptides, have demonstrated convincing inhibitory activity against fungal species including *Aspergillus flavus* (Figure 7) and show promise for transformation of plants to reduce infection of seed. In our laboratory, we (*29, 48*) reported that a synthetic lytic peptide (D4E1) gene, when transformed into tobacco, greatly enhances resistance *in planta* to *Colletotrichum destructivum.* The broad spectrum antifungal activity of the synthetic peptide D4E1 is given in Table 3 (*29*). Treatment of germinating *A. flavus* spores with tobacco leaf extracts from plants, transformed with the D4E1 gene, significantly reduced spore viability (colony forming units) relative to results obtained using extracts from non-transformed (control) plants. Similarly, in recent tests with cottonseed expressing the D4E1 gene, we demonstrated resistance to penetration of seed coats by a GFP reporter gene-containing *A. flavus* strain (*70, 71*). In addition to inhibiting the germination of *A. flavus* spores, D4E1 caused severe abnormal lytic effects on mycelial wall, cytoplasm, and nuclei. The expression of D4E1 gene in the progeny of transgenic cotton was sufficient enough to inhibit the growth *in vitro* of *Fusarium verticillioides* and *Verticillium dahliae* or *in planta* of *Thielaviopsis basicola* (*70*) and provide a good germination stand in a field infected with *Fusarium oxysporum* fsp. *vasinfectum* (*72*). Transformation of peanut with another antifungal peptide, D5C, has been reported (*73*). Although the pure D5C showed strong activity *in vitro* against *A. flavus*, it was shown that the transgenic peanut callus showed poor recovery of plants because of possible phytotoxicity of the peptide. Puterka et al. (*74*) observed that the biology and behavior of pear psylla, *Cacopsylla pyricola* Foerster, on a transgenic clone of 'Bartlett' pear, *Pyrus communis* L,, containing a synthetic antimicrobial gene, D5C1. The purpose of the original transformation was to enhance pear resistance to the bacterial disease fireblight caused by *Erwinia amylovora* (Burr.). During the conduct of the study, they observed that that the insect pest's biology and behavior were initially enhanced on transgenic pear clone. However, chronic exposure of psylla populations to transformed pear plants that express the *npt*II marker and lytic peptide genes had detrimental effects on pear psylla.

Mentag et al. (*75*) demonstrated bacterial disease resistance of transgenic hybrid poplar (*Populus tremula* L. x *Populus alba* L.) expressing the synthetic antimicrobial peptide D4E1. The transgenic poplar lines were tested for resistance to *Agrobacterium tumefaciens*, *Xanthomonas populi* pv. *populi* and *Hypoxylon mammatum* (Wahl.) Miller. One transgenic poplar line, Tr23, bearing the highest transcript accumulation for the D4E1 gene, showed a significant reduction in symptoms caused by *A. tumefaciens* and *X. populi*. However, none of the transgenic poplar lines showed a significant difference in disease response to the fungal pathogen *H. mammatum*.

Transgenic expression of a synthetic substitution analog of magainin, MSI-99 imparted disease resistance in both tobacco (*Nicotiana tabacum* L.) and banana [*Musa* spp. cv. Rasthali (AAB)]. Transgenic tobacco showed enhanced resistance to *Sclerotinia sclerotiorum*, *Alternaria alternata* and *Botrytis cinerea* where as transgenic banana plants showed resistance to *Fusarium oxysporum* f.sp. *cubense* and *Mycosphaerella musicola* (*63*). Alan et al (*76*) transformed tomato with MSI-99 and they observed that transgenic tomato plants were more inhibitory against a bacterial pathogen *Pseudomonas syringae* pv. *tomato* (bacterial speck pathogen) than against the fungal pathogens - *Alternaria solani* (early blight) and the oomycete pathogen *Phytophthora infestans* (late blight) possibly due to proteolytic degradation and lower expression of the peptide.

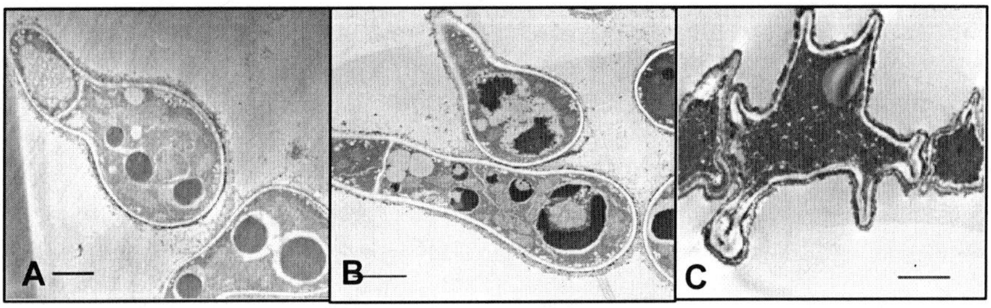

Figure 7. D4E1 effects on Aspergillus flavus spores. Transmission electron micrographs of pre-germinated A. flavus spores exposed to the antifungal peptide D4E1. A) Control in potato dextrose broth; B) Spores exposed to 10 μM D4E1 for 1h. C) Note cytoplasmic degradation due to exposure to D4E1 leading to eventual lysis at 25 μM D4E1 for 1h (K. Rajasekaran, unpublished).

Table 3. Broad spectrum antimicrobial activity of D4E1 in vitro

Phytopathogen	IC_{50} (μM)	MIC (μM)
Alternaria alternata	12.39	>25.0
Aspergillus flavus	7.75	25.0
Aspergillus flavus 70-GFP	11.01	25.0
Cercospora kikuchii	8.67	>25.0
Colletotrichum destructivum	13.02	>25.0
Claviceps purpurea	1.60	20.0
Fusarium graminearum	2.10	25.0
Fusarium moniliforme	0.88	12.5
Fusarium oxysporum	2.05	12.5
Penicillium italicum	5.92	>25.0
Phytophthora cinnamomi	nd	4.67
Phytophthora parasitica	nd	4.67
Pseudomonas syringae pv. *tabaci*	0.52	2.25
Pythium ultimum	nd	13.33
Rhizoctonia solani	nd	26.7
Thielaviopsis basicola	0.52	6.0
Verticillium dahliae	0.60	5.25
Xanthomonas campestris pv. *malvacearum*	0.19	1.25

nd = not determined ; Source: reproduced from (*29*)

In addition to nuclear transformation, we are in the process of exploring the possibility of expressing antifungal genes in plastids with the objective of higher expression and preventing transgene escape through pollen, as indicated in the previous section. For example, the antimicrobial peptide MSI-99, an analog of magainin 2, was expressed via the chloroplast genome (*18*) of tobacco. Leaf extracts from T2 generation plants showed 96% inhibition of growth against the bacterial pathogen *P. syringae* pv. *tabaci.* In addition, leaf extracts from T1 generation plants inhibited the growth of pregerminated spores of three fungal species, *A. flavus*, *Fusarium verticillioides*, and *Verticillium dahliae*, by more than 95%, compared with non-transformed control plant extracts. *In planta* assays with the bacterial pathogen *P. syringae* pv *tabaci* resulted in areas of necrosis around the point of inoculation in control leaves, whereas transformed leaves showed no signs of necrosis, demonstrating high-dose release of the peptide at the site of infection by chloroplast lysis. *In planta* assays with the fungal pathogen, *Colletotrichum destructivum*, showed necrotic anthracnose lesions in non-transformed control leaves, whereas transformed leaves showed no lesions. In addition to lytic peptide genes, a variety of other candidate antifungal genes from bacterial, plant, and mammalian sources have a good probability of being active against *A. flavus* upon transformation into plants.

Potential Problems Associated with Antimicrobial Peptides

In general, the majority of antimicrobial peptides function primarily by compromising the membrane of the target organism (lytic activity) though some also appear to function at the DNA or protein level [reviewed in Marcos et al. (*7*). With respect to their use in agriculture, many natural antimicrobial peptides have undesirable properties such a nontarget toxicity, poor activity, and susceptibility to protease degradation. Generation of hybrid peptides, rational design of peptide analogs, and synthetic peptide combinatorial chemistry have been used in an effort to increase antimicrobial peptide activity and stability while diminishing nontarget toxicity (*7, 77, 78*). Hybrid peptides and their analogs such as cecropin::mellitin have demonstrated resistance to bacterial and fungal pathogens in transgenic potato (*77, 79*) while cecropin::magainin analogs displayed increased bactericidal/tumoricidal activity without inducing hemolysis (*80*). Rational design of substitution analogs of the naturally occurring cecropin-B peptide identified, Shiva-1, an analog that retained only 46% homology to cecropin-B that conferred enhanced resistance to bacterial wilt in tobacco (*11*). Our studies on the synthetic peptide D4E1 have shown that it was more resistant to plant and fungal protease degradation than the natural peptide, cecropin-A (*24*). Additionally, introduction of a gene encoding an antimicrobial peptide into the plastid genome and its subsequent production and compartmentalization within the plastid improved antimicrobial activity while reducing proteolytic degradation (*18*). López-Garcia et al. (*81*) screened a synthetic peptide combinatorial library in a positional scanning format to identify improved versions of the hexapeptide PAF19. They identified a number of bioactive peptides with improved activity against a group of fungal phytopathogens responsible for postharvest decay in fruits yet these did not demonstrate increased, nontarget toxicity to *Escherichia coli* and *Saccharomyces cerevisiae.* In some cases substitution of just one amino acid residue could result in loss of activity against a particular fungal pathogen. Combinatorial chemistry was also used to increase antibacterial activity while maintaining low cytotoxicity of cyclic decapeptides (*82*).

Conclusion

The importance of peptides, either natural or synthetic, in control of microbial pathogens is well established from the steady stream of publications in the last decade or so. Modern technology in automated peptide synthesis and combinatorial chemistry have made the task of designing novel, environmentally benign, yet target pathogen(s)-specific potent peptides more attainable than before. In addition, availability of transgenic technology in several crops via both nuclear and/or organelle transformation provides much needed expertise in facilitating a rapid development of disease-resistant, commercially-useful germplasm or varieties of food and feed crops. In addition, identification of potent antifungal peptides from different sources will be valuable in transferring them to susceptible crops to combat fungal infection resulting in preharvest mycotoxin contamination compromising food and feed safety. More importantly, the combination of technologies presented in this review provides a means of enhancing the host-plant resistance of susceptible crop species in a relatively short time frame compared to conventional breeding.

References

1. Oerke, E.-C. *J Agric Sci.* **2006,** 141, 31-43.
2. Rajasekaran, K.; Cary, J. W.; Jacks, T. J.; Cleveland, T. E. In: Rajasekaran, K.; Jacks, T. J.; Finley, J.W. Eds. *Crop Biotechnology.* American Chemical Society Symposium Series No. 829. Washington, DC: American Chemical Society. **2002,** pp.97-117.
3. *Mycotoxins in Agriculture and Food Safety;* Sinha.K.K.; Bhatnagar, D. Eds.; Marcel Dekker, Inc.: New York, NY, **1998**.
4. *Aflatoxins and other mycotoxins: An agricultural perspective.* Council for Agricultural and Science Technology (CAST) Reports, Ames, IA. **2003**.
5. Bhatnagar, D.; Rajasekaran, K.; Brown, R. L.; Cary, J. W.; Yu, J.; Cleveland, T. E. In: Wilson, C.L. Eds.; *Microbial Food Contamination.* Boca Raton, FL: CRC Press. pp. 395-426. **2007**.
6. Bhatnagar, D.; Rajasekaran, K.; Cary, J. W.; Brown, R. L.; Yu, J.; Cleveland, T. E. In: Leslie, J. F.; Bandyopadhyay, R.; Visconti, A. Eds.; *Mycotoxins: Detection Methods, Management, Public Health, and Agricultural Trade.* CABI Publishing: Wallingford, Oxfordshire, UK, **2008,** pp. 257-276.
7. Marcos, J. F.; Munoz, A.; Perez-Paya, E.; Misra, S.; Lopez-Garcia, B. *Annu Rev Phytopathol* **2008,** 46, 273-301.
8. Jaynes, J. M.; Xanthopoulos, D.; Destefano-Beltran, L.; Dodds, J. H. *BioEssays* **1987,** 6(6), 263-270.
9. de Pace, C.; Trinca, S.; Caccia, R.; Scarascia, G.; Dodds, J. H.; Jaynes, J. M. *Molecular Methods for Potato Improvement.* Agribookstore. **1991,** 85-95.
10. Destefano-Beltran, L.; Nagpala, P.G.; Cetiner, S. M.; Denny, T.; Jaynes, J. M. In *Biotechnology in Plant Disease Control,* Chet, I. Ed.. Wiley-Liss: Wilmington, DE. **1993,** 175-189.
11. Jaynes, J. M.; Nagpala, P.; Destefano-Beltran, L.; Huang, J. H.; Kim, J.; Denny, T.; Cetiner, S. *Plant Sci* **1993,** 89, 43-53.
12. Punja, Z. K. *Can J Plant Pathol* **2006,** 28,S298–S308.
13. Lee, J.-Y.; Boman, A.; Sun, C.; Andesson, M.; Jornvall, H.; Mutt, V.; Boman, H. G. *Proc. Natl. Acad. Sci. USA.* **1989,** 86, 9159-9162.
14. Zasloff, M. *Proc. Natl. Acad. Sci. USA.* **1987,** 84, 5449-5453.
15. Haberman, E. *Science*, **1972,** 177, 314- 322.
16. Norelli, J. L.; Aldwinckle, H. S.; Destefano-Beltran, L.; Jaynes, J. M. *J Cell Biochem.* **1994,** 18A, 89.
17. Alan, A.R.; Earle, E.D. *Molec Plant-Microbe Interact*, **2002,** 15(7), 701-708.
18. DeGray, G.; Rajasekaran, K.; Smith, F.; Sanford, J.; Daniell, H. *Plant Physiol* **2001,** 127, 852-62.
19. *Chemistry and Biochemistry of the Amino Acids* Barrett, G.C. Ed. . Chapman and Hall: New York, NY. **1985**.
20. Jensen, O. N., *Curr. Opin. Chem. Biol.*, **2004**, 8, 33-41.
21. Engelman,, D. M. *Ann. Rev. Biophys. Chem.* **1986,** 15, 330-357.

22. *College Mathematics*, Nielsen, K. L. 1958, reprinted. Barnes & Noble, Inc: New York, NY. **1970**.
23. Cox, S. A. Seven Dimensional (and up) Einsteinian Hyperspherical Universe. www.geocities.com/aletawcox/ **2005.**
24. De Lucca, A. J.; Bland, J. M.; Grimm, C.; Jacks, T. J.; Cary, J. W.; Jaynes, J. M.; Cleveland, T. E.; Walsh, T. J. *Can J Microbiol* **1998,** 44, 514-20.
25. Schwab, U.; Gilligan, P.; Jaynes, J.; Henke, D. *Antimicrob Agents Chemother*. **1999,** p. 1435-1440.
26. Martin, A.; Danforth, H.D.; Jaynes, J.M.; Thornton, *J. Parasitology Research* **1999,** 85, 331-336.
27. Arlotti, J. A.; Cimino, T. S.; Nguyen, T. S.; Dhir, R.; Thomas, A.; Jaynes, J. M.; Caldwell, A. L.; Getzenberg, R. H. *Urologic Oncology* **2001,** 6, 97-102.
28. Ma, J.; Kennedy-Stoskopf, S.; Jaynes, J.; Tompkins, W. *J Virol.* **2002,** 76, 9952-9961.
29. Rajasekaran, K.; Stromberg, K.; Cary, J. W.; Cleveland, T. E. *J Agric Food Chem* **2001,** 49(6), 2799-2803.
30. *Transgenic Crops of the World – Essential Protocols.* Curtis, I. S. Ed.; Kluwer Academic Publishers: Dordrecht, The Netherlands. **2004,** 454p.
31. *Gene Transfer to Plants. Springer Lab Manual.* Potrykus, I., Spangenberg, G. Eds.;. Springer-Verlag: Berlin, Germany. **1995,** 361p.
32. *Agrobacterium – From Biology to Biotechnology.* Tzfira, T., Citovsky, V. Eds.;. Springer Science: New York, NY. **2008,** 750p.
33. *Particle Bombardment for Genetic Engineering of Plants.* Christou, P. Academic Press: San Diego, CA. **1996,** 199p.
34. Gurr, S. J.; Rushton, P. J. *Trends Biotechnol* **2005,** 23(6), 283-290.
35. Schardl, C. L.; Byrd, A. D.; Benzion, G.; Altschuler, M. A.; Hildebrand, D. F.; Hunt, A. G.. *Gene* **1987,** 61(1), 1-11.
36. Garbarino, J. E.; Belknap, W. R. *Plant Mol Biol* **1994,** 24(1), 119-27.
37. Sunilkumar, G.; Connell, J. P.; Smith, C. W.; Reddy A. S.; Rathore K. S. *Transgen Res* **2002,** 11(4), 347-59.
38. Skadsen, R. W.; Sathish, P.; Federico, M. L.; Abebe, T.; Fu , J.; Kaeppler, H. F. *Plant Mol Biol* **2002,** 49(5), 545-555.
39. Tamayo, M. C.; Rufat, M.; Bravo J. M.; San Segundo, B. *Planta* **2000,** 211(1), 62-71.
40. Yevtushenko, D. P.; Misra, S. *Plant Biotechnol J.* **2007,** 5(6), 720-734.
41. Lutz, K. A.; Azhagiri, A. K.; Tungsuchat-Huang, T.; Maliga P. *Plant Physiol.* **2007,** 145: 1201–1210.
42. Maliga, P. *Annu Rev Plant Biol.* **2004,** 55, 289-313
43. Verma, D.; Daniell, H. *Plant Physiol* **2007,** 145(4), 1129-1143.
44. Daniell, H. *Proc Natl Acad Sci U S A* **2007,** 104(17), 6879-6880.
45. Rajasekaran, K.; Bhatnagar, D.; Brown, R. L.; Chen, Z-Y.; Cary, J. W.; Cleveland, T. E. *In:* Ramasamy C, Ramanathan S, Dhakshinamoorthy M, eds.; *Perspectives of Agricultural Research and Development,* Tamil Nadu Agricultural University Centennial Issue, Coimbatore, India. **2005,** pp. 434-467.
46. Michelmore, R. W. *Curr Opin Plant Biol* **2003,** 6(4), 397-404.

47. *Fungal Disease Resistance in Plants - Biochemistry, Molecular Biology, and Genetic Engineering.* Punja, Z. K. Food Products Press: New York, NY. **2004.**
48. Cary, J. W.; Rajasekaran, K.; Jaynes, J. M.; Cleveland, T. E. *Plant Sci.* **2000,** 153, 171-180.
49. Mitsuhara, I.; Yatou, O.; Iwai, T.; Naito, Y.; Nawa, Y.; Ohashi, Y.; *Plant Biotechnol* **2006,** 23, 63-69.
50. Huffaker, A.; Pearce, G.; Ryan, C. A. *Proc Natl Acad Sci U S A* **2006,** 103, 10098-10103.
51. Kanzaki, H.; Nirasawa, S.; Saitoh, H; Ito, M.; Nishihara, M.; Terauchi, R.; Nakamura, I. *Theor Appl Genet* **2002,** 105,809-814.
52. Kazan, K.; Rusu, A.; Marcus, J. P.; Goulter, K. C.; Manners, J.M. *Mol Breed* **2002,** 10, 63-70
53. Ko, K. S.; Norelli, J. L.; Reynoird, J. P.; Boresjza-Wysocka, E.; Brown, S. K.; Aldwinckle, H. S. *Biotechnol Lett* **2000,** 22, 373-381.
54. Koo, J. C.; Chun, H. J.; Park, H. C.; Kim M. C.; Koo, Y. D.; Koo, S. C.; Ok, H. M.; Park, S. J.; Lee, S. H.; Yun, D. J.; Lim, C. O.; Bahk, J. D.; Lee, S. Y.; Cho, M. J. *Plant Mol Biol* **2002,** 50, 441-452
55. Francois, I. E. J. A.; De Bolle, M. F. C.; Dwyer, G.; Goderis, I. J. W. M.; Woutors, P. F. J.; Verhaert, P. D.; Proost, P.; Schaaper, W. M. M.; Cammue, B. P. A.; Broekaert, W. F. *Plant Physiol* **2002,** 128, 1346-1358.
56. Swathi Anuradha, T., Divya, K., Jami, S.K., Kirti, P.B. *Plant Cell Rep,* **2008,** 27, 1777-1786.
57. Rao, A. G.. *Mol Plant-Microbe Interact* **1995,** 8, 6-13,.
58. Broekaert, W. F.; Cammue, B. P. A.; Debolle, M. F. C.; Thevissen, K.; Desamblanx, G. W.; Osborn, R. W. *Crit Rev Plant Sci* **1997,** 16, 297-323.
59. Hancock, R. E. W.; Chapple, D. S. *Antimicrob Agents Chemother* **1999,** 43, 1317-1323.
60. Reddy, K. V.; Yedery, R. D.; Aranha, C. *Int J Antimicrob Agents.* **2004,** 24,536-547.
61. Bowman, H. G.; Hultmark, D. *Annu Rev Microbiol.* **1987,** 31, 103-126.
62. Li, Q. Q.; Lawrence, C. B.; Xing, H. Y.; Babbitt, R. A.; Bass, W. T.; Maiti, I. B.; Everett. N. P. *Planta* **2001,** 212, 635-639.
63. Chakrabarti, A.; Ganapathi, T. R.; Mukherjee, P. K.; Bapat, V. A. *Planta* **2003,** 216, 587-596.
64. Huang, Y.; Nordeen, R. O.; Di, M.; Owens, L. D.; McBeath, J. H. *Phytopathol* **1997,** 87, 494-499.
65. Arce, P.; Moreno, M.; Gutierrez, M.; Gebauer, M.; Dell'Orto, P.; Torres, H.; Acuna, I.; Oliger, P.; Venegas, A.; Jordana, X.; Kalazich, J.; Holuigue, L. *Amer J Potato Res* **1999,** 76, 169-177.
66. Sharma, A.; Sharma, R.; Imamura, M.; Yamakawa, M.; Machii, H. *FEBS Lett* **2000,** 484,7-11
67. Liu, Q.; Ingersoll, J.; Owens, L.; Salih, S.; Meng, R.; Hammerschlag, F. *Plant Cell Rep* **2001,** 20, 306-312.
68. Reynoird, J. P.; Mourgues, F.; Norelli, J.; Aldwinckle, H. S.; Brisset, M. N.; Chevreau, E. *Plant Sci* **1999,** 149, 23-31.

69. Ponti, D.; Mangoni, M. L.; Mignogna, G.; Simmaco, M.; Barra, D. *Biochem J* **2003,** 370,121-127.
70. Rajasekaran, K.; Cary, J. W.; Jaynes, J. M.; Cleveland, T. E. *Plant Biotechnol J.* **2005,** 3, 545-554.
71. Rajasekaran, K.; Cary, J. W.; Cotty, P.; Cleveland, T. E. *Mycopathologia.* **2008,** 165, 89-97.
72. Rajasekaran, K.; Ulloa, M.; Hutmacher, R.; Cary, J. W.; Cleveland, T. E.. *Proc 2006 Beltwide Cotton Conferences.* National Cotton Council, Memphis, TN. **2006.**
73. Weissinger, A.; Liu, Y-S.; Scanlon, S.; Murray, J.; Cleveland, T. E.; Jaynes, J.; Mirkov, E.; Moonan, F. *Proc USDA-ARS Aflatoxin Elimination Workshop,* **1999,** p. 66-68.
74. Puterka, G. J.; Bocchetti, C.; Dang, P.; Bell, R. L.; Scorza, R. *J Econ Entomol* **2002,** 95,797-802.
75. Mentag, R.; Luckevich, M.; Morency, M. J.; Seguin, A. *Tree Physiol* **2003,** 23,405-411.
76. Alan, A.R.; Blowers, A.; Earle, E.D. Plant Cell Reports, **2004,** 22(6), 388-396.
77. Osusky, M.; Zhou, G.; Osuska, L.; Hancock, R. E.; Kay, W. W.; Misra, S. *Nat Biotechnol* **2000,** 18, 1162-1166.
78. Blondelle, S. E.; Lohner, K. *Biopolymers* **2000,** 55, 74-87.
79. Osusky M, Osuska L, Hancock RE, Kay WW, Misra S. *Transgen Res* **2004,** 13,181-190.
80. Shin, S. Y.; Lee, S. H.; Yang, S. T.; Park, E. J.; Lee, D. G.; Lee, M. K.; Eom, S. H.; Song, W. K.; Kim, Y.; Hahm, K. S.; Kim, J. I. *J Pept Res* **2001,** 58, 504-514.
81. Lopez-Garcia, B.; Perez-Paya, E.; Marcos, J. F. *Appl Environ Microbiol* **2002,** 68, 2453-2460.
82. Monroc, S.; Badosa, E.; Besalu, E.; Planas, M.; Bardaji, E.; Montesinos, E.; Feliu, L. *Peptides* **2006,** 27, 2575-2584.

Chapter 10

Phytotoxicity of trichothecenes

S. P. McCormick

Mycotoxin Research Unit, USDA-ARS-NCAUR, Peoria IL 61604

Many plant pathogenic species of the genus *Fusarium* produce trichothecenes, a large group of sesquiterpene epoxides that are inhibitors of eukaryotic protein synthesis. Although some *Fusarium* trichothecenes are virulence factors in plant disease, the phytotoxicities of many trichothecenes have only recently been investigated. Two test systems have been used to evaluate the structural features of trichothecenes that may impact phytotoxicity. Detached leaves of *Arabidopsis thaliana* plants and cultures of the unicellular plant *Chlamydomonas reinhardtii* were both treated with solutions of trichothecenes from a library of natural and synthetically modified trichothecenes. Results of these studies suggest that trichothecenes with a C-3 acetoxy group are generally less phytotoxic than those with a C-3 hydroxyl group but that other structural features may be important. Isotrichodermol, the earliest trichothecene precursor of the mycotoxins T-2 toxin and deoxynivalenol, was toxic in both assays.

Trichothecenes are oxygenated sesquiterpene mycotoxins (Figure 1) produced by species of *Fusarium*, *Myrothecium*, *Trichothecium*, *Trichoderma*, and *Stachybotrys*. *Fusarium* trichothecenes all have an oxygen function at C-3 and can have additional oxygen functions at C-4, C-7, C-8 and C-15.

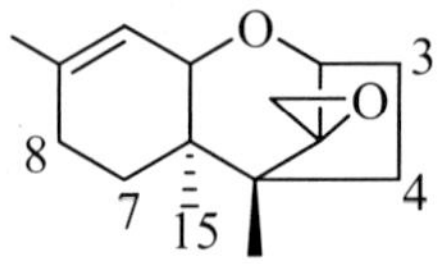

Figure 1. General trichothecene structure. Numbers indicate carbons positions with oxygenation in Fusarium.

Fusarium Trichothecene Biosynthesis

The biosynthetic pathways of T-2 toxin in *Fusarium sporotrichioides* and deoxynivalenol or nivalenol in *F. graminearum* involve a sequence of oxygenation and esterification reactions controlled by up to 15 genes, most of which are localized in a 25 kB cluster (*1, 2*). Trichodiene, a hydrocarbon product of farnesyl pyrophosphate, is formed by the sesquiterpene cyclase, trichodiene synthase (*3*). Trichodiene (Figure 2) is oxygenated to form isotrichotriol (Figure 2) which then cyclizes to form isotrichodermol (3-hydroxytrichothecene) (Figure 2). The first four oxygenation steps are controlled by a single gene, *Tri4* (*4*). The pathway then proceeds through a series of oxygenation, esterification, and deacetylation steps to produce more complex trichothecenes such as 4,15-diacetoxyscirpenol (4,15 DAS), T-2 toxin, or nivalenol (Figure 2). One important feature of trichothecene biosynthesis in *Fusarium* is the early addition of an acetyl group at C-3. This acetyl group remains in place through the remaining oxygenations and esterifications (*5*).

Although a relatively large number of trichothecenes have been reported and structurally characterized, many of these compounds are minor products of large scale fermentations. A relatively small number of trichothecenes and related compounds have been examined for toxicity in animal or plant systems. Mutant strains of *F. sporotrichioides* and *F. graminearum,* generated during the characterization of trichothecene biosynthesis (*5, 6, 7*), have afforded an efficient means of generating a library of trichothecenes in quantities that are sufficient for phytotoxicity screens *(8, 9)*. Additional derivatives have been prepared with biotransformation and synthetic modification.

Trichothecene resistance

Trichothecenes are inhibitors of protein synthesis in eukaryotes. Some trichothecene-producing fungi, such as *Myrothecium* and *Trichothecium*, have altered ribosomal proteins that render them less susceptible to their own toxins (*10, 11*). Yeast strains have also been identified that have an altered ribosomal protein L3 and are resistant to trichothecene mycotoxins (*12*). A modified L3 has been genetically engineered into tobacco to increase resistance to deoxynivalenol (*13*)

Research to determine how *Fusarium* protects itself from trichothecenes identified *Tri101*, an acetyltransferase gene that controls the addition of a C-3 acetyl group (*14*). Gene disruption of *Tri101* resulted in the accumulation of isotrichodermol and indicated that the gene also controlled a key step in trichothecene biosynthesis (5). This acetyl group is removed by the *Tri8* esterase as a final step in e.g. T-2 toxin biosynthesis (*7*). The acetyl group protects the fungus from its own toxin during biosynthesis and can be thought of an off/on switch for toxicity.

Since trichothecenes have been identified as virulence factors in wheat head scab (15), a strategy for improving resistance to the fungal disease is to express genes for trichothecene resistance in plants. *Tri101* has been engineered into tobacco (*16*), wheat (*17*), barley (*18*), and rice (*19*) in an attempt to introduce resistance to the toxin and thereby increase resistance to *Fusarium.* Engineering wheat and barley with *Fusarium sporotrichioides Tri101* has been reviewed *(20)*. Recent work has shown significant differences in substrate specificity for *Tri101* proteins from *F. sporotrichioides* and *F. graminearum.* With deoxynivalenol as the substrate, the K_m values for *F. graminearum* TRI101 and F. sporotrichioides TRI101 were 11.7 μM and 1463 μM, respectively *(21)*. This suggests that trichothecene resistance could be greatly improved by expressing the *F. graminearum Tri101* ortholog in plants.

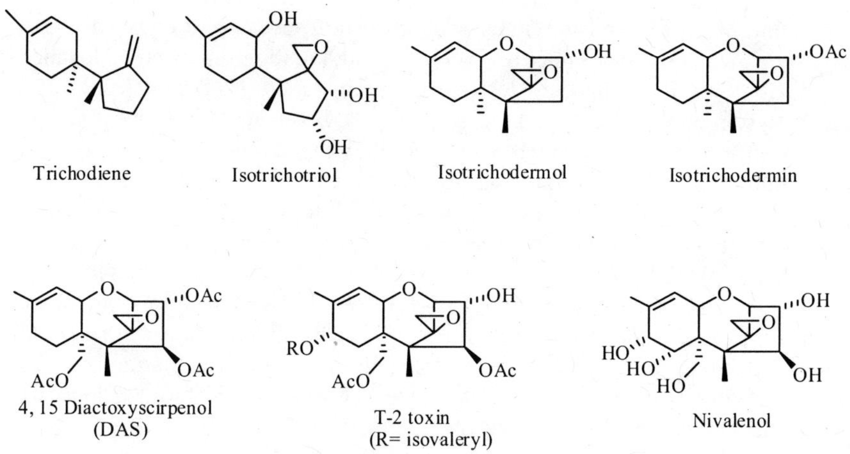

Figure 2. Structures of trichothecenes and the trichothecene precursors trichodiene ands isotrichotriol.

Phytotoxicity assays

Toxicity studies on mycotoxins, including trichothecenes, have mostly focused on animals or animal cell lines. The identification of trichothecenes as virulence factors has increased interest in the phytotoxicity of these compounds. Trichothecenes are known to be inhibitors of protein synthesis but additional effects have been noted including effects that include inhibition of mitochondrial function, electron transport, changes in membrane fluidity, reduced coleoptile growth, changes in seed germination, root and shoot growth, leaf chlorosis and necrosis, bleaching and degradation of chlorophyll (*22*). Bioassays have used whole plants, cell suspension culture and callus culture. Most bioassays have been limited by the availability of trichothecenes so often only a few compounds are tested. Many bioassays that were originally developed for work with environmental pollutants or herbicides should be adaptable for studies with mycotoxins.

Chlamydomonas bioassay

Chlamydomonas reinhardtii is a unicellular green alga that has been used as a model system for photosynthetic and other studies. It can be grown easily on a defined medium in the lab and has a rapid doubling time. It is amenable to transformation (*23, 24*) which may make it a viable system for screening genes for toxin resistance. *Chlamydomonas* sp. have been previously used in aquatic toxicology to assess the deleterious effects of metals and other pollutants *(25)*. The algae have also been used to screen herbicides *(26)*.

C. reinhardtii was used to assess the relative phytotoxicity of a group of trichothecenes (*8*) with either a C-3 hydroxy or C-3 acetoxy group. Cultures were initiated with 1 X 10^5 cells/ml on a high salt, high acetate medium (*27*) containing 80 μM of an individual trichothecene. Cells were counted after 8 days of growth and the number of doublings and doubling time were calculated. Of the fourteen trichothecenes tested, five compounds, isotrichodermol, 4,15-diacetoxyscirpenol (4,15-DAS), T-2 toxin, deoxynivalenol (DON), and 3-decalonectrin, inhibited growth and cell doubling of liquid cultures. One interesting finding from this survey was that the simple trichothecene, isotrichodermol, was as toxic as 4,15-DAS or T-2 toxin to *C. reinhardtii.* After 8 days, cultures grown in the presence of isotrichodermol had only 3 X 10^5 cells/ml compared to over 500 X 10^5 cells/ml in control cultures (*8*). Five C-3 acetoxy analogs, isotrichodermin, 3,4,15-TAS, 3-ADON, 3-acetyl T-2 toxin and calonectrin, were significantly less toxic. These data support the role that the C-3 acetyl group may have in controlling toxicity. Four other compounds tested, 8-hydroxy isotrichodermol, 8-hydroxy isotrichodermin, 3,15-didecalonectrin and 15-decalonectrin (*8*) were less phytotoxic (230-236 X 10^5 cells/ml after 8 days). This suggested that additional structural feature may lower toxicity.

Effects of C-7 and C-8 oxygenation

Two other sets of data comparisons suggest that additional hydroxylation decreases phytotoxicity. 8-hydroxyisotrichodermol and 3,15-didecalonectrin were less toxic than isotrichodermol. Similarly, 8-hydroxyisotrichodermin and 15-decalonectrin were less toxic than isotrichodermin (*8*).

In order to see if other trichothecene substitution patterns are correlated with phytotoxicity, some additional isotrichodermol derivatives were tested with

Chlamydomonas reinhardtii. *C. reinhardtii* cells were counted and cultures started with 1 X 10^5 cells/ml on a liquid high salt, high acetate medium as before (*8, 27*) but were supplemented with a higher concentration (100 μM final concentration) of the trichothecene to be tested and grown for a shorter time period – 5 or 6 days rather than 8 days. Table I shows the growth, number of doublings and doubling time for isotrichodermol and four related compounds. The 6 day incubation was adequate to see differences in toxicity. This screen identified three phytotoxic compounds, isotrichodermol (3OH), 7-hydroxy isotrichodermol (3,7-diOH) and 8-keto isotrichodermol (3OH 8keto), that strongly inhibited the growth of *C. reinhardtii.* 8-hydroxy isotrichodermol and 7,8-dihydroxyisotrichodermol (3,7,8 triOH) were less phytotoxic in this assay. The results indicated that addition of a C-7 hydroxyl or C-8 keto function did not affect toxicity but that C-8 hydroxylation decreased toxicity.

Table I. Effect of trichothecenes on *C. reinhardtii* growth after 6 days

Toxin	Cells (X 10^5)	Doublings	Doubling time (hr)
Control	75.6	6.2	23.1
3-OH	3.1	1.6	94.4
3,7 diOH	2.5	1.3	112.3
3,8 diOH	22.3	4.4	33.6
3,7,8 triOH	41.6	5.4	26.8
3OH 8keto	2.7	1.4	110.9

Phytotoxicity of *Fusarium graminearum* mycotoxins

The initial screen of fourteen trichothecenes with *Chlamydomonas reinhardtii* indicated that deoxynivalenol (DON) caused moderate growth inhibition (79 X 10^5 cells/ml compared to 560 X 10^5 cells/ml for control cultures) (*8*). 3-ADON had growth rates similar to control cultures indicating that C-3 acetylation is an off/on switch for this toxicity.

Four additional deoxynivalenol and nivalenol compounds were tested against *C. reinhardtii.* The results of this study are shown in Table II and Table III. 15-ADON treated cultures had a growth rate similar to that of deoxynivalenol. Both compounds have a free C-3 hydroxyl group. Nivalenol (NIV) was the most toxic compound tested in this group, severely inhibiting growth to less than two doublings after six days. Acetylation of nivalenol to form 3,15-diANIV reduced this toxicity.

Table II. Effect of trichothecenes on *C. reinhardtii* growth after 5 days

Toxin	Cells ($X\ 10^5$)	Doublings	Doubling time (hr)
Control	47.8	5.6	21.7
DON	28.2	4.6	27.1
3ADON	49.2	5.6	21.4
15ADON	24.6	4.3	29.6
3,15diADON	38.3	5.4	22.1

Table III. Effect of trichothecenes on *C. reinhardtii* growth after 6 days

Toxin	Cells ($X\ 10^5$)	Doublings	Doubling time (hr)
Control	71.9	6.2	23.4
Nivalenol	3.2	1.6	100.6
3,15-diANIV	46.1	5.5	26.3

Arabidopsis thaliana model system

The second bioassay developed for evaluating trichothecene phytotoxicity used the small mustard plant *Arabidopsis thaliana. A. thaliana* is an attractive model system due to its short generation time as well as the availability of the complete genome sequence. In addition, *A. thaliana* is susceptible to infection by *Fusarium graminearum* (*28, 29*). A thorough study of the effects of three trichothecenes, 4,15-DAS, DON, and T-2 toxin on *A. thaliana* plants grown on toxin-amended agar indicated that both 4, 15-DAS and DON inhibited root growth (*30*). T-2 toxin caused stunted shoot growth and morphological changes to the leaves including reddening from anthocyanins. 4,15-DAS also inhibited seed germination, an effect also observed with tobacco seeds (*16*).

The *A. thaliana* detached leaf assay is a four week, labor-intensive process (*9*). Briefly, seeds of Columbia (Col-4) ecoptype were surface disinfested with sodium hypochlorite and water. Seeds were then sown in a grid pattern on the surface of a Petri plates filled with Murashige-Skoog mineral medium adjusted to pH 5.9 in agar, and incubated 3 days in the dark at 4 °C. Plates and seeds were then incubated for an additional 11 days under fluorescent light at 25 °C. Agar blocks containing plants were cut from the Petri dish and transferred to 24 well plates and the plants were incubated for an additional week. Leaves with petioles were sliced from three-week old plants with a scalpel, floated in distilled water and then transferred individually to a well of 96 well microplates containing a solution of test solution in water. The treated leaves were scored for chlorosis after an additional week (*9*).

The study looked at 24 trichothecenes ranging from trichodiene and isotrichotriol (Figure 2) to more complex oxygenated and acetylated compounds

(*9*). Compounds were tested at five concentrations (0 to 100 μM) and dose response, LD_{50} and ED_{50} were calculated. This screen identified six relatively non-toxic compounds that showed little or not chlorosis at 100 μM and had an LD50, the concentration at which 50% of the leaves were dead, over 100 μM: trichodiene, isotrichotriol, 3,15-didecalonectrin, 15-decalonectrin, scirpentriol and nivalenol (Figure 2). The most toxic compounds in this assay, with LD_{50} under 10 μM, were isotrichodermol, calonectrin, 15-acetoxyscirpenol, 4,15-DAS, HT-2 toxin, T-2 toxin, 3-acetyl T-2 toxin, 15-ADON and 3,15-diacetylnivalenol (*9*).

Arabidopsis thaliana screen of toxin library

Time and labor were a limiting factor in the number of trichothecenes originally tested with *A. thaliana* detached leaves. Additional compounds were screened in a similar manner but at a single concentration, 100 μM. Each leaf was scored after one week in a toxin solution as in the earlier study. The number of leaves that had more than 50% chlorosis (rating 3) or were dead were counted and this sum was used to calculate a %dead figure. This combined figure was previously used to calculate LD_{50} (*9*). Two experiments with 48 leaves/compound were completed for each trichothecene or *Fusarium* metabolite. While this streamlined assay may miss subtle differences in toxicity, it was easy to distinguish between compounds that were toxic (>80% dead) and non-toxic (<10% dead) at 100 μM.

In addition to trichodiene and isotrichotriol, two other trichodiene derivatives, isotrichodiol and isotrichotetraol, and three non-trichothecene *Fusarium* metabolites, zearalenone, butenolide and culmorin were non-toxic to *A. thaliana* leaves.

T-2 toxin was the most phytotoxic trichothecene tested (LD_{50} of 0.5 μM) in the first bioassay (*9*) and it caused 100% dead leaves in the second assay. Other neosolaniol-based trichothecenes (oxygenation at C-3, C-4, C-8, C-15), neosolaniol, 8-butyryl neosolaniol, 8-propionyl neosolaniol, HT-2 toxin, 3-acetyl T-2 toxin, as well as 4-deoxy T-2 toxin, caused substantial chlorosis and greater than 80% dead leaves.

4,15-DAS was also highly phytotoxic (LD_{50} of 1.5 μM) in first assay (*9*) and killed all leaves tested at 100 μM in the second assay. 3,15-DAS caused more than 80% dead leaves in the second assay. Of the compounds tested with oxygenation at C-3, C-4, and C-15, only scirpentriol was relatively non-toxic (LD_{50} > 100 μM) (*9*).

Effect of additional hydroxylation

The first screen of 24 trichothecenes indicated that increased oxygenation was associated with lower phytotoxicity. Isotrichodermol was moderately toxic in the first bioassay with an LD_{50} of 9 μM but 3,15-didecalonectrin and scirpentriol had reduced toxicity (*9*). In the second assay at 100 μM, isotrichodermol killed 60% of the leaves. Four additional hydroxylated derivatives of isotrichodermol were tested at 100 μM, 7-hydroxyisotrichodermol, 8-hydroxyisotrichodermol, 7,8-dihydroxyisotrichodermol, and 4,8- dihydroxy-isotrichodermol, and each killed fewer than 10% of the leaves. 8-ketoisotricho- dermol caused 14 % dead leaves.

C-3 oxygenation

All *Fusarium* trichothecenes have a C-3 oxygen function as a result of the four oxygenations controlled by *Tri4.* Three *Trichothecium roseum* trichothecene metabolites that lack C-3 oxygenation were tested. Trichothecene (no extraskeletal oxygens) and trichothecolone (C4-OH, C-8 keto) were non-toxic (0 % dead). The *T. roseum* mycotoxin trichothecin (C-4 butyryloxy, C-8 keto) however, was phytotoxic, causing 97.5% dead leaves. These results indicate that the trichothecene C-3 oxygen function is not required for phytotoxicity of *Arabidopsis* leaves.

Detoxification of trichothecenes

Acetylation of isotrichodermol to form isotrichodermin resulted in a decrease in phytotoxicity – from an LD_{50} of 9 μM to LD_{50} of 16 μM (*9*). Similar decreases in toxicity were observed between 4,15-DAS and 3,4,15-TAS, and 15-ADON and 3,15-diADON. In the second assay, isotrichodermin killed 40% of the leaves tested compared to 60% with isotrichodermol.

Both *Chlamydomonas* and *Arabidopsis* assays indicated that the C-3 acetyl group can reduce the phytotoxicity of some trichothecenes. However, this acetyl group can be removed by esterases in the fungus (*7*). There is ample evidence that plant esterases can remove C-3 and C-15 acetyl groups. For example, *Fusarium* strains can be characterized as 3-ADON producers or 15-ADON producers in culture, but *Fusarium*-infected grains are primarily contaminated with deoxynivalenol rather than one of the acetylated derivatives (*31*).

Two other derivatives of C-3 hydroxy trichothecenes were produced and tested with *A. thaliana* detached leaves. O-methyl derivatives of isotrichodermol and 4,15-DAS were prepared with methyl iodide. Neither 3-O-methylisotrichiodermol nor 3-O-methyl 4,15 DAS caused any chlorosis of detached leaves at 100 μM (0% dead).

A soil bacterium has been isolated that can convert deoxynivalenol into 3-keto-deoxynivalenol (*32*). This compound had reduced immunosuppressive activity when compared to deoxynivalenol but the phytotoxicity of the 3-keto derivative was not determined. 3-keto derivatives of isotrichodermol and 15-ADON were prepared using a modification of the Swern oxidation method (*33*, *34*). Both 3-keto products were completely non-toxic at 100 μM to *A. thaliana* leaves (0% dead).

Comparison of *Chlamydomonas* and *Arabidopsis* bioassays

There are some noticeable differences in the phytotoxicity of individual trichothecenes between the *Chlamydomonas* and *Arabidopsis* bioassays. C-3 acetylation significantly decreased the toxicity of five compounds in *Chlamydomonas* (*8*). Only C-3 acetylation of two of these compounds, 4,15-DAS and T-2 toxin, caused a large decrease in toxicity in *Arabidopsis* (*9*). These differences may be due to differences in uptake or in the relative metabolism of the compounds tested to less or to more toxic products. For example, glycosylation of a trichothecene C-3 hydroxyl group by a plant glycosyltransferase or removal of a C-3 acetyl group by a plant esterase could change which toxin was actually tested and at what concentration. Treated algal cultures and leaves were not examined for metabolism of the trichothecenes tested.

There was also a significant difference between the two assays for one pair of compounds, nivalenol and its acetylated derivative, 3,15-diacetylnivalenol. Nivalenol was not at all toxic to *A. thaliana* leaves (LD_{50} > 100) but 3,15-diacetylnivalenol was quite toxic (LD_{50} of 4.3 μM) (*8*). These compounds had the opposite effects on *C. reinhardtii* cultures. Nivalenol was quite toxic (Table III) but the 3,15-diacetyl derivative caused much less growth inhibition.

Targets for resistance

Although *Fusarium Tri101* has been the main focus for introducing trichothecene resistance into plants, other routes to trichothecene detoxification may lead to improved resistance. Glycosylation of the C-3 hydroxyl group has been correlated with moderate scab resistant lines derived from of Sumai-3 wheat and a glucosylransferase has been isolated from *Arabidopsis thaliana* that can detoxify deoxynivalenol (*35*). Other possible detoxification proteins are C-3 oxidase *(32)*, epoxide reductase *(36),* and epoxide hydrolase *(37).*

An alternate strategy for improving disease resistance is by chemically blocking, with plant metabolites or herbicides, trichothecene biosynthesis. The relatively simple trichothecene, isotrichodermol, was shown to be phytotoxic in both *Chlamydomonas* and *A. thaliana* bioassays. A single gene, *Tri4*, controls the conversion of trichodiene to form this toxic product (*4*). This suggests that inhibitors that block the TRI4 enzyme would effectively block the production of a phytotoxin and could decrease the virulence of the fungus. A screen of plant shikimates identified a number of compounds that block trichothecene T-2 toxin biosynthesis in *F. sporotrichioides (38).* To obtain the intermediate and TRI4 substrate trichodiene, inhibition of this enzyme by xanthotoxin was used (*39*). A yeast screening system for TRI4 inhibitors, using a synthetic trichodiene analog, identified flavone, xanthotoxin, and two other furanomcoumarins as strong inhibitors of TRI4 activity (*40*).

Summary

This paper compared two bioassays for measuring the relative phytotoxicity of trichothecenes. Both *Arabidopsis* and *Chlamydomonas* are amenable to transformation and both phytotoxicity assays could be adapted to assess toxin resistance genes. Heterologous expression of genes in *Chlamydomonas* has been problematic due to its G-C rich codon bias *(41)*. One goal of *A. thaliana* studies was to develop a model system for screening *Arabidopsis* ecotypes (*28, 42*) for new sources of resistance to trichothecenes, and to identify which toxin to use in such a screen. The detached leaf assay may not be the most efficient method for screening for phytotoxicity or trichothecene resistance in *A. thaliana,* but has the advantage of focusing on a single symptom of phytotoxicity, leaf chlorosis. Germinating and growing whole plants on toxin-containing media may be a more tractable assay if sufficient amounts of trichothecenes are available. Isotrichodermin (Figure 2) may be the best trichothecene for an initial screen of ecotypes. Using this compound will increase the likelihood that novel resistance genes can be found rather than acetyltransferase or glycosyltransferase genes.

The bioassays also indicated that the simple trichothecenes, isotrichodermol and isotrichodermin, are phytotoxic. Chemicals that target the TRI4 enzyme will block biosynthesis of the toxic trichothecenes and may increase plant resistance to *Fusarium* diseases in plants.

References

1. Desjardins, A.E. *Fusarium* mycotoxins: chemistry, genetics, and biology. The American Pathological Society, St. Paul, Minnesota, **2006**, 260 pp.
2. Kimura, M.; Tokai, T.; Takahashi-Ando, N.; Ohsato, S.; Fujimura, M. Molecular and Genetic studies of trichothecene biosynthesis: pathways, genes, evolution. *Biosci. Biotechnol. Biochem.* **2007**, 71, 2105-2123
3. Hohn, T.M.; VanMiddlesworth, F. Purification and characterization of the sesquiterpene cyclase trichodiene synthase from *Fusarium sporotrichioides.* Arch. Biochem. Biophys. **1986,** 251, 756-761.
4. McCormick, S.P.; Alexander, N.J.; Proctor, R.H. *Fusarium Tri4* encodes a multifunctional oxygenase required for trichothecene biosynthesis. *Can. J. Microbiol.* **2006**, 52, 636-642.
5. McCormick, S.P.; Alexander, N.J.; Trapp, S.E.; Hohn, T.M. Disruption of TRI101, the gene encoding trichothecene 3-O-acetyltransferase, from *Fusarium sporotrichioides. Appl. Environ. Microbiol.* **1999**, 65, 5252-5256.
6. McCormick, S.P.; Hohn, T.M.; Desjardins, A.E. Isolation and characterization of Tri3, a gene encoding 15-O-acetyltransferase from *Fusarium sporotrichioides. Appl. Environ. Microbiol.* **1996**, 62-353-359.
7. McCormick, S.P.; Alexander, N.J. *Fusarium Tri8* Encodes a Trichothecene C-3 Esterase. *Appl. Environ. Microbiol.* **2002**, 68, 2959-2964.

8. Alexander, N.J.; McCormick, S.P.; Ziegenhorn, S.L. Phytotoxicity of selected trichothecenes using *Chlamydomonas reinhardtii* as a model system. *Natural Toxin*, **1999**, 7, 265-269.
9. Desjardins, A.E.; McCormick, S.P.; Appell, M. Structure-activity relationships of trichothecenes toxins in an *Arabidopsis thaliana* leaf assay. *J. Agric. Food Chem.* **2007**, 55:6487-6492.
10. Hobsen, A.N.; Cundliffe, E. Ribosomal resistance to the 12,13-epoxytrichothecene antibiotics in the producing organism *Myrothecium verrucaria. Biochem. J.* **1980**, 190, 765-770.
11. Iglesias, M.; Ballesta, J.P. Mechanism of resistance to the antibiotic trichothecin in the producing fungi. *Eur. J. Biochem.* **1994**, 223, 447-453.
12. Fernandez-Lobato, M.; Cannon, M.; Mitlin, J.A.; Mount, R.C.; Jimenez, A. Characterization of *Saccharomyces cerevisiae* strains displaying high-level or low-level resistance to trichothecene antibiotics. *Biochem. J.* **1990**, 267, 709-713.
13. Harris, L.J.; Gleddie, S.C. A modified *Rpl3* gene from rice confers tolerance of the *Fusarium graminearum* mycotoxin deoxynivalenol to transgenic tobacco. *Physiol. Mol. Plant Pathol.* **2001**, 58, 173-181.
14. Kimura, M.; Kaneko, I.; Komiyama, M.; Takatsuki, A.; Koshino, H.; Yoneyama, K.; Yamaguchi, I. Trichothecene 3-O-acetyltransferase protects both the producing organism and transformed yeast from related mycotoxins. Cloning and characterization of *Tri101*. *J. Biol. Chem.* **1998**, 273, 1654-1661.
15. Proctor, R.H.; Hohn, T.M.; McCormick, S.P. Reduced virulence of *Gibberella zeae* caused by disruption of a trichothecene toxin biosynthetic gene. *Mol. Plant-Microbe Interact.* **1995**, 8, 593-601.
16. Muhitch, M. J.; McCormick, S. P.; Alexander, N. J.; Hohn, T. M. Transgenic expression of the *TRI101* or *PDR5* gene increases resistance of tobacco to the phytotoxic effects of the trichothecene 4,15-diacetoxyscirpenol. *Plant Sci.* **2000**, 157, 201-207.
17. Okubara, P.A.; Blechl, A. E.; McCormick, S. P.; Alexander, N.J.; Dill-Macky, R.; Hohn, T.M. Engineering deoxynivalenol metabolism in wheat through the expression of a fungal trichothecene acetyltransferase gene. *Theor. Appl. Genet.* **2002**, 106, 74-83.
18. Manoharan, M.; Dahleen, L. S.; Hohn, T. M.; Neate, S. M.; Yu, X.; Alexander, N. J.; McCormick, S. P.; Bregitzer, P.; Schwarz, P. B.; Horsley, R. D. Expression of 3-OH trichothecene acetyltransferase in barley (*Hordeum vulgare* L.) and effects on deoxynivalenol. *Plant Sci.* **2006**, *171*, 699-796.
19. Ohsato, S.; Ochiai-Fukuda, T.; Nishiuchi, T.; Takahashi-Ando, N.; Koizumi, S.; Hamamoto, H.; Kudo, T.; Yamaguchi, I.; Kimura, M. Transgenic rice plants expressing trichothecene 3-O-acetyltransferase show resistance to the *Fusarium* phytotoxin deoxynivalenol. *Plant Cell Rep.* **2006**, 26, 531-538.
20. Alexander, N.J. The TRI101 story: engineering wheat and barley to resist Fusarium head blight. World Mycotoxin J. **2008**, 1, 31-37.
21. Garvey, G.S.; McCormick, S.P.; Rayment, I. Structural and functional characterization of the TRI101 trichothecene 3-O-acetyltransferase from

Fusarium sporotrichioides and Fusarium graminearum: kinetic insights to combating Fusarium head blight. *2008*, *J. Biol. Chem.* 283, 1660-1669.

22. McLean, M. The phytotoxicity of *Fusarium* metabolites: an update since 1989. *Mycopathologia* **1996**, 133, 163-179.
23. Kindle, K.L. High-frequency nuclear transformation of *Chlamydomonas reinhardtii*. *Proc. Natl. Acad. Sci. U.S.A.* **1990**, 87, 1228-1232.
24. Shimogawara, K.; Fujiwara, S.; Grossman, A.; Usua, H. High-efficiency transformation of *Chlamydomonas reinhardtii* by electroporation. *Genetics* **1998**, 148, 1821-1828.
25. Geoffroy, L.; Gilbin, R.; Simon, O.; Floriani, M.; Adam, C.; Pradines, C.; Cournac, L.; Garnier-Laplace, J. Effect of selenate on growth and photosynthesis of *Chlamydomonas reinhardtii*. *Aquat. Toxicol.* **2007**, 83, 149-158.
26. Hess, F.D. A *Chlamydomonas* algal bioassay for detecting growth inhibitor herbicides. *Weed Science* **1980**, 28, 515-520.
27. Harris, E.H. The *Chlamydomonas* Sourcebook. San Diego, CA, Academic Press **1989,** 780 pp.
28. Urban, M.; Daniels, S.; Mott, E.; Hammond-Kosack, L. *Arabidopsis* is susceptible to the cereal ear blight fungal pathogens *Fusarium graminearum* and *Fusarium culmorum*. *Plant J.* **2002**, 32, 961-973.
29. Cuzick, A.; Urban, M.; Hammond-Kosack, K. *Fusarium* graminearum gene deletion mutants map1 and tri5 reveal similarities and differences in the pathogenicity requirements to cause disease on *Arabidopsis* and wheat floral tissue. *New Phytologist* **2008**, 177:990-1000.
30. Masuda, D.; Ishida, M.; Yamaguchi, K.; Yamaguchi, I.; Kimura, M.; Nishiuchi, T. Phytotoxic effects of trichothecenes on the growth and morphology of *Arabidopsis thaliana*. *J. Exp. Bot.* **2007**. 58, 1617-1626.
31. Ward, T.J.; Clear, R.M.; Rooney, A.P.; O'Donnell, K.; Gaba, D.; Patrick, S.; Starkey, D.E.; Gilbert, J.; Geiser, D.M., Nowicki, T.W. An adaptive evolutionary shift in Fusarium head blight pathogen populations is driving the rapid spread of more toxigenic *Fusarium graminearum* in North America. *Fung. Genet.* **2008**, 45, 473-484.
32. Shima, J. S.; Takase, S.; Takahashi, Y.; Iwai, Y.; Fujimoto, H.; Yamazaki, M.; Ochi, K. Novel detoxification of the trichothecene mycotoxin deoxynivalenol by a soil bacterium isolated by enrichment culture. *Appl. Environ. Microbiol.* **1997**, 63, 3825-3830.
33. Savard, M.E.; Blackwell, B.A.; Greenhalgh, R. An ^{1}H nuclear magnetic resonance study of derivatives of 3-hydroxy-12,13-epoxytrichothec-9-enes. *Can. J. Chem.* **1987**, 65, 2254-2262.
34. Ohsugi, S.; Nishide, K.; Oono, K.; Okuyama, K.; Fudesaka, M.; Kodamaa, S.; Nodea, M. New odorless method for the Corey–Kim and Swern oxidations utilizing dodecyl methyl sulfide (Dod-S-Me). *Tetrahedron* **2003**, 59, 8394-8398.
35. Lemmens, M.; Scholz, U.; Berthiller, F,; Dall'Asta, C.; Koutnik, A.; Schuhmacher, R.; Adam, G.; Buerstmayr, H.; Mesterházy, A.; Krska, R.; Ruckenbauer, P. The ability to detoxify the mycotoxin deoxynivalenol colocalizes with a major quantitative trait locus for Fusarium head blight resistance in wheat. *Mol. Plant-Microbe Interact.* **2005,** 18, 1318-1324.

36. Fuchs, E.; Binder, E.M.; Heidler, D.; Krska, R. Structural characterization of metabolites after the microbial degradation of type A trichothecenes by bacterial strain BBSH 797. *Food Addit. Contam.* **2002**, 18, 379-386.
37. Weijers, C.A.G.M. Enantioselective hydrolysis of aryl, alicyclic and aliphatic epoxides by *Rhodotorula glutinis*. *Tetrahedron: Asymmetry* **1997**, 8, 639-647.
38. Desjardins, A.E.; Plattner, R.D.; Spencer, G.F. Inhibition of trichothecene toxin biosynthesis by naturally occurring shikimate aromatics. *Phytochemistry* **1988**, 27, 767-771.
39. Hesketh, A.R.; Gledhill, L.; Marsh, D.C.; Bycroft, B.W.; Dewick, P.M.; Gilbert, J. Biosynthesis of trichothecene mycotoxins: identification of isotrichodiol as a post-trichodiene intermediate. *Phytochemistry* **1991**, 30, 2237-2243.
40. Takahashi-Ando, N.; Ochiai, N.; Tokai, T.; Ohsato, S.; Nishiuchi, T.; Yoshida, M.; Fujimura, M.; Kimura, M. A screening system for inhibitors of trichothecene biosynthesis: hydroxylation of trichodiene as a target. *Biochem. Lett.* **2008**, 30, 1055-1059.
41. Franklin, S.E.; Mayfield, S.P. Prospects for molecular farming in the green alga *Chlamydomonas reinhardtii*. Curr. Opin. Plant Biol. **2004**, 7, 159-165.
42. Chen, X.; Steed, A.; Harden, C.; Nicholson, P. Characterization of *Arabidopsis thaliana-Fusarium graminearum* interactions and identification of variation in resistance among ecotypes. *Mol. Plant Pathol.* **2006**, 7, 391-403.

Chapter 11

Identification of Maize Breeding Markers through Investigations of Proteins Associated with Aflatoxin-Resistance

Robert L. Brown[1], Zhi-Yuan Chen[2], Thomas E. Cleveland[1], Abebe Menkir[3], Ahmad Fakhoury[4]

[1]USDA, ARS, SRRC, New Orleans, LA 70124
[2]Department of Plant Pathology and Crop Physiology, Louisiana State University Agricultural Center, Baton Rouge, LA 70803
[3]International Institute of Tropical Agriculture, Ibadan, Nigeria
[4]Southern Illinois University, Carbondale, IL 62901

The goal of a collaborative research project between International Institute of Tropical Agriculture (IITA) in Nigeria and ARS-Southern Regional Research Center (SRRC) in New Orleans is to develop maize inbred lines with resistance against aflatoxin contamination by *Aspergillus flavus*. A second goal is to identify gene markers in these lines to facilitate their use in U.S. breeding and African national programs. To accomplish this, comparative proteomics of near-isogenic lines varying in aflatoxin accumulation is being employed to identify kernel resistance associated proteins with (RAPs). A number of RAPs have been identified and several further characterized through physiological and biochemical investigations conducted to determine a potential role in resistance and, therefore, fitness as breeding markers. Three RAPs, a trypsin inhibitor, pathogenesis-related protein and glyoxalase I were investigated as well, using RNAi gene silencing and plant transformation. Results of proteome and characterization studies are discussed.

Screening for Resistance to Aflatoxin Accumulation

A number of approaches to eliminating aflatoxins from susceptible crops such as corn, cottonseed, peanut, and treenuts have been advanced, however, the best and most widely explored strategy is the development of preharvest host resistance. This is because *A. flavus* infects affected crops prior to harvest (*1*). Effective, reliable, and rapid screening techniques are indispensable prerequisites to breeding for resistance to aflatoxin accumulation in maize (*2, 3*). Brown et al. (*4*) developed a rapid laboratory-based kernel-screening assay (KSA) that creates higher and more uniform levels of infection and aflatoxin production and allows differentiation of resistant and susceptible maize genotypes. This assay provides consistent ranking of maize genotypes in different tests and the results seem to be correlated with resistance levels expressed by maize genotypes in field trials (*4*).

This rapid assay employs a very simple and inexpensive procedure (*4*). Kernels screened by the KSA are usually incubated for seven days in 100% humidity, at a temperature of 31°C that favors *A. flavus* growth and aflatoxin production. Aflatoxin amounts in kernels from KSA experiments can be obtained two weeks after experiments are initiated. KSA experiments confirmed GT-MAS:gk resistance to aflatoxin production and demonstrated that the resistance in otherwise viable kernels is maintained, when the pericarp barrier is breached (*5*). Penetration through the pericarp barrier was achieved by wounding the kernel with a hypodermic needle down to the endosperm, prior to inoculation. Wounding allows the fungus to freely access the seed without physical barriers, thereby facilitating differentiation between stable resistance mechanisms in operation, and physical attributes of the seed imparting temporary resistance. The results of this study indicated the presence of two levels of resistance: at the pericarp and at the subpericarp level. The former was supported by KSA studies which demonstrated a role for pericarp waxes in kernel resistance (*6, 7, 8*) and highlighted quantitative and qualitative differences in pericarp wax between GT-MAS:gk and susceptible genotypes (*9*). The KSA also confirmed sources of resistance among 31 inbreds tested in Illinois field trials (*4, 10*).

The KSA has several advantages as compared to traditional breeding techniques (*4*): 1) it can be performed and repeated several times throughout the year and outside of the growing season; 2) it requires few kernels; 3) it can detect/identify different kernel resistance mechanisms expressed; 4) it can dispute or confirm field evaluations (e.g. identify escapes); and, 5) relationships between laboratory findings and inoculations in the field have been demonstrated. The KSA has proven be a valuable complement to standard breeding practices in the evaluation of germplasm for aflatoxin-resistance. This was most recently demonstrated through the registration and release of six aflatoxin-resistant maize inbreds developed through a collaboration between IITA-Nigeria (International Institute of Tropical Agriculture) and the Southern Regional Research Center (SRRC) of USDA-ARS in New Orleans (*11*). Field trials, however, are irreplaceable for confirmation of resistance.

Developing resistance to fungal infection in wounded as well as intact kernels would go a long way toward solving the aflatoxin problem (*12*). Studies demonstrating subpericarp (wounded-kernel) resistance in corn kernels have led to research to identify subpericarp resistance mechanisms. When kernels of susceptible genotypes were allowed to imbibe water at 100% humidity at 31°C for 3 days prior to being inoculated in the KSA protocol, certain kernel proteins became elevated, and aflatoxin levels were drastically and significantly reduced compared to unimbibed controls (*13*). These observations suggest that even susceptible kernels contain proteins capable of being induced during imbibition that can inhibit growth and/or fungal elaboration of aflatoxins. A recent investigation into maize kernel resistance (*14*), however, determined that both constitutive and induced proteins are required for resistance to aflatoxin production. It demonstrated that one major difference between resistant and susceptible genotypes is the relatively high level of constitutively expressed antifungal proteins in resistant lines compared to susceptible lines. A function of these high levels of constitutive proteins may be to delay fungal invasion, and subsequent aflatoxin formation, until other antifungal proteins can be synthesized to form an active defense system.

Identification of Resistance-Associated Proteins (RAPs)

In a previous study, when imbibed susceptible kernels, for example, showed decreased aflatoxin levels, this was paralleled by increases in ribosome inactivating protein (RIP) and zeamatin (*15*). Both zeamatin and RIP have been shown to inhibit *A. flavus* growth *in vitro* (*15*). In another study, two kernel proteins were identified from a resistant corn inbred (Tex6) which may contribute to resistance to aflatoxin contamination (*16*). One protein, 28 kDa in size, inhibited *A. flavus* growth, while a second, over 100 kDa in size, primarily inhibited toxin formation. Recently, the antifungal protein was identified as an endochitinase and was shown to be a major contributor to Tex6 resistance (*17*). When a commercial maize hybrid was inoculated with aflatoxin and nonaflatoxin-producing strains of *A. flavus* at milk stage, one induced chitinase and one ß-1,3-glucanase isoform were detected in maturing infected kernels, while another isoform was detected in maturing uninfected kernels (*18*).

In another investigation, an examination of kernel protein profiles of 13 maize genotypes revealed that a constitutively-produced 14 kDa trypsin inhibitor protein (TI) is present at relatively high concentrations in seven resistant maize lines, but at low concentrations or is absent in six susceptible lines (*19*). The mode of action of TI against fungal growth may be partially due to its inhibition of fungal -amylase, limiting *A. flavus* access to simple sugars (*20*) required not only for fungal growth, but also for toxin production (*21*). TI also demonstrated antifungal activity against other mycotoxigenic species (*22*). The identification of these proteins may provide markers for plant breeders, and may facilitate the cloning and introduction of antifungal genes through genetic engineering into other aflatoxin-susceptible crops.

Using Comparative Proteomics to Identify RAPs

To increase protein reproducibility, spot resolution and detection sensitivity by 10 to 20 fold and, thus, enhance ability to identify more constitutively-expressed RAPs, a proteomics approach was employed. Kernel proteins from several resistant and susceptible genotypes were compared using large format 2-D gel electrophoresis, and over a dozen such protein spots, either unique or 5-fold up-regulated in resistant lines, were identified, isolated from preparative 2-D gels and analyzed using ESI-MS/MS after in-gel digestion with trypsin (*23, 24*). These proteins can be grouped into three categories based on their peptide sequence homology: (1) storage proteins, such as globulins (GLB1, GLB2), and late embryogenesis abundant proteins (LEA3, LEA14); (2) stress-responsive proteins, such as aldose reductase (ALD), glyoxalase I (GLX1) and heat shock proteins, and (3) antifungal proteins, including TI. In total, approximately 17 proteins upregulated in resistant versus susceptible lines have been identified using comparative proteomics (Table I).

Table I. Resistance-associated Proteins Identified in Maize Kernels using Proteomics

Antifungal	Stress-responsive	Storage
Zeamatin	Peroxiredoxin 1	Globulin I
TI (14 kDa)	Aldose reductase	Globulin II
PR-10	Glyoxalase I	LEA III
TI (10 kDa)	Anionic Peroxidase	LEA 14
β-1,3-glucanase	Cold Regulated Protein	
RIP	Water stress inducible	
	Heat shock	

SOURCE: Table adapted from Reference 25.

No investigation has been conducted, thus far, to determine the possible direct involvement of stress-related proteins in host fungal resistance. However, increased temperatures and drought, which often occur together, are major factors associated with aflatoxin contamination of corn kernels (*26*). Possession of unique or higher levels of hydrophilic storage or stress-related proteins, such as the aforementioned, may put resistant lines in an advantageous position over susceptible genotypes in the ability to synthesize proteins and defend against pathogens under stress conditions. Further studies including physiological and biochemical characterization, genetic mapping, plant transformation using RAP genes, RNAi gene silencing experiments (*27*) and marker-assisted breeding should clarify the roles of stress-related RAPs in kernel resistance.

The screening of progeny generated through the collaborative project between IITA-Nigeria and USDA-ARS-SRRC facilitated the identification of near-isogenic lines from the same backcross differing significantly in aflatoxin accumulation (Table II), and proteome analysis of these lines is being conducted

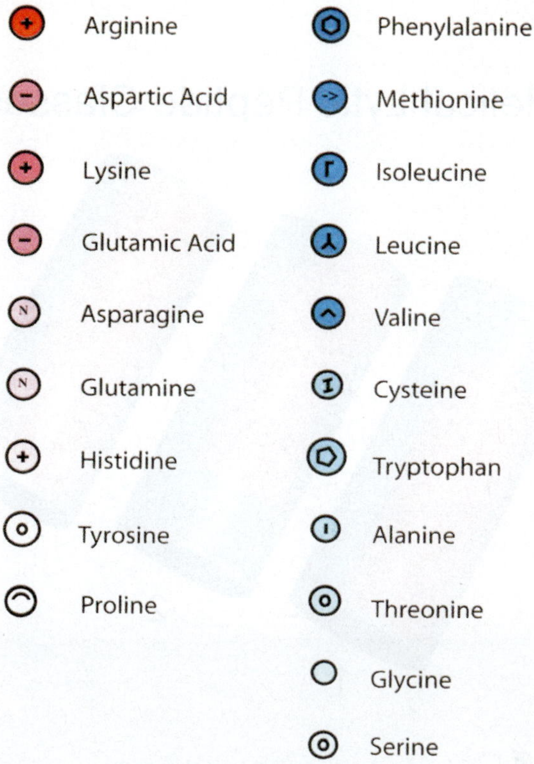

Figure 9.1. Amino acids in lytic peptides. The symbols represent a molecular alphabet

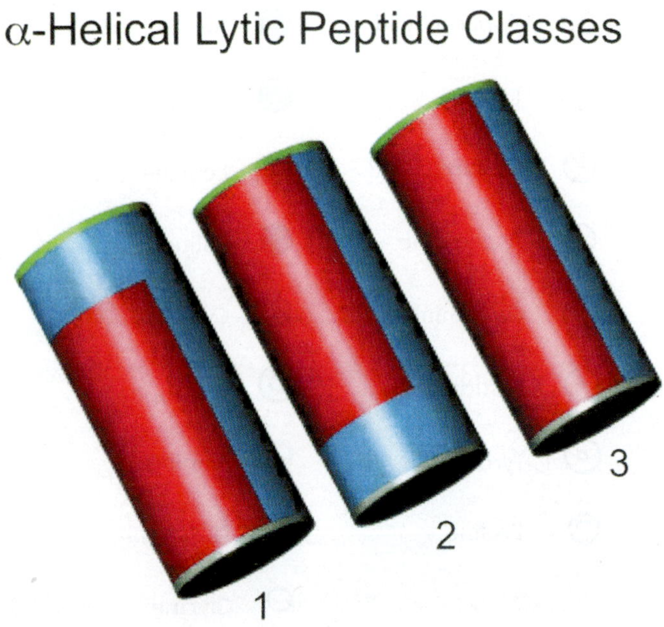

Figure 9.2. α-helical lytic peptide classes

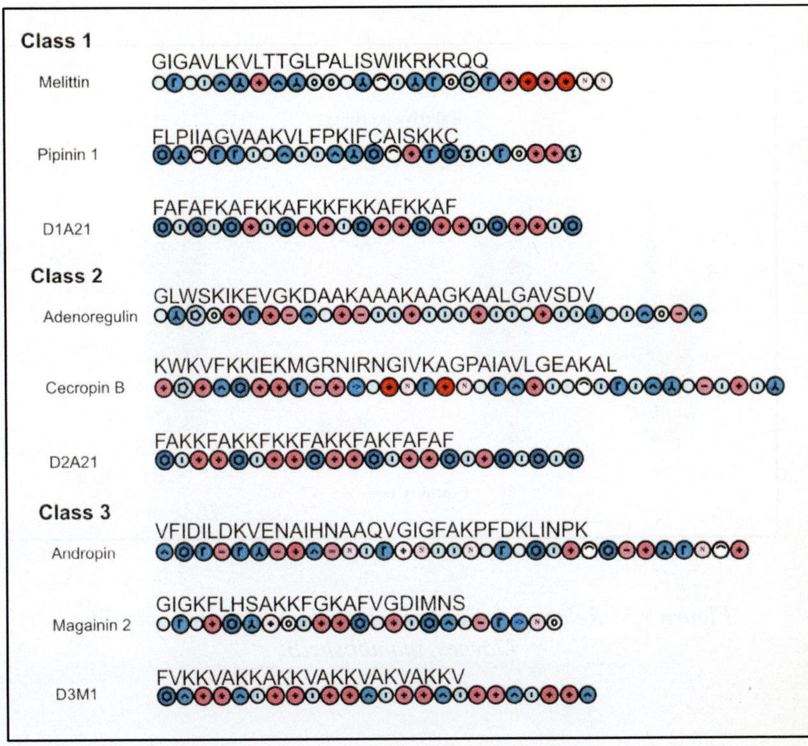

Figure 9.3. Natural lytic peptides.

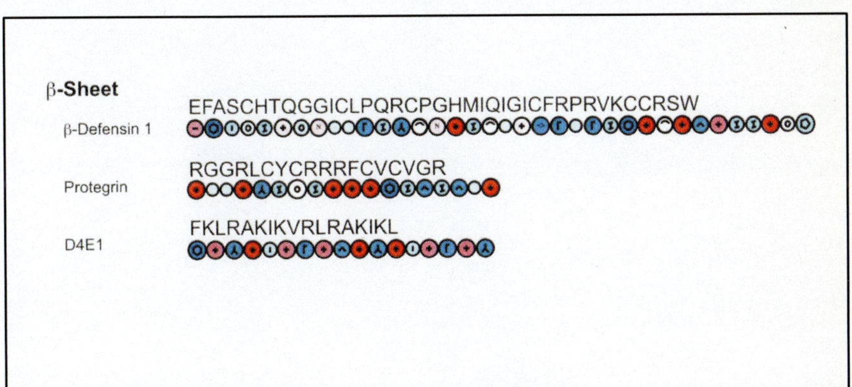

Figure 9.4. β-sheet peptides

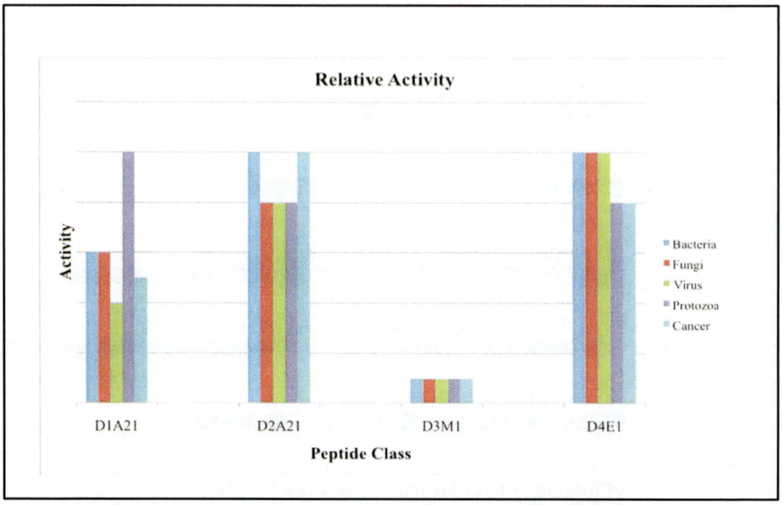

Figure 9.5. Relative Activity of selected synthetic lytic peptides (Jaynes, unpublished)

(*27, 28*). Investigating corn lines sharing close genetic backgrounds should enhance the identification of RAPs clearly without the confounding effects experienced with lines of diverse genetic backgrounds.

Table II. Near-Isogenic Maize Pairs Varying in Aflatoxin Levels via KSA

Line	Pedigree	AFB_1(ppb)
1	(GT-MAS:gk x KU1414SR x GT-MAS:gk)-8-1-2-3-B*6	1263 b
2	(GT-MAS:gk x KU1414SR x GT-MAS:gk)-8-1-2-4-B*6	11 e
25	1368 x GT-MAS:gk-8-1-1-4-B*4	5 h
31	1368 x GT-MAS:gk-8-1-1-3-B*3	1984 bc
86	1368 x MI82-23-2-1-3-B*4	156 f
87	1368 x MI82-23-2-1-5-B*4	1226 de
111	GT-MAS:gk x Babangoyo-1-1-3-B*6	3710 c
112	GT-MAS:gk x Babangoyo-1-1-1-B*2	276 fg

NOTE: Percentage of genetic similarity between pairs is as follows: Lines 1 and 2 =93.75%; Lines 25 and 31=87.5%; Lines 86 and 87=87.5%; Lines 111 and 112=75%.

Further Characterization of RAPs

Of the constitutively-expressed proteins identified through proteomics, several have been further investigated to understand their potential involvement in resistance. Among those investigated are: 1) aldose reductase (ALD), 2) glyoxalase I (GLX-I), 3) pathogenesis related protein 10 (PR-10), 4) peroxiredoxin antioxidant (PER1), 5) cold-regulated protein (ZmCORp), and 6) trypsin inhibitor-10kDa (ZmTIp).

Aldose reductase, which is reported to have a role in plant stress tolerance, is found constitutively-produced at higher levels in kernel embryo tissue of resistant versus susceptible maize genotypes (*24*). However, after *A. flavus'* infection, ALD activity was found to be higher in highly susceptible lines than in resistant lines. Interestingly, lines determined to be intermediately susceptible varied from highly susceptible lines in ALD production.

Glyoxalase I, also isolated from kernel embryo, is involved in the conversion of cytotoxic methylglyoxal (MG) into D-lactate, along with GLX II, and is suggested to be important to plant stress tolerance (*29*). Higher GLX-I activity was observed in maize kernels of resistant genotypes than in susceptibles both constitutively and after *A. flavus'* infection. However, infection significantly increased MG levels in two of three susceptible lines. MG was also shown to induce aflatoxin production in vitro; its mode of action may be to stimulate expression of the aflatoxin regulatory gene *aflR*.

During an investigation of PR-10, which was isolated from kernel endosperm (*30*), it was discovered that during kernel development, *pr-10* expression increased fivefold between 7 and 22 days after pollination, and was induced upon *A. flavus* infection in the resistant but not the susceptible genotype. It was also shown that PR-10 had ribonuclease and antifungal activities. Leaf extracts of transgenic tobacco plants expressing *pr-10* also demonstrated RNase activity and inhibited *A. flavus* growth.

PER1 protein, also produced in the endosperm, demonstrated peroxidase activity *in vitro*. Also, *per1* expression was significantly higher in a resistant genotype versus a susceptible one during the late stages of development, and was significantly induced upon *A. flavus* infection (31).

ZMCORp protein has a sequence similar to cold-regulated protein, however, it exhibited lectin-like hemagglutination activity against fungal conidia and sheep erythrocytes (*32*). When tested against *A.flavus*, it was shown to inhibit germination of conidia by 80% and to decrease mycelial growth by 50% when germinated conidia were incubated with the protein. Quantitative real-time RT-PCR revealed ZmCORp to be expressed 50% more in kernels of a resistant maize line versus a susceptible.

ZmTIp, a 10 kDa trypsin inhibitor, had an impact on fungal growth, but not as great as previously investigated TIs (*33*).

Of the six proteins discussed above, GLX I, PR-10 and also the 14 kDa TI are being investigated as to impact on aflatoxin formation through RNAi gene silencing (34). Data, while still being evaluated, indicate an important role for PR-10 in aflatoxin-resistance, as transgenic maize plants where *pr-10* expression is significantly decreased also accumulate significantly more aflatoxin compared to controls.

Conclusions

A great deal of progress has been made towards controlling aflatoxin contamination of maize through the development of germplasm resistant to the growth of aflatoxigenic species, and or biosynthesis of toxins by these species. Therefore, the identification of resistance traits in maize can, using marker-assisted breeding, facilitate a more rapid development of resistant, commercially-useful germplasm. Genetic engineering provides a tool especially useful for transferring resistance genes identified in maize into crops with little natural genetic diversity (e.g. cotton), and for testing the efficacy of the putative resistance genes.

Comparative proteome analysis of maize kernels has identified categories of proteins other than the expected antifungals that are associated with aflatoxin-resistance. The relationships among plant stresses such as drought, high temperature and fungal infection may indicate a connection between kernel stress tolerance and its ability to resist *A. flavus* colonization.

References

1. Lillehoj, EB. The aflatoxin in maize peoblem: The historical perspective. In *Proceedings of the CIMMYT Aflatoxin in Maize Workshop*, **1987**, pp 13-32.
2. Brown, R.L.; Chen, Z.-Y.; Cleveland, T.E.; Russin, J.S. Advances in the development of host resistance in corn to aflatoxin contamination by *Aspergillus flavus. Phytopathol.* **1999**, *89*, 113-117.
3. Gorman, D.P.; Kang, M.S. Preharvest aflatoxin contamination in maize: resistance and genetics. *Plant Breed.* **1991**, *107*, 1-10.
4. Brown, R.L.; Cleveland, T.E.; Payne, G.A.; Woloshuk, C.P.; Campbell, K.W.; White, D.G. Determination of resistance to aflatoxin production in maize kernels and detection of fungal colonization using an *Aspergillus flavus* transformant expressing *Escherichia coli* ß-glucuronidase. *Phytopathol.* **1995**, *85,* 983-989.
5. Brown, R.L.; Cotty, P.J.; Cleveland, T.E.; Widstrom, N.W. Living maize embryo influences accumulation of aflatoxin in maize kernels. *J. Food Prot.* **1993**, *56*, 967-971.
6. Gembeh, S.V.; Brown, R.L.; Grimm, C.; Cleveland, T.E. Identification of chemical components of corn kernel pericarp wax associated with resistance to *Aspergillus flavus* infection and aflatoxin production. *J. Agric. Food Chem.* **2001**, *49*, 4635-4641.
7. Guo, B.Z.; Russin, J.S.; Cleveland, T.E.; Brown, R.L.; Widstrom, N.W. Wax and cutin layers in maize kernels associated with resistance to aflatoxin production by *Aspergillus flavus*. *J. Food Prot.* **1995**, *58*, 296-300.
8. Guo, B.Z.; Russin, J.S.; Cleveland, T.E.; Brown, R.L.; Damann, K.E. Evidence for cutinase production by *Aspergillus flavus* and its possible role in infection of corn kernels. *Phytopathol.* **1996**, *86,* 824-829.
9. Russin, J.S.; Guo, B.Z.; Tubajika, K.M.; Brown, R.L.; Cleveland, T.E.; Widstrom, N.W. Comparison of kernel wax from corn genotypes resistant or susceptible to *Aspergillus flavus*. *Phytopathol.* **1997**, *87*, 529-533.
10. Campbell, K.W.; White, D.G. Evaluation of corn genotypes for resistance to Aspergillus ear rot, kernel infection, and aflatoxin production. *Plant Dis.* **1995**, *79*, 1039-1045.
11. Menkir, A.; Brown, R.L.; Bandyopadhyay, R.; Cleveland, T.E. Registration of six tropical germplasm lines with resistance to aflatoxin contamination. *J. Plant Regist.* **2008,** *2, 246-250.*
12. Payne, G.A. Aflatoxin in maize. *Crit. Rev. Plant Sci.* **1992**, *10*, 423-440.
13. Guo, B.Z.; Russin, J.S.; Brown, R.L.; Cleveland, T.E.; Widstrom, N.W. Resistance to aflatoxin contamination in corn as influenced by relative humidity and kernel germination. *J. Food Prot.* **1996**, *59*, 276-281.
14. Chen, Z.–Y.; Brown, R.L.; Cleveland, T.E.; Damann, K.E.; Russin, J.S. Comparison of constitutive and inducible maize kernel proteins of

genotypes resistant or susceptible to aflatoxin production. *J. Food Prot.* **2001**, *64*, 1785-1792.

15. Guo, B.Z.; Chen, Z.-Y.; Brown, R.L.; Lax, A.R.; Cleveland, T.E.; Russin, J.S.; Mehta, A.D.; Selitrennikoff, C.P.; Widstrom, N.W. Germination induces accumulation of specific proteins and antifungal activities in corn kernels. *Phytopathol.* **1997**, *87*, 1174-1178.
16. Huang, Z.; White, D.G.; Payne, G.A. Corn seed proteins inhibitory to *Aspergillus flavus* and aflatoxin biosynthesis. *Phytopathol.* **1997**, *87*, 622-627.
17. Moore, K.G.; Price, M.S.; Boston, R.S.; Weissinger, A.K.; Payne, G.A. A chitinase from Tex6 maize kernels inhibits growth of *Aspergillus flavus*. *Phytopathol.* **2004**, *94*, 82-87.
18. Ji, C.; Norton, R.A.; Wicklow, D.T.; Dowd, P.F. Isoform patterns of chitinase and β-1,3-glucanase in maturing corn kernels (*Zea mays* L.) associated with *Aspergillus flavus* milk stage infection. *J. Agric. Food Chem.* **2000**, *48*, 507-511.
19. Chen, Z.-Y.; Brown, R.L.; Lax, A.R.; Guo, B.Z.; Cleveland, T.E.; Russin, J.S. Resistance to *A. flavus* in corn kernels is associated with a 14 kDa protein. *Phytopathol.* **1998**, *88*, 276-281.
20. Chen, .Z-Y.; Brown, R.L.; Russin, J.S.; Lax, A.R.; Cleveland, T.E. A corn trypsin inhibitor with antifungal activity inhibits *Aspergillus flavus* α-amylase. *Phytopathol.* **1999**, *89*, 902-907.
21. Woloshuk, C.P.; Cavaletto, J.R.; Cleveland, T.E. Inducers of aflatoxin biosynthesis from colonized maize kernels are generated by an amylase activity from *Aspergillus flavus*. *Phytopathol.* **1997**, *87*, 164-169.
22. Chen, Z.-Y.; Brown, R.L.; Lax, A.R.; Cleveland, T.E.; Russin, J.S. Inhibition of plant pathogenic fungi by a corn trypsin inhibitor over-expressed in *Escherichia coli*. *Appl. Environ. Microbiol.* **1999**, *65*, 1320-1324.
23. Chen, Z.-Y.; Brown, R.L.; Damann, K.E.; Cleveland, T.E. Proteomics analysis of kernel embryo and endosperm proteins of corn genotypes resistant or susceptible to *Aspergillus flavus* infection. *Proc. USDA-ARS Aflatoxin Elimination Workshop* **2000**, pp 88.
24. Chen, Z.-Y.; Brown, R.L.; Damann, K.E.; Cleveland, T.E. Identification of unique or elevated levels of kernel proteins in aflatoxin-resistant maize genotypes through proteome analysis. *Phytopathol.* **2002**, *92*, 1084-1094.
25. Cary, J.W.; Rajasekaran, K.; Yu, J.; Brown, R.L.; Bhatnagar, D.; Cleveland, T.E. Transgenic approaches for pre-harvest control of mycotoxin contamination. *World Mycotoxin J.* **2009** (in press).
26. Payne, G.A. Process of contamination by aflatoxin-producing fungi and their impact on crops. In *Mycotoxins in Agriculture and Food Safety*; Sinha, K.K.; Bhatnagar, D. , Eds.; Marcel Dekker: New York, NY, **1998**; pp 279-306.
27. Brown, R.L.; Chen, Z.-Y.; Menkir, A.; Cleveland, T.E. Using biotechnology to enhance host resistance to aflatoxin contamination of corn. *African J. Biotechnol.* **2003**, *2*, 557-562.
28. Menkir, A.; Brown, R.L.; Bandyopadhyay, R.; Chen, Z.-Y.; Cleveland, T.E. A USA-Africa collaborative strategy for identifying,

characterizing, and developing maize germplasm with resistance to aflatoxin contamination. *Mycopathol.* **2006**, *162*, 225-232.

29. Chen, Z.-Y.; Brown, R.L.; Damann, K .E.; Cleveland, T.E. Identification of a maize kernel stress-related protein and its effect on aflatoxin accumulation. *Phytopathol.* **2004**, *94,* 938-945.
30. Chen, Z.-Y.; Brown, R.L.; Rajasekaran, K.; Damann, K.E.; Cleveland, T.E. Evidence for involvement of a pathogenesis-related protein in maize resistance to *Aspergillus flavus* infection and aflatoxin production. *Phytopathol.* **2006**, *96*, 87-95.
31. Chen, Z.-Y.; Brown, R.L.; Damann, K.E.; Cleveland, T.E. Identification of maize kernel endosperm proteins associated with resistance to aflatoxin contamination by *Aspergillus flavus*. *Phytopathol.* **2007**, *97,* 1094-1103.
32. Baker, R.; Brown, R.L.; Cleveland, T.E.; Chen, Z.-Y.; Fakhoury, A. A maize lectin-like protein with antifungal activity against *Aspergillus flavus. J. Food Prot.* **2009**, *72*, 120-127.
33. Baker, R.; Brown, R.L.; Cleveland, T.E.; Chen, Z.-Y.; Fakhoury, A. A maize trypsin inhibitor (ZmTIp) with limited activity against *Aspergillus flavus*. *J. Food Prot.* **2009**, *72*, 185-188.
34. Chen, Z.-Y.; Brown, R.L.; Damann, K.; Cleveland, T.E. The effect of PR-10 expression on host resistance in maize against Aspergillus flavus infection and aflatoxin production. *Proceedings of the USDA-ARS Aflatoxin and Fumonisin Elimination and Fungal Genomics Workshop* **2006**, pp. 55.

Chapter 12

Advances in Understanding the Biosynthesis of Fumonisins

Ryan Gerber[1], Lili Lou[1], Justin Huffman[1], Xiangcheng Zhu[1,3], Ting Lin[1,4], Lucy Q Li[1], Isis Arreguin[1], Robert A.E. Butchko[2], Robert H. Proctor[2], and Liangcheng Du[1*]

[1]Department of Chemistry, University of Nebraska, Lincoln, NE 68588
[2]Mycotoxin Research Unit, U. S. Department of Agriculture-ARS, Peoria, IL 61604
[3]Current address: Department of Chemical Engineering and Bioengineering, Zhejiang University, Hangzhou, 310027, China
[4]Current address: School of Life Science, Xiamen University, Xiamen 361005, China

Fumonisins are a group of economically important, polyketide-derived mycotoxins. Since the cloning of the fumonisin polyketide synthase (PKS) gene from *Fusarium verticillioides* in 1999, significant advances have been made in understanding the molecular mechanisms for fumonisin biosynthesis. A cluster of 17 genes (*FUM*) required for fumonisin biosynthesis, regulation, and possibly resistance has been extensively studied using gene disruption, domain swapping, and heterologous expression approaches. Genetic manipulations of the fumonisin PKS gene have shed light on the mechanism by which highly-reducing PKSs control polyketide structure. Heterologous expression and examination of reactions catalyzed by the corresponding enzymes has reveled the role of some *FUM* genes in fumonisin biosynthesis. Together, these efforts have established a general biosynthetic pathway for fumonisins and revealed several fascinating enzymatic reactions. In spite of the progress, many questions remain. Future efforts should focus on addressing the molecular mechanistic details of assembly, release, and tailoring of the fumonisin polyketide chain.

Fumonisins are mycotoxins produced by multiple species of the filamentous fungus *Fusarium*, most notably *F. verticillioides* (synonym *F. moniliforme*, teleomorph *Gibberella moniliformis*, synonym *G. fujikuroi* Mating Population A). *F. verticillioides* has received widespread attention due to its worldwide occurrence on maize (*Zea mays*) and its ability to cause ear, stalk, and seedling rot of this important crop (*1*). To date, at least 28 fumonisins have been isolated from fungi and can be classified into four groups, fumonisin A, B, C, and P series (*2*). B-series fumonisins are the most abundant, and fumonisin B_1 (FB_1) constitutes approximately 70% of the total fumonisin content found in naturally contaminated maize and is the most toxic fumonisin analogue (*3, 4*). Fumonisins cause several fatal diseases in animals, including pulmonary edema in swine, leukoencephalomalacia in horses, and cancer in rats and mice (*3, 5, 6*). In humans, fumonisins are suspected to cause esophageal cancer (*3*) and may play a role in liver cancer (*7*) and neural tube defects (*8*). Fumonisin-induced diseases are probably caused by inhibition of sphingolipid biosynthesis through competitive inhibition of the enzyme sphinganine *N*-acyltransferase, also known as ceramide synthase (*9*). Fumonisins, along with AAL-toxins, are known as sphinganine-analog mycotoxins (SAMT) due to their structural similarity to the early precursors of sphingolipid pathways (Figure 1) (*10, 11*). The International Agency for Research on Cancer classified FB_1 as a Group 2B carcinogen (probable carcinogen of humans). The Joint FAO/WHO Expert Committee on Food Additives recommended a maximum tolerable daily intake of 2 mg/kg body weight for FB_1.

During the past decade, our understanding of the molecular mechanism of fumonisin biosynthesis has advanced considerably. A cluster of 17 fumonisin biosynthetic genes (*FUM*) has been cloned. The function of many of the genes has been investigated using genetic and biochemical approaches, and much of the fumonisin biosynthetic pathway has been elucidated. These studies have revealed that *F. verticillioides* employs several unusual, sometimes unprecedented, mechanisms to assemble the complex structure of fumonisins. An improved understanding of these mechanisms will be useful in the development of rational approaches toward mycotoxin elimination in agriculture and the food industry. The purpose of this review is to summarize the recent advances in the genetics and biochemistry of fumonisin biosynthesis.

Chemical Structure and Biosynthetic Genes

With the exception of the C-series, fumonisins consist of a linear 20-carbon backbone decorated with 3-5 hydroxyls, two methyl groups, one amino group, and two tricarballylic esters (Figure 1). The biosynthetic origins of fumonisins have been determined. Specifically, carbons 3-20 of B-series fumonisins have been shown, through isotope feeding experiments, to be derived from acetate, while the amino group and C-1 and C-2 are derived from alanine (*12, 13*). The methyl groups at C-12 and C-16 are derived from methionine (*14*). The hydroxyls on C-5, C-10, C-14, and C-15 are from molecular oxygen, while the hydroxyl at C-3 is derived from the carbonyl group of acetate (*10*). It is generally believed that the tricarballylic esters originate from the citric acid cycle (*12*).

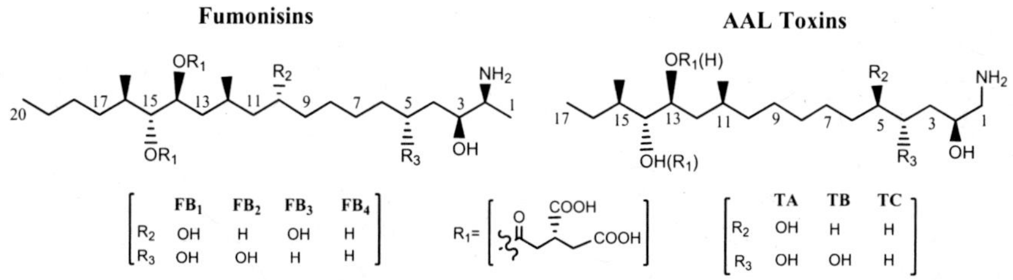

Figure 1. Chemical structure of fumonisins and AAL-toxins.

A cluster of genes required for fumonisin biosynthesis was identified by Proctor et al. after they cloned a polyketide synthase (PKS) gene, *FUM1* (formerly *FUM5*), and subsequently sequenced a 75-kb region adjacent to the PKS gene from *F. verticillioides* (*15-17*). They initially identified 15 contiguous genes that were co-expressed in *F. verticillioides*, including *FUM1*, *FUM6-8, FUM3* (*FUM9*), *FUM10-11, FUM2* (*FUM12*), *FUM13-19.* More recently, Brown et al. identified two additional genes (*FUM20* and *FUM21*) in the cluster by characterization of over 87,000 expressed sequence tags (ESTs) from *F. verticillioides* (*18, 19*). The ESTs correspond to almost 11,000 unique genes that may represent ~80% of the genes in the *F. verticillioides* genome. *FUM20* is a small gene (encoding 40 amino acids) and its function is not yet known. *FUM21* is predicted to encode a Zn(II)2Cys6 transcription factor and strains in which *FUM21* has been deleted are blocked in fumonisin production and expression of other *FUM* genes. Thus, the *FUM21* protein is most likely a pathway-specific (or narrow-domain) transcription factor. Four previously identified regulatory genes (*FCK1*, *FCC1*, *PAC1*, and *ZFR1*) that are not located in the cluster and affect *FUM* gene expression and fumonisin production (*20-23*) are either broad-domain regulators (*PAC1* and *ZFR1*) or global regulators (*FCK1* and *FCC1*).

Although the *FUM* gene cluster is proposed to consist of 17 genes, several of the *FUM* genes (*FUM15* – *FUM18*) are not essential for fumonisin production. Disruption of any of these genes does not result in any significant change in fumonisin production in the mutants. *FUM15* is predicted to encode a cytochrome P450 monooxygenase. The fumonisin backbone contains several hydroxyl groups, but none of the hydroxyls were affected by a disruption of the gene (Proctor, unpublished data). This suggests that *FUM15* might be involved in a hydroxylation at a yet-to-be-identified position of fumonisin carbon backbone. *FUM16* and *FUM10* are predictd to encode fatty acid-CoA synthetases. Disruption of *FUM10* resulted in production of non-esterified fumonisins (hydrolyzed HFB_3 and HFB_4), whereas disruption of *FUM16* did not change the fumonisin production phenotype (*24*). *FUM17* and *FUM18* are predicted to encode ceramide synthases, and were proposed to be involved in resistance to fumonisins, because fumonisins inhibit ceramide synthase and a similar gene (*Asc*-1) in tomato confers resistance to FB_1 and AAL toxin (*25*).

FUM19 is similar to ABC transporter genes and probably also involved in the resistance mechanism (*16*). Proctor et al. analyzed at least 5 *FUM19* mutants and found that they all exhibited a small change in the ratio of FB_3 and FB_1, while transformants that retained the wild-type *FUM19* did not exhibit a change in the ratio. This phenotype was observed in two idependent experiments, suggesting that *FUM19* plays a subtle role in the final pattern of fumonisin production. Because they are clustered and co-expressed with other *FUM* genes that are required for normal fumonisin production, it is plausible that *FUM17, FUM18, FUM19* could play a role in fumonisin biosynthesis or provide resistance to the toxins but that other genes in the genome can compensate for the loss of function in the mutants. For example, the genome sequence database for *F. verticillioides* indicates that the fungus has three ceramide synthase genes in addition to *FUM17* and *FUM18* and as many as 20 ABC transporter genes in addition to *FUM19*. In addition, *FUM11* (encoding a tricarboxylate transporter) and *FUM20* also appeared non-essential (*18, 24*). *FUM11* mutants produced the wild-type complement of FB_1, FB_2, FB_3 and FB_4, but they also produced half hydrolyzed and keto half hydrolyzed fumonisins at much higher levels than the wild type. So, *FUM11* and *FUM19* seem to belong to the same type for their role in fumonisin biosynthesis, i.e. non-essential but relevant.

The *FUM* genes have also been examined in at least three other fungal species, *F. proliferatum*, *F. oxysporum* and *Aspergillus niger*. Prior to the description of *FUM20* and *FUM21* in *F. verticillioides*, Waalwijk et al. identified the entire 15-gene *FUM* cluster in *F. proliferatum* with the same gene order and orientation as *F. verticillioides* (*26*). The sequences outside the cluster were highly dissimilar except for a 2-kb region just upstream of *FUM1*, which is most likely the regulatory gene *FUM21* that was subsequently described (*19*). The *FUM* genes within the two gene clusters share an identity of 77-89% at the protein level. More recently, Proctor et al. sequenced the *FUM* gene cluster in *F. oxysporum* O-1890 (*27*). The cluster also has an identical gene order and orientation as the other two *Fusarium* species, except for an extra ORF (*CPM1*) located downstream to *FUM19*. In general, there is little variability in *FUM* clusters within the *Fusarium* species. However, the gene cluster from *Aspergillus niger* exhibits marked differences from the *Fusarium* clusters (*28*). The *Aspergillus* cluster does not contain *FUM11*, *2*, *16*, *17*, and *18*. The absence of *FUM2* is consistent with the fact that *A. niger* produces FB_2, but not FB_1 and FB_3. The *FUM2* protein catalyzes fumonisin C-10 hydroxylation, and FB_2 does not have this hydroxyl but FB_1 and FB_3 have it (*29*). The absence of *FUM11, 16, 17*, and *18* in *A. niger* also supports the hypothesis that they are not essential for fumonisin production. Curiously, this cluster contains a number of ORFs that are not present in the *Fusarium* clusters, including genes for another PKS, 3-ketoreductase, alkaline phytoceramidase, MFS transporter, and four genes with unknown function (*28*). It is also worth noting that a distinct population of *F. verticillioides* that occurs on banana lacks the *FUM* cluster except for a 754-bp remnant of *FUM21* and a 905-bp remnant of *FUM19* (*30*). The banana strains do not produce fumonisins. However, co-transformation of two cosmids that together contain the entire *FUM* gene cluster from a maize strain of the fungus restored fumonisin production in a banana strain. This is the first direct demonstration that the *FUM* cluster, which was defined mainly based on sequence analysis, is indeed sufficient for fumonisin production. The study also found that fumonisin production is required for development of foliar disease symptoms on maize seedlings, which is an indication of a physiological function of the mycotoxins.

Biosynthetic Mechanism

Polyketide Chain Assembly

B-series fumonisins contain a 20-carbon chain, in which C-3 to C-20 are derived from acetate (*12*). All evidence supports that the 18-carbon portion is a polyketide. The PKS gene, *FUM1*, was the the first gene identified in the cluster (*15*). *FUM1* encodes for a typical modular PKS, consisting of seven domains, β-ketoacyl synthase (KS), acyltransferase (AT), dehydratase (DH), methyltransferase (MT), β-ketoacyl reductase (KR), enoylreductase (ER), and acyl carrier protein (ACP) (Figure 2). In fungi, this type of domain organization is typical of so-called Highly-Reducing PKSs (HR-PKSs), which are responsible for synthesizing polyketides in which β-carbonyl functions are fully or almost fully reduced (*11, 31*). The presence of the MT domain in the PKS suggests that the two methyl groups on the C-12 and C-16 are incorportated during, rather than after, polyketide chain assembly. It is also consistent with the feeding experiment that showed the methyl groups are from methionine (via *S*-adenosylmethionine) rather than from methylmalonate (*14*). Yu et al. showed that a point mutation on the active site of the MT domain resulted in production of demethylated intermediates of fumonisins, providing experimental support for this biosynthetic mechanism (*32*).

Earlier gene disruption experiments have shown that *FUM1* is absolutely essential for fumonisin production (*15*). Since fungal PKS are single-modular, iteratively used enzymes, no biosynthetic intermediate is expected in the gene disruption mutants. In order to understand how the PKS controls the size of final polyketide chain, the Du group adopted the domain-swapping approach that has been successfully used in the studies of bacterial multi-modular PKS (*33*). A group of HR-PKS that have the same domain organization but synthesize very different polyketide chains were used, including *FUM1* for the 18-carbon chain of fumonisins from *F. verticillioides* (*15*), *ALT1* for the 16-carbon AAL-toxins from *Alternaria alternata* f. sp. *Lycopersici* (*34*), *LovF* for the 4-carbon side chain of lovastatin from *Aspergillus terreus* (*35*), and *PKS1* for the 35 to 45-carbon chain(s) of T-toxins from *Cochliobolus heterostrophus* (*36*). First, the KS domain of *FUM1* was replaced with the KS domain of *PKS1* (*37*). The *F. verticillioides* strain containing the hybrid *FUM1* was able to produce fumonisins, showing that the foreign KS domain can function and interact with the six domains of Fum1p. In another experiment, the *FUM1*-KS domain was replaced with the KS domain of *LovF*, which encodes lovastatin diketide synthase (LDKS) (*35*). *F. verticillioides* strains containing the KS domain of *LovF* produced aromatic compounds, dihydroisocoumarins (*38*). Together, the KS-swapping experiments suggest that KS alone is not sufficient to control the structure of polyketide products, although the KS domain does play a role in this process. Interestingly, the dihydroisocoumarins produced by the mutants are known to have antimalarial, antituberculosis, and antifungal activities (*39*). This opens a possible new means to control the mycotoxin-producing *F. verticillioides* in agriculture.

More recently, Zhu et al. introduced the AAL-toxin PKS gene, *ALT1*, from the tomato pathogen *Alternaria alternata* f. sp. *Lycopersici* into *F. verticillioides* strain 57-7-7, which does not produce fumonisins due to a point mutation that introduces a premature stop codon in the *FUM1* coding region (*40*) (Figure 2). The transformant produced FB series as well as the 3-keto analogs. The results showed that the C16-synthesizing *ALT1* is able to support the C18 fumonisin biosynthesis in *F. verticillioides*, suggesting that the final size of the polyketides is not determined by the PKSs alone. Unlike bacterial PKS and other fungal PKS, fungal HR-PKS do not have a thioesterase/cyclase (TE/CLC) domain for the release of the polyketide chain that is covalently attached to the ACP domain of the PKS (*11*). This suggests that this group of polyketides involves a distinct mechanism for chain releasing. The data suggest that the distinct chain releasing mechanism is probably the key factor determining the size of final products.

Polyketide Chain Release

Precursor feeding experiments showed that C-1 and C-2 of fumonisins are from alanine (*13*), while C3 to C20 from acetate (*12*). These data imply that the C3-20 polyketide chain is released via introduction of a new carbon-carbon bond. This mechanism is very different from the known TE/CLC-catalyzed chain releasing mechanism, in which a nitrogen or an oxygen atom is the nucleophile to attack the carbonyl group of acyl-*S*-ACP that leads to the formation of an amide bond or an ester bond in the products (*11*). The nucleophilic nitrogen (an amino) or oxygen (a hydroxyl) is often part of the biosynthetic intermediate that is to be released and results in the formation of lactam or lactone, respectively. The release of fumonisin polyketide is unique in that it uses a carbon nucleophile and results in extension of the carbon chain and introduction of an amino group into the final product. This process is not catalyzed by a TE/CLC-like domain but is most likely catalyzed by a pyridoxal 5′-phosphate (PLP)-dependent enzyme.

Within the *FUM* cluster, *FUM8* is predicted to encode a 2-oxoamino synthases (class II α-aminotransferases), a group of PLP-dependent enzymes that catalyze the condensation of amino acids and acyl-CoA thioester substrates (*17*). These include serine-palmitoyltransferase, 5-aminolevulinate synthase, 8-amino-7-oxononanoate synthase, and 2-amino-3-ketobutyrate CoA ligase. Therefore, it is likely that the *FUM8* protein (Fum8p) catalyzes a similar reaction between alanine and the dimethyl stearoyl-*S*-PKS (Figure 2). The known 2-oxoamino synthases are either heterodimers or homodimers. Serine-palmitoyltransferase, for example, consists of two separate subunits. In the yeast *Saccharomyces cerevisiae* these subunits are designated as Lcb1p and Lcb2p, where Lcb2p is the catalytic subunit that contains the active site lysine required for PLP binding (*41*). Fum8p is unique in that it is a natural fusion of the two subunits; its N-terminal half is a homolog of Lcb2p and its C-terminal half is a homolog of Lcb1p. Although the PLP-dependent chain releasing mechanism has been proposed (*11*), the experimental evidence is still lacking. Fum8p is not only the key factor to determine the polyketide chain size but also the primary factor to determine whether the fungus produces B-series or C-series fumonisins (*27*). The chemical structure of FCs is identical to FBs, except that FCs are one carbon shorter, lacking the C-1 of FBs. *F. verticillioides* and most other *Fusarium* species that have been examined produce predominately FBs. However, *F. oxysporum* O-1890 produces predominantly FCs (*42*). Complementation of *F. verticllioides FUM8* mutants with the *FUM8* orthologue from *F. oxysporum* O-1890 restored fumonisin production, but the complemented mutants produced predominantly FCs rather than FBs (*27*). This suggests that the amino acid (alanine versus glycine) substrate specificity of Fum8p determines whether *Fusarium* produces B or C-series fumonisins.

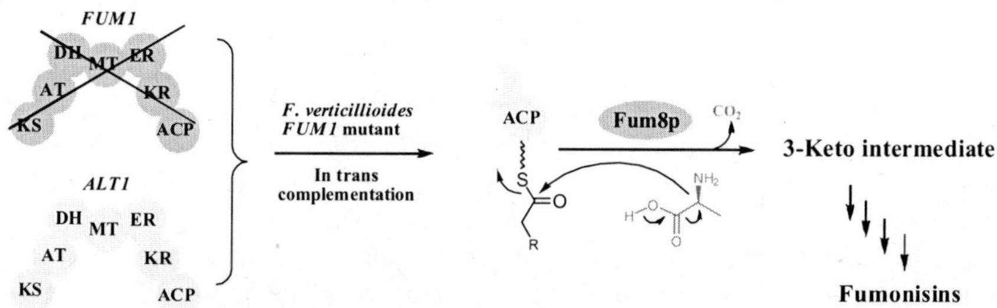

Figure 2. Complementation of fumonisin biosynthesis in a FUM1 mutant of F. verticillioides with the AAL-toxin-synthesizing PKS gene, ALT1.

Tailoring of the Polyketide Chain

The polyketide chain of fumonisins undergoes several modification during the formation of B or C-series fumonisins. These tailoring steps include the oxidoreductions and esterifications. One question has been whether the tailoring steps take place prior to or after the polyketide chain release. Current experimental evidence supports that the chain release precedes the tailoring steps (*43*). In vitro studies have also shown that the tailoring enzymes use free substrates, rather than PKS-bound substrates (*44-46*).

The condensation between the polyketide acyl-*S*-Fum1p and alanine results in a 3-keto intermediate (Figure 3). The reduction of this ketone group is the first step of the oxidoreductions. The first evidence for *FUM13* to be responsible for this reduction step came from the gene disruption experiment carried out by Butchko et al. (*47*). This gene was predicted to encode a short chain dehydrogenase/reductase, which catalyzes the dehydrogenation or reduction of various substrates like alcohols and aromatic compounds (*48*). *FUM13* disruption mutants exhibited a 90% reduction in B-series fumonisin production and produced two new fumonisin analogues not produced by the wild type. MS and NMR analysis of the analogues indicated that they were 3-keto form of FB_3 and FB_4 (*47*). However, the presence of the low level of normal fumonisins in the mutants is peculiar. One possibility is that an additional 3-ketoreductase gene present in the fungus may have compensated *FUM13*'s function. Yi et al. expressed the gene in *E. coli* and used the purified enzyme Fum13p to show the 3-keto reduction activity of the enzyme (*45*). The results showed that Fum13p is an NADPH-dependent ketoreductase. The in vitro study provided direct evidence for the function of *FUM13*. The enzyme is believed to stereospecifically reduce the 3-keto group to produce a 3*S* configuration at C-3, as found in fumonisins. Recently, Gelderblom et al. isolated two stereoisomers of FB_3 and FB_4 (*49*). The relative and absolute stereochemistry of the *3-epi* fumonisins (*2S*, *3R*) were established by NMR and chemical methods. Naturally isolated samples of FB_3 and FB_4 contained 10-40% of the epimers. Thus, it is likely that Fum13p could catalyze the reduction at both the *Si* and *Re* face of the 3-keto functionality.

One of the interesting structural features of fumonisins is the vicinal diol at C-14 and C-15. Previously, it was suggested that the vicinal diol could be introduced by hydrolysis of an epoxide (*50*). However, this would require an epoxide hydrolase, which is not present in the *FUM* cluster (*16*). It is more likely that the two hydroxyls are introduced by one or a combination of P450 monooxygenases, such as the enzyme encoded by *FUM6*. Seo et al. previously showed that *FUM6* disruption mutants did not produce any detectable fumonisin intermediate (*17*), suggesting that the 3-keto intermediate is not stable, possibly due to its linear, sphanganine-like structure that could easily be redirected into other metabolic pathways. Bojja et al. used a co-culture approach to restore fumonisin production in the mutants (*43*). The disruption mutants of *FUM1* or *FUM8* did not accumulate any intermediate. However, a co-culture of either the *FUM1* mutant or the *FUM8* mutant with the *FUM6* mutant led to fumonisin production. The results suggest that the *FUM6* mutant produced a fumonisin intermediate that can be further metabolized by biosynthetic enzymes present in the *FUM1* or *FUM8* mutants. Several potential intermediates accumulated in the early stage of co-cultures were also detected by LC-MS. These metabolites had the general carbon skeleton of fumonisins with 1-4 hydroxyls. The data suggest that the timing of these hydroxylation steps appears flexbile, as the metabolites had up to 4 hydroxyls but without any of the tricarballylic esters. Together, the current data support that *FUM6* is the gene responsible for the vicinal diol formation (Figure 3). However, the vicinal di-hydroxylation reaction remains to be demonstrated using heterologously expressed Fum6p.

Besides *FUM6*, there are two other genes, *FUM2* and *FUM15*, in the *FUM* cluster that encode P450 monooxygenases (*16*). As discussed above, *FUM15* is not essential for fumonisin biosynthesis, and its function remains unclear. Proctor et al. showed that *FUM2* disruption mutants produced only FB_2 and FB_4, which lack the C-10 hydroxyl (*51*). Thus, the *FUM2*-encoded P450 monooxygenase catalyzes hydroxylation of the fumonisin backbone at C-10. Further evidence for the function of *FUM2* resulted from analysis of a natural variant of *F. verticillioides* that produces only FB_2 and FB_4, a C-10 hydroxylation deficient phenotype. Desjardins et al determined by meiotic analysis that this variant phenotype resulted from a nonfunctional allele at a single locus (*52*). Introduction of a wild-type copy of the *FUM2* gene into an *F. verticillioides* strain with the phenotype restored production of C-10-hydroxylated fumonisins (FB_1 and FB_3) and thereby confirmed that the *FUM2*–encoded monooxygenase (Fum2p) catalyzes fumonisin C-10 hydroxylation (*51*). In spite of the genetic evidence, when C-10 hydroxylation occurs relative to other reactions in the fumonisin biosynthetic pathway is not clear (Figure 3). This can only be demonstrated by studying the substrate specificity and catalytic efficiency of the heterologously expressed Fum2p.

The structure of fumonisins includes two tricarballylic functions esterified to the vicinal diol at on C-14 and C-15. Tricarballylic esters are rare in natural products. Gene disruption experiments showed that four genes, *FUM7, FUM10, FUM11,* and *FUM14*, are involved in the formation of the tricarballylic esters (*24, 46*). *FUM7* mutants produced tetradehydro-fumonisins, where a double bond is present on each of the tricarballylic esters. The exact nature of the compounds, including the stereo- (*cis*/*trans*) and regiochemistry of the double bond, is not clear. *FUM7*-encoded enzyme is an iron-containing alcohol dehydrogenase similar to maleylacetate reductase that catalyzes the reversible reduction of a double to a single bond (*53*). This combined with the phenotype of *FUM7* mutants suggests that *FUM7* catalyzes the reduction of a double bond in a tricarboxylic acid precursor, such as aconitic acid. However, feeding purified tetradehydro-FB_1 to cultures of *FUM gene* mutants did not result in FB_1 production (*24*). This indicates that tetradehydro-fumonisins are not biosynthetic intermediates but instead are shunt products of the *FUM7* mutant. Thus, the exact function of *FUM7* remains to be demonstrated.

FUM11 mutants produced the wild-type complement of fumonisins as well as the half-hydrolyzed and keto half-hydrolyzed homologues of fumonisins (*24*). Compared to the wild type, the amount of the hydrolyzed homologues in the mutants was significantly increased, from 1.2% to 15-26% for the half-hydrolyzed and from 2.1% to 6-10% for the keto half-hydrolyzed. The results indicate that *FUM11* is not essential but important for a complete esterification of fumonisins. The gene was predicted to encode a tricarboxylate transporter. The fumonisin production phenotype of *FUM11* disruption mutants is consistent with impaired transportation of a tricarboxylate precursor(s) from mitochondria to the cytoplasm, where fumonisin biosynthesis presumably takes place, and this in turns results in a limited pool of available precursors for esterification to the fumonisin backbone.

The *FUM10* and *FUM14* disruption mutants exhibited the same phenotype; they produced hydrolyzed FB_3 (HFB_3) and hydrolyzed FB_4 (HFB_4) (*24, 46*). This production phenotype indicate that the two genes are directly involved in the esterification. This has been confirmed for *FUM14* by in vitro studies using purified Fum14p (*46*). *FUM10* is predicted to encode an acyl-CoA synthethase or the adenylation (A) domain of nonribosomal peptide synthetases (NRPS) (*16*), whereas *FUM14* was predicted to encode a peptidyl carrier protein (PCP) and condensation (C) domain (*46*). Together with the *FUM7*-encoded reductase (Re) domain, these genes may encode a "stand-alone" complex analogous to the A-PCP-C-Re domains of an NRPS. In the putative complex, the A domain would recognize and activate a specific tricarboxylate substrate as acyl-adenylate, which would be loaded onto the PCP domain as an acyl thioester. The reduction catalyzed by the Re domain would likely take place prior to the esterification, such as at the PCP-bound stage. Then, the C domain would transfer the tricarballylic moiety from acyl-*S*-PCP to the diol of hydrolyzed fumonisin (most likely HFB_4) to produce FB_4. This NRPS complex is unusual in two aspects. First, typical NRPS minimally contain three domains, A, PCP, and C, which usually reside on the same protein (*54, 55*). Thus, Fum10p-7p-14p would constitute a "stand-alone" NRPS complex. Second, NRPSs typically catalyze amide bond (C-N) formation, whereas Fum14p catalyzes ester bond (C-O) formation (*46*).

FUM3 disruption mutants produced only FB_3 and FB_4, fumonisins that lack a C-5 hydroxyl group (*56*). This phenotype was the same as that described previously for a natural variant of *F. verticillioides* (*52*). Meiotic and molecular genetic analyses deomonstrated that the genetic lesion in the natural variant corresponded to a nonfunctional allele of *FUM3* (*56*). The sequence of *FUM3* indicates that it encodes a dioxygenase with similarity to 2-ketoglutarate-dependent dioxygenases. The deduced amino acid sequence of Fum3p includes a HRD motif at positions 146-148, which is consistent with the conserved HxD motif found in 2-ketoglutarate-dependent dioxygenases. This HxD motif is believed to be involved in iron binding, as histidine and aspartic acid take part in holding a ferrous ion in the crystal structure of the enzyme isopenicillin *N* synthase, which also uses dioxygen as a cosubstrate (*57, 58*). *FUM3* was expressed in the yeast *S. cerevisiae* and the purified enzyme was shown to convert FB_3 to FB_1. The C-5 hydroxylase activity requires the presence of 2-ketoglutarate, iron(II), ascorbic acid and catalase (*44*). Thus, *FUM9* was demonstrated to encode a 2-ketoglutarate-dependent dioxygenase required for the C-5 hydroxylation of the fumonisin backbone. Interestingly, although the 3-*epi*-FB_3 and 3-*epi*-FB_4 have been isolated, the corresponding 3-epimers of FB_1 and FB_2 have never been detected (*49*). This suggests that the 3*R* configuration in these epimers are not recognized by Fum3p and thus 3-*epi*-FB_3 and 3-*epi*-FB_4 could not be converted to 3-*epi*-FB_1 and 3-*epi*-FB_2 hav, respectively.

In summary, the functions of all genes, except *FUM20,* in the *FUM* cluster have been examined by gene disruption analysis. Several (*FUM3*, *FUM13*, *FUM14*) have also been examined by *in vitro* activity of heterologously expressed and purified enzymes. These efforts have led to a relatively clear biosynthetic pathway for fumonisins, as outlined in Figure 3.

Final Remarks

Although significant progress has been made in the past decade, there are still many unknowns regarding the biosynthetic mechanisms for fumonisins. First, the molecular mechanistic details of enzymes (Fum1p and Fum8p) that control the structure of the final polyketide chain are not clear. Although the genetic and biochemical studies have revealed some insights, the ultimate answers may have to come from studies of the 3-dimensional structure of the two large multi-domain proteins as well as their interactions. Second, it is unclear how a single cytochrome P450 monooxygenase (Fum6p) introduces both hydroxyl groups of the vicinal diol. The vicinal diol is not only a very interesting structural feature, but also an important functional group for the activity of fumonisins. The study of its biosynthetic mechanism could potentially lead to an understanding of an unusual enzymatic reactions useful in

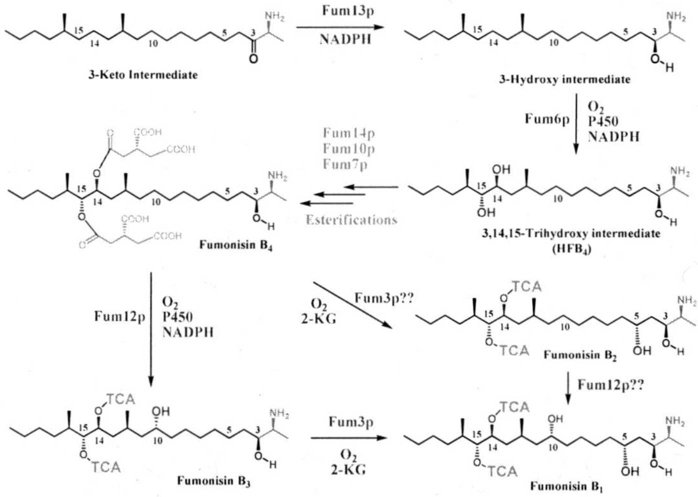

Figure 3. Proposed biosynthetic pathway for the modifications of fumonisin carbon backbone. The conversion of FB_4 to FB_2 and then to FB_1 has not been established. TCA, tricarballylic acid; 2-KG, 2-ketoglutarate.

designing inhibitors of fumonisin production. Third, the reactions catalyzed by the "stand-alone" NRPS complex (Fum10p-7p-14p) need to be further determined. Among all the unusual structural features of fumonisins, the tricarballylic esters are most intriguing. The biochemical studies have revealed that Fum14p is an unusual NRPS catalyzing a new reaction. But how the three enzymes interact and work together to add tricarballylic groups to the carbon backbone of fumonisins requires further investigations. Last, the self-protection mechanism by which fumonisin-producing fungi avoid the toxic effects of the mycotoxins is still unknown. Although *FUM17*, *FUM18* and *FUM19* have been proposed candidates for self-protection genes (*16*), the chemical and biochemical basis by which these genes confer protection is not clear.

Acknowledgement

Research in the Du laboratory has been supported by NSF (MCB-0614916), Nebraska Research Initiatives (NRI), and NSFC (#30428023). Isis Arreguin was supported by an NSF/DOE REU program at the Nebraska Redox Biology Center. The research was performed in facilities renovated with support from NIH (RR015468-01).

References

1. Miller, J. D. Factors that affect the occurrence of fumonisin. *Environmental Health Perspectives, Supplement 2* **2001**, *109*, 321-324.
2. Rheeder, J. P.; Marasas, W. F.; Vismer, H. F. Production of fumonisin analogs by *Fusarium* species. *Applied and Environmental Microbiology* **2002**, *68*, 2101-2105.
3. Marasas, W. F. Discovery and occurrence of the fumonisins: a historical perspective. *Environmental Health Perspectives, Supplement 2* **2001**, *109*, 239-243.
4. Nelson, P. E.; Desjardins, A. E.; Plattner, R. D. Fumonisins, mycotoxins produced by *Fusarium* species: biology, chemistry, and significance. *Ann Rev Phytopath* **1993**, *31*, 233-252.
5. Gelderblom, W. C.; Jaskiewicz, K.; Marasas, W. F.; Thiel, P. G.; Horak, R. M.; Vleggaar, R.; Kriek, N. P. Fumonisins-novel mycotoxins with cancer-promoting activity produced by *Fusarium moniliforme*. *Appl Environ Microbiol* **1988**, *54*, 1806-1811.
6. Marasas, W. F.; Kellerman, T. S.; Gelderblom, W. C.; Coetzer, J. A.; Thiel, P. G.; van der Lugt, J. J. Leukoencephalomalacia in a horse induced by fumonisin B_1 isolated from *Fusarium moniliforme*. *Onderstepoort Journal of Veterinary Research* **1988**, *55*, 197-203.
7. Ueno, Y.; Iijima, K.; Wang, S. D.; Sugiura, Y.; Sekijima, M.; Tanaka, T.; Chen, C.; Yu, S. Z. Fumonisins as a possible contributory risk factor for primary liver cancer: a 3-year study of corn harvested in Haiman, China by HPLC and ELISA. *Food Chem Toxicol* **1997**, *35*, 1143-1150.
8. Missmer, S. A.; Suarez, L.; Felkner, M.; Wang, E.; Merrill Jr., A. H.; Rothman, K. J.; Hendricks, K. A. Exposure to fumonisins and the occurrence of neural tube defects along the Texas-Mexico border. *Environmental Health Perspectives* **2006**, *114*, 237-241.
9. Wang, E.; Norred, W. P.; Bacon, C. W.; Riley, R. T.; Merrill Jr., A. H. Inhibition of sphingolipids biosynthesis by fumonisins, implications for diseases associated with *Fusarium moniliforme*. *J Biol Chem* **1991**, *266*, 14486-14490.
10. Caldas, E. D.; Sadilkova, K.; Ward, B. L.; Jones, A. D.; Winter, C. K.; Gilchrist, D. G. Biosynthetic Studies of Fumonisin B1 and AAL Toxins. *J Agric Food Chem* **1998**, *46*, 4734-4743.
11. Du, L.; Zhu, X.; Gerber, R.; Huffman, J.; Lou, L.; Jorgenson, J.; Yu, F.; Zaleta-Rivera, K.; Wang, Q. Biosynthesis of sphinganine-analog mycotoxins. *J Ind Microbiol Biotechnol* **2008**, *35*, 455-464.
12. Blackwell, B. A.; Edwards, O. E.; Fruchier, A.; ApSimon, J. W.; Miller, J. D. NMR structural studies of fumonisin B1 and related compounds from *Fusarium moniliforme*. *Adv Exp Med Biol* **1996**, *392*, 75-91.
13. Branham, B. E.; Plattner, R. D. Alanine is a precursor in the biosynthesis of fumonisin B1 by *Fusarium moniliforme*. *Mycopathologia* **1993**, *124*, 99-104.
14. Plattner, R. D.; Shackelford, D. D. Biosynthesis of labeled fumonisins in liquid cultures of *Fusarium moniliforme*. *Mycopathologia* **1992**, *117*, 17-22.

15. Proctor, R. H.; Desjardins, A. E.; Plattner, R. D.; Hohn, T. M. A polyketide synthase gene required for biosynthesis of fumonisin mycotoxins in *Gibberella fujikuroi* mating population A. *Fungal Genet Biol* **1999**, *27*, 100-112.
16. Proctor, R. H.; Brown, D. W.; Plattner, R. D.; Desjardins, A. E. Co-expression of 15 contiguous genes delineates a fumonisin biosynthetic gene cluster in *Gibberella moniliformis*. *Fungal Genet Biol* **2003**, *38*, 237-249.
17. Seo, J. A.; Proctor, R. H.; Plattner, R. D. Characterization of four clustered and coregulated genes associated with fumonisin biosynthesis in *Fusarium verticillioides*. *Fungal Genet Biol* **2001**, *34*, 155-165.
18. Brown, D. W.; Cheung, F.; Proctor, R. H.; Butchko, R. A.; Zheng, L.; Lee, Y.; Utterback, T.; Smith, S.; Feldblyum, T.; Glenn, A. E.; Plattner, R. D.; Kendra, D. F.; Town, C. D.; Whitelaw, C. A. Comparative analysis of 87,000 expressed sequence tags from the fumonisin-producing fungus *Fusarium verticillioides*. *Fungal Genet Biol* **2005**, *42*, 848-861.
19. Brown, D. W.; Butchko, R. A.; Busman, M.; Proctor, R. H. The *Fusarium verticillioides FUM* gene cluster encodes a Zn(II)2Cys6 protein that affects *FUM* gene expression and fumonisin production. *Eukaryot Cell* **2007**, *6*, 1210-1218.
20. Bluhm, B. H.; Woloshuk, C. P. Fck1, a C-type cyclin-dependent kinase, interacts with Fcc1 to regulate development and secondary metabolism in *Fusarium verticillioides*. *Fungal Genet Biol* **2006**, *43*, 146-54.
21. Flaherty, J. E.; Pirttila, A. M.; Bluhm, B. H.; Woloshuk, C. P. PAC1, a pH-regulatory gene from *Fusarium verticillioides*. *Appl Environ Microbiol* **2003**, *69*, 5222-5227.
22. Flaherty, J. E.; Woloshuk, C. P. Regulation of fumonisin biosynthesis in *Fusarium verticillioides* by a zinc binuclear cluster-type gene, ZFR1. *Appl Environ Microbiol* **2004**, *70*, 2653-2659.
23. Shim, W. B.; Woloshuk, C. P. Regulation of fumonisin B(1) biosynthesis and conidiation in *Fusarium verticillioides* by a cyclin-like (C-type) gene, FCC1. *Appl Environ Microbiol* **2001**, *67*, 1607-1612.
24. Butchko, R. A.; Plattner, R. D.; Proctor, R. H. Deletion analysis of *FUM* genes involved in tricarballylic ester formation during fumonisin biosynthesis. *J Agric Food Chem* **2006**, *54*, 9398-9404.
25. Brandwagt, B. F.; Kneppers, T. J.; Nijkamp, H. J.; Hille, J. Overexpression of the tomato Asc-1 gene mediates high insensitivity to AAL toxins and fumonisin B1 in tomato hairy roots and confers resistance to *Alternaria alternata* f. sp. *lycopersici* in *Nicotiana umbratica* plants. *Mol Plant Microbe Interact* **2002**, *15*, 35-42.
26. Waalwijk, C.; van der Lee, T.; de Vries, I.; Hesselink, T.; Arts, J.; Kema, G. H. Synteny in toxigenic *Fusarium* species: The fumonisin gene cluster and the mating type region as examples. *Eur J Plant Pathol* **2004**, *110*, 533-544.
27. Proctor, R. H.; Busman, M.; Seo, J. A.; Lee, Y. W.; Plattner, R. D. A fumonisin biosynthetic gene cluster in *Fusarium oxysporum* strain O-1890 and the genetic basis for B versus C fumonisin production. *Fungal Genet Biol* **2008**, *45*, 1016-1026.

28. Baker, S. E. Aspergillus niger genomics: past, present and into the future. *Med Mycol* **2006**, *44 Suppl 1*, S17-21.
29. Frisvad, J. C.; Smedsgaard, J.; Samson, R. A.; Larsen, T. O.; Thrane, U. Fumonisin B2 production by *Aspergillus niger*. *J Agric Food Chem* **2007**, *55*, 9727-9732.
30. Glenn, A. E.; Zitomer, N. C.; Zimeri, A. M.; Williams, L. D.; Riley, R. T.; Proctor, R. H. Transformation-mediated complementation of a *FUM* gene cluster deletion in *Fusarium verticillioides* restores both fumonisin production and pathogenicity on maize seedlings. *Mol Plant Microbe Interact* **2008**, *21*, 87-97.
31. Cox, R. J. Polyketides, proteins and genes in fungi: programmed nano-machines begin to reveal their secrets. *Org Biomol Chem* **2007**, *5*, 2010-2026.
32. Yu, F.; Zhu, X.; Du, L. Developing a genetic system for functional manipulations of *FUM1*, a polyketide synthase gene for the biosythesis of fumonisins in *Fusarium verticillioides*. *FEMS Microbiol Letters* **2005**, *248*, 257-264.
33. Staunton, J.; Weissman, K. J. Polyketide biosynthesis: a millennium review. *Nat Prod Rep* **2001**, *18*, 380-416.
34. Akamatsu, H.; Otani, H.; Kodama, M. Characterization of a gene cluster for host-specific AAL-toxin biosynthesis in the tomato pathotype of *Alternaria alternata. Fungal Genet. Newsl. (2003, Suppl. The 22nd Fungal Genetics Conference, p121)* **2003**.
35. Kennedy, J.; Auclair, K.; Kendrew, S. G.; Park, C.; Vederas, J. C.; Hutchinson, C. R. Modulation of polyketide synthase activity by accessory proteins during lovastatin biosynthesis. *Science* **1999**, *284*, 1368-1372.
36. Yang, G.; Rose, M. S.; Turgeon, B. G.; Yoder, O. C. A polyketide synthase is required for fungal virulence and production of the polyketide T-toxin. *Plant Cell* **1996**, *8*, 2139-2150.
37. Zhu, X.; Yu, F.; Bojja, R. S.; Zaleta-Rivera, K.; Du, L. Functional replacement of the ketosynthase domain of *FUM1* for the biosynthesis of fumonisins, a group of fungal reduced polyketides. *J Ind Microbiol Biotech* **2006**, *33*, 859-868.
38. Zhu, X.; Yu, F.; Li, X. C.; Du, L. Production of dihydroisocoumarins in *Fusarium verticillioides* by swapping ketosynthase domain of the fungal iterative polyketide synthase Fum1p with that of lovastatin diketide synthase. *J Am Chem Soc* **2007**, *129*, 36-37.
39. Kongsaeree, P.; Prabpai, S.; Sriubolmas, N.; Vongvein, C.; Wiyakrutta, S. Antimalarial dihydroisocoumarins produced by Geotrichum sp., an endophytic fungus of *Crassocephalum crepidioides*. *J Nat Prod* **2003**, *66*, 709-711.
40. Zhu, X.; Vogeler, C.; Du, L. Functional Complementation of Fumonisin Biosynthesis in *FUM1*-Disrupted *Fusarium verticillioides* by the AAL-Toxin Polyketide Synthase Gene ALT1 from *Alternaria alternata* f. sp. *Lycopersici*. *J Nat Prod* **2008**, *71*, 957-960.
41. Hanada, K. Serine palmitoyltransferase, a key enzyme of sphingolipid metabolism. *Biochim Biophys Acta* **2003**, *1632*, 16-30.

42. Seo, J. A.; Kim, J. C.; Lee, Y. W. Isolation and characterization of two new type C fumonisins produced by *Fusarium oxysporum*. *J Nat Prod* **1996**, *59*, 1003-1005.
43. Bojja, R. S.; Cerny, R. L.; Proctor, R. H.; Du, L. Determining the biosynthetic sequence in the early steps of the fumonisin pathway by use of three gene-disruption mutants of *Fusarium verticillioides*. *J Agric Food Chem* **2004**, *52*, 2855-2860.
44. Ding, Y.; Bojja, R. S.; Du, L. Fum3p, a 2-ketoglutarate-dependent dioxygenase required for C-5 hydroxylation of fumonisins in *Fusarium verticillioides*. *Appl Environ Microbiol* **2004**, *70*, 1931-1934.
45. Yi, H.; Bojja, R. S.; Fu, J.; Du, L. Direct evidence for the function of *FUM13* in 3-ketoreduction of mycotoxin fumonisins in *Fusarium verticillioides*. *J Agric Food Chem* **2005, *53*,** 5456-5460.
46. Zaleta-Rivera, K.; Xu, C.; Yu, F.; Butchko, R. A.; Proctor, R. H.; Hidalgo-Lara, M. E.; Raza, A.; Dussault, P. H.; Du, L. A bidomain nonribosomal peptide synthetase encoded by *FUM14* catalyzes the formation of tricarballylic esters in the biosynthesis of fumonisins. *Biochem* **2006**, *45*, 2561-2569.
47. Butchko, R. A.; Plattner, R. D.; Proctor, R. H. *FUM13* encodes a short chain dehydrogenase/reductase required for C-3 carbonyl reduction during fumonisin biosynthesis in *Gibberella moniliformis*. *J Agric Food Chem* **2003**, *51*, 3000-3006.
48. Jornvall, H.; Persson, B.; Krook, M.; Atrian, S.; Gonzalez-Duarte, R.; Jeffery, J.; Ghosh, D. Short-chain dehydrogenases/reductases. *Biochem* **1995**, *34*, 6003-6013.
49. Gelderblom, W. C.; Sewram, V.; Shephard, G. S.; Snijman, P. W.; Tenza, K.; van der Westhuizen, L.; Vleggaar, R. Structure and natural occurrence of stereoisomers of the fumonisin B series mycotoxins. *J Agric Food Chem* **2007**, *55*, 4388-4394.
50. Morisseau, C.; Ward, B. L.; Gilchrist, D. G.; Hammock, B. D. Multiple epoxide hydrolases in *Alternaria alternata* f. sp. *lycopersici* and their relationship to medium composition and host-specific toxin production. *Appl Environ Microbiol* **1999**, *65*, 2388-2395.
51. Proctor, R. H.; Plattner, R. D.; Desjardins, A. E.; Busman, M.; Butchko, R. A. Fumonisin production in the maize pathogen *Fusarium verticillioides*: genetic basis of naturally occurring chemical variation. *J Agric Food Chem* **2006**, *54*, 2424-2430.
52. Desjardins, A. E.; Plattner, R. D.; Proctor, R. H. Linkage among genes responsible for fumonisin biosynthesis in *Gibberella fujikuroi* mating population A. *Appl Environ Microbiol* **1996**, *62*, 2571-2576.
53. Seibert, V.; Thiel, M.; Hinner, I. S.; Schlomann, M. Characterization of a gene cluster encoding the maleylacetate reductase from *Ralstonia eutropha* 335T, an enzyme recruited for growth with 4-fluorobenzoate. *Microbiol* **2004**, *150*, 463-472.
54. Walsh, C. T.; Gehring, A. M.; Weinreb, P. H.; Quadri, L. E.; Flugel, R. S. Post-translational modification of polyketide and nonribosomal peptide synthases. *Cur Opin Chem Biol* **1997**, *1*, 309-315.

55. Marahiel, M. A.; Stachelhaus, T.; Mootz, H. D. Modular peptide synthetases involved in nonribosomal peptide synthesis. *Chem Rev* **1997**, *97*, 2651-2674.
56. Butchko, R. A.; Plattner, R. D.; Proctor, R. H. *FUM9* is required for C-5 hydroxylation of fumonisins and complements the meitotically defined Fum3 locus in *Gibberella moniliformis*. *Appl Environ Microbiol* **2003**, *69*, 6935-6937.
57. Shibasaki, T.; Mori, H.; Chiba, S.; Ozaki, A. Microbial proline 4-hydroxylase screening and gene cloning. *Appl Environ Microbiol* **1999**, *65*, 4028-4031.
58. Roach, P. L.; Clifton, I. J.; Fulop, V.; Harlos, K.; Barton, G. J.; Hajdu, J.; Andersson, I.; Schofield, C. J.; Baldwin, J. E. Crystal structure of isopenicillin *N* synthase is the first from a new structural family of enzymes. *Nature* **1995**, *375*, 700-704.

Chapter 13

Regulation of *Aspergillus flavus* aflatoxin biosynthesis and development

Jeffrey W. Cary[1], Leanne Szerszen[2], and Ana M. Calvo[2]

[1]ARS, U.S. Department of Agriculture, Southern Regional Research Center, New Orleans, LA 70124
[2]Department of Biological Sciences, Northern Illinois University, DeKalb, IL 60115

The filamentous fungus *Aspergillus flavus* produces a family of potent mutagenic and carcinogenic polyketide-derived compounds collectively known as aflatoxins. These secondary metabolites contaminate a number of oilseed crops during growth of the fungus and this can result in severe negative economic and health impacts. The biosynthesis and regulation of these toxins represent one of most studied areas of all the fungal secondary metabolites. Much of the information obtained on the AF biosynthetic genes and regulation of AF biosynthesis was obtained through studies using *A. flavus* and *A. parasiticus* and also the model fungus *Aspergillus nidulans* that produces sterigmatocystin (ST), the penultimate precursor to AF. There has long been anecdotal evidence of a genetic linkage between production of secondary metabolites and fungal morphogenesis however the exact mechanism of this relationship was not clear. A breakthrough in the genetic mechanisms governing AF production and *A. flavus* development was made upon the discovery in *A. nidulans* of a G-protein-mediated signaling pathway that regulated both ST biosynthesis and asexual conidiation. Further studies in *A. nidulans* and *A. flavus* and also of the fungus-host plant interaction have identified a number of genetic factors that link secondary metabolism and morphological differentiation processes in *A. flavus* as well as filamentous fungi in general. The focus of this review is to provide an overview of research that characterized the genes involved in the biosynthesis and regulation of AF in *A. flavus*, how environmental and nutritional factors control expression of these genes, and the role of global regulators in AF production, fungal development and virulence. The impact of fungal whole genome sequence and microarray technology in the identification of novel genes involved in the regulation of AF production and development as well as virulence are also discussed.

The Aflatoxin Biosynthetic Gene Cluster

Linkage of AF pathway genes was first evidenced in an *A. parasiticus* cosmid clone that harbored both the *aflD* (=*nor-1*) and *aflM* (=*ver-1*) genes (*1*) and soon thereafter transcriptional mapping of overlapping cosmid clones of *A. parasiticus* and *A. flavus* established that the genes involved in AF biosynthesis were present on a single gene cluster (*2, 3*). Four structurally related forms of AFs occur in nature depending on the presence of the dihydro-bisfuran rings (B_1 and G_1) or the tetrahydro-bisfuran rings (B_2 and G_2). In general, *A. parasiticus* produces both B and G AFs while *A. flavus* only produces the B forms. In *A. parasiticus* and *A. flavus* the enzymes and regulatory proteins required for AF production are encoded by at least 29 genes that are clustered in a 70-kb DNA region. *Aspergillus nidulans* has a similar biosynthetic cluster required for the production of the mycotoxin ST, an intermediate in AF formation, that spans about 60 kb and contains homologs of many of the AF biosynthetic genes (*4*). Both pathways contain a gene that encodes a positive-acting, pathway-specific transcriptional regulator, AflR, that is responsible for the co-regulation of many if not all of the AF/ST biosynthetic genes (*5, 6*). Ehrlich and collaborators (*7, 8*) sequenced and compared the AF gene clusters from a number of *Aspergillus* species including *A. nomius*, two sclerotial morphotypes of *A. flavus*, and an unnamed *Aspergillus* species from West Africa that produces both B and G AFs. They found that the gene order (Fig. 1) for all of the isolates was the same as that for the *A. parasiticus* AF cluster. Twenty two of these genes were homologous to genes in the *A. nidulans* ST cluster, however the order and direction of transcription of some of the ST genes differed from that of the AF gene cluster. A comparison of the AF and ST gene clusters is provided by Ehrlich et al. (*7*). Clusters varied in length from 66.1 kb for the small (S) sclerotial morphotype and 66.5 kb for the large (L) morphotype of *A. flavus*. Sclerotia are survival structures that remain dormant in adverse environments and then germinate under favorable conditions. The difference in cluster size between the sclertotial morphotypes is due in large part to deletions of portions of the coding and intergenic regions of the *aflF* (=*norB*) and *aflU* (=*cypA*) genes. Loss of *aflU* function results in loss of the ability to produce G AFs. Deletions are not only limited to the *aflF-aflU* gene region. An examination of 38 nonaflatoxigenic *A. flavus* field isolates showed 8 distinct patterns of large (≥40 kb) deletions involving the AF gene cluster some extending from within the cluster to the end of the chromosome (*9*). The *A. nomius* cluster was found to be

68.4 kb in length. In all the AF gene clusters analyzed, the *aflF* gene represented the proximal end of the cluster while, *aflY* (=*hypA*), a gene of unknown function was at the distal end. New evidence suggests that the *nadA* gene, adjacent to *aflY* and once believed to be part of a sugar utilization gene cluster, is actually involved in biosynthesis of G AFs and therefore constitutes the distal end of the cluster (*10*). Of the 29 genes identified in the pathway, only 7 (*aflF, aflT, hypB1, aflE* (=*norA*), *hypE, hypB2*, and *aflY*) have yet to have the function of their protein product determined experimentally.

Other *Aspergillus* section *Flavi* species such as the nonaflatoxigenic *A. oryzae* and *A. sojae* are generally recognized as safe (GRAS) and are used in food fermentations. Both fungi harbor an AF gene cluster however specific defects and/or deletions in the *aflR* gene and other AF pathway structural genes result in lack of AF production (*11-14*). Both the AF clusters of *A. flavus/parasiticus* and *A. oryzae* and the ST cluster of *A. nidulans* are located in the subtelomeric regions of their respective chromosomes (*15*). In general, there appears to be a preference for the location of secondary metabolic gene clusters to the subtelomeric regions. Subtelomeric regions are known to be active regions for intra-molecular recombinations, insertions or deletions, and translocations (*15*). Therefore, the proximity of the AF cluster to the telomere may facilitate the rapid reorganization and evolution of these genes in a species-specific fashion. Carbone et al. (*15*) have identified specific gene modules that exist in the AF and ST clusters. They postulate that these modules have arisen from gene duplications that retain the pre-duplicated gene's function or the duplicated copy's function may augment a specific pathway function. Additionally, the duplicated copy may evolve a completely new function. They hypothesized that it is possible that genes may become separated after their duplication and that differences in gene order between AF and ST clusters may be the result of gene reorganization in an ST-producing ancestor.

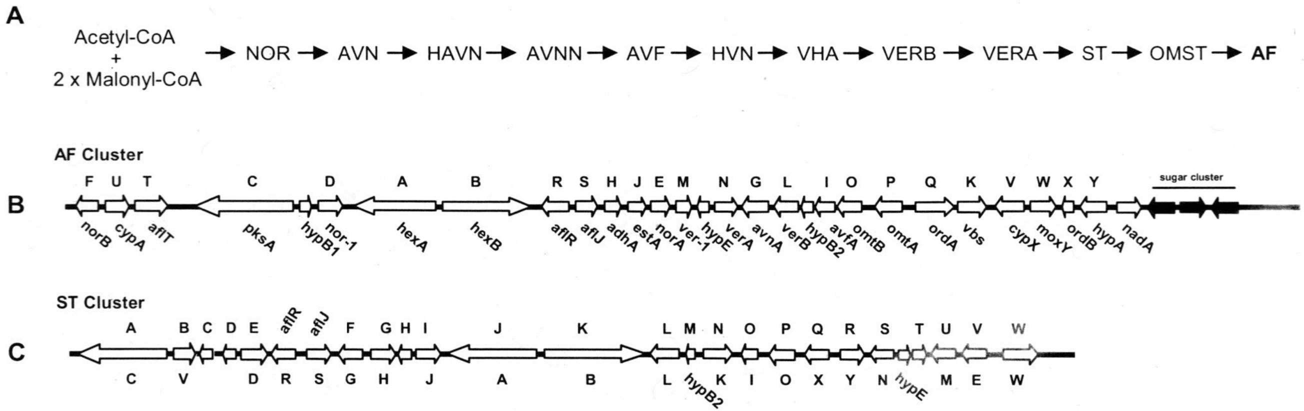

Figure 1. (A) Metabolic precursors of aflatoxin: NOR, norsolorinic acid; AVN, averantin; HAVN, hydroxyaverantin; AVNN, averufanin; AVF, averufin; HVN, hydroxyversicolorone; VHA, versicolorone hemiacetal acetate; VERB, versicolorin B; VERA, versicolorin A; ST, sterigmatocystin; OMST, O-methylsterigmatocystin; AF, aflatoxin. Schematic representation of the (B) aflatoxin and (C) sterigmatocystin biosynthetic gene clusters. The direction of transcription is indicated by arrows. In the AF cluster the gene designations above the schematic represent the afl nomenclature while those below represent the enzymatic function nomenclature. In the ST cluster the gene designations above represent the stc nomenclature and those below indicate the AF cluster homolog.

Transcriptional Regulation of Aflatoxin Biosynthesis and Development

Aflatoxin Pathway-Specific Regulators

A positive regulatory gene, *aflR*, encodes a sequence-specific, Gal4-type C6-zinc binuclear cluster DNA binding protein that is required for transcriptional activation of the AF structural genes (*16, 17*). AflR also regulates the expression of *A. nidulans* ST biosynthetic genes (*5*). The AflR protein binds to the palindromic sequence 5'-TCGN_5CGR-3' within the promoter region of AF structural genes (*5, 6*). Interestingly, a study by Price et al. (*18*) using *A. flavus* EST microarray transcription profiling studies identified two additional genes (*niiA* and *hlyC*) located well outside of the AF gene cluster that were upregulated under AF-conducive conditions. In these genes, the consensus AflR binding sites were observed, however, they were located from 1.8 to 2.3 kb upstream of the translational start sites. A number of studies have also shown that elements upstream of the *aflR* coding region may serve as binding sites for proteins that negatively regulate *aflR* expression (*19-21*). In addition *aflR* influences *A. flavus* morphogenesis. Loss of *aflR* does not result in loss of spore or sclerotial formation, however, it does impact the numbers of spores or sclerotia being formed (*22*).

Another gene that appears to play a role in the regulation of AF production is *aflS* (=*aflJ*). This gene is adjacent to and divergently transcribed from *aflR*. AflS does not demonstrate any significant homology to other genes or deduced protein sequences present in the genome databases (*23*). Though the exact function of AflS has not been determined, it is required for production of wild-type levels of AF and it has been shown to interact with AflR (*23-25*). Interestingly, though *aflS* overexpression increased AF production its overexpression did not result in elevated transcription of mid- to late-AF pathway intermediate genes such as *aflM* (=*ver-1*) or *aflP* (=*omtA*). However, transformants expressing *aflR* and *aflS* produced five times more *aflC* (=*pksA*) transcripts and four times more *aflD* (=*nor-1*) transcripts than strains expressing only *aflR* (*25*). This would suggest that *aflS* modulates the regulation of early genes in the AF biosynthetic pathway.

Global Regulatory Factors

Early studies by Kale et al. (*26*) suggested a genetic connection between AF production and development in *A. parasiticus*. This observation was extended by the characterization of a G-protein-mediated signal transduction pathway in the model fungus *A. nidulans* that regulates both asexual conidiation and ST production (*27*). This topic has been covered extensively in previous reviews (*28-30*). In general, G-protein-coupled receptors activated by a ligand transmit the signal to two downstream signaling pathways: a G-protein-cAMP-dependent kinase (*28*) and/or mitogen-activated protein kinase pathway, affecting several cell functions, including morphogenesis and mycotoxin biosynthesis (*31, 32*)

VeA

The *veA* gene is well known as a regulator of light-dependent morphogenesis in aspergilli. *A. nidulans* cultures exposed to light develop asexually forming abundant air-borne conidia, while in the dark sexual development is favored resulting in the formation of fruiting bodies called cleistothecia (*33*). Although the *veA* gene product does not present homology with any other proteins of known function, *veA* has been found in the genome of many filamentous fungi, being particularly conserved in Ascomycetes (Calvo et al., unpublished). Deletion of *veA* blocks cleisthothecia production (*34*). It is also known that the *veA* gene plays a global regulatory role in the synthesis of secondary metabolites, including mycotoxins (*34-38*). Studies have demonstrated that *veA* controls the transcription of genes necessary for the synthesis of AF in *A. flavus* and *A. parasiticus* and ST in *A. nidulans*. VeA was shown to be required for the expression of *aflR* and subsequent activation of AF/ST biosynthetic genes (*35-38*). Importantly, the *A. flavus* and *A. parasiticus* deletion mutants were also unable to produce sclerotia, further supporting the hypothesis that sclerotia could be sterile vestiges of cleistothecia (Fig. 2) (*35, 36, 39*).

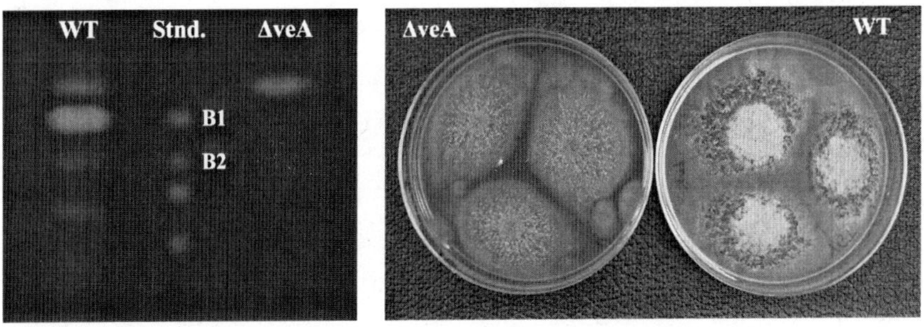

Figure. 2. Analysis of A. flavus veA mutant (ΔveA). Left panel: TLC analysis of wild-type (WT) and ΔveA strains showing lack of AF production in the mutant. Right panel: YGT agar cultures of WT and ΔveA strains following 5 d growth. Note abundant sclerotia (dark structures) in WT and absence in ΔveA.

These findings on the role of VeA in the regulation of secondary metabolism and morphogenesis, together with the fact that *veA* has only been found in fungi (phylogenetic studies in Calvo's lab, unpublished), suggest that *veA* or *veA*–dependent genes could serve as targets for development of strategies to decrease the detrimental effects of mycotoxin contamination in our food supplies (*40, 41*). The *veA* gene also regulates the synthesis of additional secondary metabolites such as cyclopiazonic acid and aflatrem toxins in *A. flavus* and the antibiotic penicillin in *A. nidulans* that were reduced or *veA* deletion mutants (*34, 36*).

As mentioned above, VeA function is light-dependent. Interestingly, Stinnett et al. (*42*) showed that *A. nidulans* VeA migration to the nucleus is light-dependent. While in the dark VeA is located mainly in the nuclei, under light VeA is found abundantly in the cytoplasm. Blue light (440–500 nm) had a similar effect to that observed with white light, preventing an efficient accumulation of VeA in the nuclei (*42*). The effect of exposure to red light (625–740 nm) was similar but more moderate in comparison with blue and white light. Blumenstein et al. (*43*) reported that a phytochrome-like protein called FphA acts as a red-light sensor in *A. nidulans* and represses sexual development and ST production. Previous studies suggested a possible interaction of VeA with light-responsive proteins (*42, 44*) and it was recently demonstrated that FphA protein interacts with VeA as part of a protein complex in the nucleus (*45*). These studies indicated that red and blue-light perception occurs in an integrative way in the nuclear VeA-protein complex affecting morphogenesis and secondary metabolism. Furthermore, VeA abundancy in the nucleus was negatively affected by FphA in the light. This, together with the fact that nuclear concentration of VeA is also reduced by blue light (*42*) provide further evidence that both, red- and blue-light sensing systems are influencing VeA nuclear localization and consequently toxin production. The fact that the VeA complex includes proteins that respond to external stimuli, such as light, and proteins that affect the expression of secondary metabolic gene clusters, such as LaeA, further elaborated below, suggested a possible scaffold role for VeA. A scaffold-like role for VeA in the nucleus has been identified. In *A. nidulans* nuclei, VeA was also shown to physically interact with VelB (a velvet-like protein), which is expressed during sexual development (*46*). VeA is believed to bridge VelB to LaeA. Deletion of either *velB* or *veA* resulted in defects in both ST and cleistothecial production.

Our lab has demonstrated that VeA also plays a role in *A. flavus* virulence (Duran, Cary and Calvo, submitted). Virulence of *A. flavus* on peanut and maize seeds was reduced in the absence of the *veA* gene product. Generation of air-borne asexual conidia was reduced in viable or non-viable peanut seeds and in viable maize seed. Production of AF and sclerotia in peanut and maize seed was completely blocked when infected with the *A. flavus veA* mutant (Δ*veA*). *In planta* inoculated cotton bolls examined 3 weeks post-inoculation (soon after boll opening) showed that conidiation was decreased in bolls inoculated with the Δ*veA* strain and spread [as determined by intercarpellary membrane damage (*47*)] of the Δ*veA* strain to locules adjacent to the inoculated locule was less than observed with the wild-type *veA* strain (data not shown, Duran, Cary and Calvo, submitted). As observed in peanuts and maize, no AF was produced in seed harvested from cotton bolls that had been inoculated with the Δ*veA* strain while AF was present in seed from wild-type *veA* inoculated bolls (data not shown, Duran, Cary and Calvo, submitted).

LaeA

The nuclear regulator LaeA has been shown to govern production of multiple secondary metabolites in *Aspergillus* species (*48-51*). LaeA is a positive regulator of *aflR* expression, which in turn activates the expression of AF/ST genes. It was shown to be a negative regulator of *veA* expression in *A.*

flavus and *laeA* deletion mutants demonstrated decreased AF production and no sclerotia were formed (*48*). This would suggest that both *veA* and *laeA* are required for AF/ST production and sclerotial formation through the formation of a regulatory protein complex in the nucleus and that regulation of *veA* by *laeA* may represent an internal mechanism to balance stoichiometry of this complex (*48*). As observed with *A. flavus veA* mutants, *A. flavus laeA* mutants also were less able to colonize peanut and maize seed *in vitro*. Interestingly, an *A. nidulans laeA* mutant was preferentially attacked by the fungivorous springtail *Folsomia candida* indicating that the arthropod could sense the presence of a secondary metabolite(s) being produced by the wild-type fungus (*52*).

The LaeA deduced amino acid sequence indicated that this protein can function as a methyltransferase, as it harbors a S-adenosyl methionine binding domain (*53*). ST cluster expression analysis showed that regulation by *laeA* is spatially limited to the cluster genes, not affecting genes adjacent to the cluster (*50, 53-55*) This was also found to be the case for the *A. nidulans* terrequinone A cluster (*54*). Based on preliminary evidence, it is hypothesized that LaeA may influence chromatin structure at cluster loci thus controlling transcriptional activation of cluster genes (*55, 56*).

The Effect of Histone Acetylation on Toxin Biosynthesis

AF biosynthesis is influenced by a number of nutritional and environmental factors (see below). Stimulation of AF biosynthesis by high levels of glucose and cAMP has been demonstrated and explained, at least in part, by production of elevated levels of AflR that in turn increases transcription of AF biosynthetic genes. However, studies have suggested that other proteins are needed to assist in the binding of AflR to the promoters of AF biosynthetic genes and also for optimal transcriptional activity. Roze et al. (*57*) identified a binding site (CRE1), unique from the AflR binding site (AflR1), in the promoter of the *aflD* gene, which served as a cAMP-response element. Mutation in CRE1 or AflR1 caused up to a three-fold decrease in cAMP-mediated stimulation of *aflD* promoter activity. They demonstrated that the CRE1 site was required for binding of a 32 kDa protein (CRE1bp) and hypothesized that CRE1bp interacts with AflR to assist in its binding to the *aflD* promoter. Several CRE1-like binding sites are found in different AF gene promoters, so this interaction may extend beyond that of just the *aflD* gene.

Importantly, CRE1 has been shown to recruit histone acetyltransferase (HAT) to promoter regions, leading to acetylation of histones, in particular histone H4, which has been demonstrated to enhance transcriptional activation (*58-60*). In a recent study, Roze et al. (*61*) demonstrated a positive correlation between the initiation and spread of histone H4 acetylation in AF gene promoters, leading to the establishment of AF pathway gene expression and AF accumulation. AF gene transcription is enhanced by AflR-binding to the promoter regions as access is made available by changes in chromatin conformation. Previous studies have shown that when AF gene promoter-reporter fusions were integrated outside of the AF cluster, promoter activity was greatly decreased supporting the model of specific regulation of cluster expression (*62*). The role of epigenetic regulation of secondary metabolic gene clusters was further supported by recent evidence that deletion of the histone deacetylase (HDAC) gene, *hdaA*, in *A. nidulans* resulted in activation of transcription of genes of the ST and penicillin biosynthetic gene clusters (*63*). Interestingly, while they showed that ST and PN production could be restored in a Δ*laeA* Δ*hdaA* double mutant, levels of these metabolites were not as high as in the Δ*hdaA* alone. This suggested that LaeA and HdaA operate through different mechanisms.

Effect of Environmental Factors on Aflatoxin Production and Fungal Development

Role of Nitrogen, Carbon, and pH

Production of AF is also under the control a number of global regulatory networks that respond to environmental and nutritional cues. These include responses to nutritional factors such as carbon and nitrogen sources and environmental factors such as pH, light, oxidative stress and temperature. Globally-acting regulatory proteins such as AreA involved in nitrogen signaling, CreA involved in carbon signaling, and PacC involved in pH mediated signaling either positively or negatively influence AF production.

In general, nitrate inhibits AF production while ammonium salts are conducive (*64*). However, nitrate has been shown to enhance ST production in *A. nidulans* while ammonium-based media repressed ST (*65*). Nitrogen metabolism in fungi is regulated by the globally-acting transcription factor AreA (*66*). Over-expression of the *aflR* gene in *A. parasiticus* resulted in release of nitrate inhibition on AF biosynthesis indicating that AreA imparts its control on toxin synthesis either directly or indirectly via *aflR* (*67*). Electrophoretic mobility shift assays (EMSAs) indicated that the *A. parasiticus* AreA binds within the *aflR-aflS* intergenic region and a number of putative AreA GATA binding sites are present within this region (*21*). Analysis of the effects of nitrate on AF production and expression of *aflR* and *aflS* in a number of *A. flavus* strains indicated variability in nitrogen regulation and this variability could often be found to correspond to differences in the number of GATA sites near the *aflS* tsp (*68*). In addition, nitrogen source can also influence formation of sclerotia in *A. flavus*. Studies of *A. flavus* growth on agar media containing either nitrate or ammonium as the sole nitrogen source showed that sclerotial development occurred with nitrate but not with ammonium [see Genomics section and (*69*)].

AF biosynthesis is induced by simple sugars such as glucose and sucrose that are either present or generated by fungal hydrolytic enzymes during invasion of seed tissues (*70, 71*). When molasses was added to three commonly used growth media, conidial production was stimulated while AF production was reduced in a small sclerotial *A. flavus* isolate whereas sclerotial formation increased or decreased depending on the medium (*72*). There is no evidence for the involvement of carbon catabolite repression by CreA in regulation of AF production as in most cases glucose stimulates AF production. No putative CreA binding sites have been identified in the promoters of AF pathway genes. A sugar utilization gene cluster has been identified distal to the *nadA* gene of the AF cluster in *A. flavus* as well as a number of other *Aspergillus* section *Flavi* species but it does not flank the ST cluster of *A. nidulans* (*7, 73*). Though not experimentally proven to be linked to AF production, the fact that this cluster is conserved among section *Flavi* species suggests that a higher-order chromatin structure encompassing both the AF and sugar cluster could be important for expression of genes in the clusters (*7*).

AF and ST production, in general, is greatest in acidic medium and tends to decrease as the pH of the medium increases (*74*). An atypical, West African strain of *A. flavus* was identified that produced less AF in acidic medium (*75*). The strain is designated *A. flavus* S_{BG} as it produced both AFB1 and AFG1 whereas most *A. flavus* strains only produce the B AFs. Interestingly, the changes in AF production between the S_B and S_{BG} strains did not correlate well with changes in *aflR* expression indicating that pH may be exerting its effects on other cellular metabolic processes that in turn regulate AF biosynthesis. Response to changes in pH is regulated by the globally-acting transcription factor PacC that is post-translationally modified by a pH-sensing protease (*76*). A number of putative PacC binding sites have been identified in the promoters of AF biosynthetic genes and could be involved in negatively regulating AF biosynthesis during growth at alkaline pH (*6, 77*). Fungal development also appears to respond to changes in pH as sclerotial production was found to be reduced by 50% at pH 4.0 or less while AF production was at its maximal (*78*).

Role of Plant Metabolites

A number of plant-based metabolites have been shown to reduce AF production as well as alter fungal development (*79*). These include volatile aldehydes (*80-82*), flavanoid compounds (*83*), neem leaf extracts and jasmonic acid (*84, 85*). Volatile aldehydes, jasmonic acid, and methyl jasmonate are all biologically-active end products of the LOX pathways in plants. LOX pathway metabolic precursors (i.e. oxylipins) have been shown have significant effects on fungal development and toxin production (see below) and these effects are more than likely mediated by the above described end products of the LOX pathways that can function as signaling molecules in transduction pathways that regulate a number of biological processes (*86*).

Ethylene and CO_2 treatment have been shown to reduce AF production in *A. parasiticus* (*87, 88*). Treatment with ethylene reduced AF accumulation in a dose-dependent manner with a 10-fold reduction observed when *A. parasiticus* was treated with 146 ppm ethylene. CO_2 at 0.1% also reduced AF accumulation about 5- fold however adding more CO_2 (0.7 or 3.0%) reversed the inhibitory effect. Treatment of infected peanut seeds with differing concentrations of ethylene and/or CO_2 inhibited AF accumulation up to 5-fold. Other volatile compounds described to affect AF production are 2-ethyl-1-hexanol, and 2-buten-1-ol. 2-ethyl-1-hexanol stimulates AF production in *A. parasiticus* (*89*). However, 2-buten-1-ol showed a dose-dependent up-regulatory or down-regulatory effect not only on AF gene transcription and AF accumulation, but also on production of asexual spores. Both 2-ethyl-1-hexanol and 2-buten-1-ol were found to be produced by *A. parasiticus* and at a higher level by *A. nidulans*. Molyneux et al. (*90*) showed that AF production is markedly decreased by the presence of natural antioxidants in tree nuts such as hydrolysable tannins, flavanoids, and phenolic acids. They hypothesized that AF biosynthesis is stimulated by oxidative stress on the fungus and that compounds that act as antioxidants such as tannins and caffeic acid can suppress AF production.

One of the most in-depth areas of study on the molecular genetics of fungal responses to plant metabolites are those that have looked at a group of metabolites known collectively as oxylipins. Oxylipins are hormone-like molecules that have been implicated as signaling molecules for cross-kingdom communication in plant-pathogen interactions (*91*). In plants, linoleic (18:2) and linolenic acid (18:3) can be converted by the action of LOX enzymes to 13*S*-hydroperoxy linoleic acid (13*S*-HPODE) and 9*S*-hydroperoxy linoleic acid (9*S*-HPODE) and 13*S*-hydroperoxy linolenic acid (13*S*-HPOTE) and 9*S*-hydroperoxy linolenic acid (9*S*-HPOTE) respectively. These plant oxylipins closely resemble the fatty acid-derived compounds known as psi factors produced by aspergilli that can also influence fungal development and toxin production (*92*). Both *A. flavus* and *A. nidulans* development is affected by the presence of 13*S*-HPODE and 9*S*-HPODE in a dose-dependent manner (*93*). Incorporation of pure 13*S*-HPODE to cultures of either *A. parasiticus* or *A. nidulans* was shown to repress AF and ST production respectively while pure 9*S*-HPODE increased toxin production (*94*). Interestingly, it appears that growth of the fungus on natural substrates such as maize or peanut seed results in altered expression of seed LOX genes thus leading to changes in the levels of plant oxylipins (*95, 96*). Postulating that plant oxylipins mimic or interfere with biological activities of endogenous fungal oxylipins, Brodhagen et al. (*97*) looked at the ability of a maize oxylipin biosynthetic gene (ZmLOX3) to substitute functionally for *A. nidulans ppo* genes that encode dioxygenases involved in synthesis of fungal psi factors. The maize *ZmLOX3* gene was introduced into wild-type *A. nidulans* and a Δ*ppoAC* strain (reduced in production of oxylipins psiBα and psiBβ, conidia, and ST) and they observed increased production of conidia and ST in both strains. They also observed that peanut seed *pnlox2-3* expression was decreased upon infection by *A. nidulans Δppo* mutants compared to levels expressed upon infection by a wild-type strain. These two experiments suggest that oxylipin cross-talk in the host seed-*Aspergillus* interaction may be reciprocal. Though not reported for *A. flavus*, research has shown that maize mutants lacking function of *ZmLOX3* had decreased levels of susceptibility to several fungal pathogens including the fumonisin B1 producer, *Fusarium verticillioides* (*98*). These results support the hypothesis that a specific plant 9-LOX isoform mediates susceptibility of maize to fungal pathogens.

Genomics

A. flavus EST and whole genome libraries are invaluable as tools for the identification of genes involved in AF biosynthesis and morphogenesis, as well as fungal pathogenesis/virulence and comparative genomics. Information gleaned from analysis of genome databases and microarray experiments will provide a better understanding of the mechanisms governing AF production and morphogenesis as well as the *A. flavus*-host plant interaction. A number of *Aspergillus* genomes have been sequenced, including *A. flavus* (*99*), *A. oryzae* (*100*), *A. nidulans* (*101*), and *A. niger* (*102*). Both *A. flavus* EST and two whole genome libraries have been generated and microarrays developed for gene profiling experiments (*103, 104*). Numerous studies have been performed using EST microarrays for profiling AF and developmental genes in *A. flavus* (*37, 105-107).* Our lab used EST microarray technology to identify genes differentially expressed in *A. flavus* wild-type *veA* and *veA* mutant strains (*37*). Microarray analysis identified 136 genes that were differentially expressed between the two strains including 27 genes that demonstrated a significant difference in expression both between strains and over time. Of the 136 genes we were able to identify subgroups of genes that exhibited expression profiles similar to those expected for genes involved in AF biosynthesis or sclerotial formation. Guo et al. (*108*) utilized EST libraries generated from developing peanut (*Arachis hypogaea* L.) seeds at three reproduction stages from a resistant (resistant to *Aspergillus* infection with reduced AF contamination) and a susceptible (susceptible to *Aspergillus* infection with high AF contamination) peanut genotype challenged by *A. parasiticus* and drought stress in the field. A number of resistance-related genes with significant up-regulation were identified from the two libraries.

A. oryzae is a close relative of *A. flavus* yet it is not a plant pathogen nor does it produce AF. Yu et al. (*103*) have begun a genome-wide comparison between *A. flavus* and *A. oryzae* in hopes of identifying *A. flavus* genes that are involved in pathogenesis and AF biosynthesis. Preliminary studies indicate that over 95% of the annotated genes are shared between these two fungal species with fewer than 300 genes being unique to each species. In addition to 11,823 *A. flavus* genes, the *A. flavus* whole genome microarrays contain the *A. oryzae* unique genes in addition to genes of interest from maize, *Fusarium* spp., mouse and human genomes. We have performed an experiment using the TIGR *A. flavus* whole genome microarrays to look at differential expression of genes between the wild-type and mutant *veA* strains (Cary and Calvo, unpublished results). Results identified a number of genes that demonstrated a significant difference in expression both between strains and over time. Of interest was the differential expression of a number of genes involved in nitrogen metabolism, in particular the nitrate reductase (*niaD*) gene. Microarray analysis showed that *niaD* expression was significantly lower in the *veA* mutant compared to the wild-type. Based on these results an *A. flavus* strain harboring a wild-type or mutant copy of *niaD* were examined for AF production and conidial and sclerotial formation following growth for 5 days on yeast extract-glucose (YGT) agar plates (Fig. 3). Microscopic examination of sclerotia showed that the *niaD* mutant produced fewer and smaller sclerotia than the wild-type strain (Fig. 3A and B). Analysis of conidial formation and AF production showed that the *niaD* mutant produced about 10-fold more conidia than the wild-type but less AF (Fig. 3C). These results indicate that nitrogen metabolism plays a role in both secondary metabolism and development in *A. flavus* and that expression of some of the genes involved are *veA*-dependent.

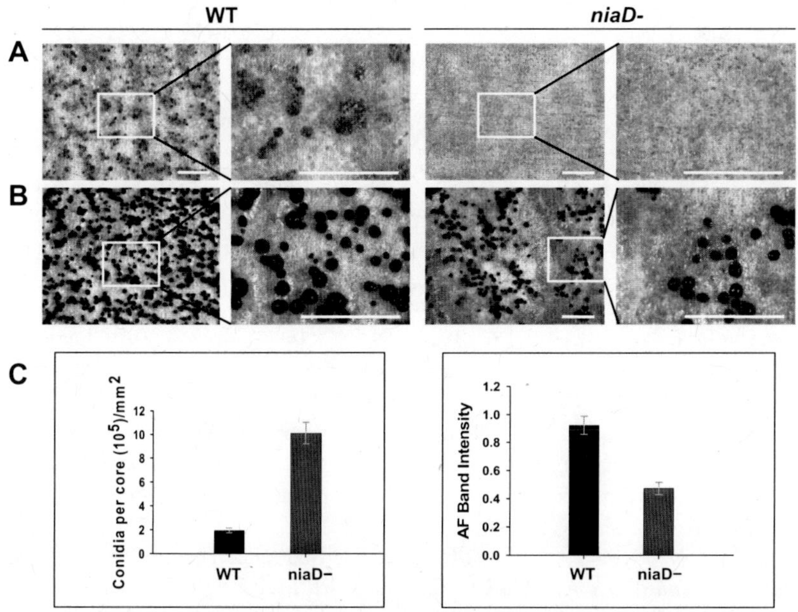

Figure 3. Microscopic examination of sclerotial formation and determination of conidial and AF production of A. flavus wild-type (WT) and niaD− strains. (A) Micrographs of A. flavus WT and niaD− sclerotia after 5-day growth on YGT agar. (B) Same as (A) except mycelia were removed by washing agar surface with 70% ethanol. Black, spherical structures are sclerotia. Bar = 2 mm. (C) Conidial counts and densitometric determination of AF production from TLC plates. Standard error is shown.

Concluding Remarks and Future Prospects

The advent of technologies for the rapid sequencing and annotation of fungal genomes and screening of microarray libraries has provided researchers with additional tools to identify genes involved in the biosynthesis of AF and development in *A. flavus* as well as its ability to invade plants tissues. The use of comparative genomics between *A. flavus* and *A. oryzae*, a close relative of *A. flavus* that is not a plant pathogen and does not produce AF, should be useful in identifying genes specifically involved in AF production and virulence. Elucidation of the role of plant metabolites and environmental factors such as oxylipins and oxidative stress in AF production and fungal development has opened up new avenues of research on the signaling pathways present in both the fungus and the host plant and how they may ultimately provide new clues into the ability of the fungus to invade crops, produce AF and survive in the field.

References

1. Skory, C. D.; Chang, P. K.; Cary, J.; Linz, J. E. Isolation and characterization of a gene from *Aspergillus parasiticus* associated with the conversion of versicolorin A to sterigmatocystin in aflatoxin biosynthesis. *Appl. Environ. Microbiol.* **1992**, *58*, 3527-3537.
2. Trail, F.; Mahanti, N.; Rarick, M.; Mehigh, R.; Liang, S. H.; Zhou, R.; Linz, J. E. Physical and transcriptional map of an aflatoxin gene cluster in *Aspergillus parasiticus* and functional disruption of a gene involved early in the aflatoxin pathway. *Appl. Environ. Microbiol.* **1995**, *61*, 2665-2673.
3. Yu, J.; Chang, P. K.; Ehrlich, K. C.; Cary, J. W.; Bhatnagar, D.; Cleveland, T. E.; Payne, G. A.; Linz, J. E.; Woloshuk, C. P.; Bennett, J. W. Clustered pathway genes in aflatoxin biosynthesis. *Appl. Environ. Microbiol.* **2004**, *70*, 1253-1262.
4. Brown, D. W.; Yu, J. H.; Kelkar, H. S.; Fernandes, M.; Nesbitt, T. C.; Keller, N. P.; Adams, T. H.; Leonard, T. J. Twenty-five coregulated transcripts define a sterigmatocystin gene cluster in *Aspergillus nidulans*. *Proc Natl Acad Sci U S A* **1996**, *93*, 1418-22.
5. Fernandes, M.; Keller, N. P.; Adams, T. H. Sequence-specific binding by *Aspergillus nidulans* AflR, a C6 zinc cluster protein regulating mycotoxin biosynthesis. *Mol. Microbiol.* **1998**, *28*, 1355-1365.
6. Ehrlich, K. C.; Montalbano, B. G.; Cary, J. W. Binding of the C6-zinc cluster protein, AFLR, to the promoters of aflatoxin pathway biosynthesis genes in *Aspergillus parasiticus*. *Gene* **1999**, *230*, 249-257.
7. Ehrlich, K. C.; Yu, J.; Cotty, P. J. Aflatoxin biosynthesis gene clusters and flanking regions. *J. Appl. Microbiol.* **2005**, *99*, 518-527.
8. Ehrlich, K. C.; Yu, J.; Cotty, P. J. Comparison of *Aspergillus* gene clusters from five *Aspergillus* species. *Fungal Genet Biol* **2004**, *submitted.*
9. Chang, P. K.; Horn, B. W.; Dorner, J. W. Sequence breakpoints in the aflatoxin biosynthesis gene cluster and flanking regions in nonaflatoxigenic *Aspergillus flavus* isolates. *Fungal Genet Biol* **2005**, *42*, 914-23.
10. Cai, J.; Zeng, H.; Shima, Y.; Hatabayashi, H.; Nakagawa, H.; Ito, Y.; Adachi, Y.; Nakajima, H.; Yabe, K. Involvement of the *nadA* gene in formation of G-group aflatoxins in *Aspergillus parasiticus*. *Fungal Genet Biol* **2008**, *45*, 1081-93.
11. Chang, P. K. Lack of interaction between AFLR and AFLJ contributes to nonaflatoxigenicity of *Aspergillus sojae*. *J. Biotechnol.* **2004**, *107*, 245-253.
12. Chang, P. K.; Ehrlich, K. C.; Hua, S. S. Cladal relatedness among *Aspergillus oryzae* isolates and *Aspergillus flavus* S and L morphotype isolates. *Int J Food Microbiol* **2006**, *108*, 172-7.
13. Lee, Y. H.; Tominaga, M.; Hayashi, R.; Sakamoto, K.; Yamada, O.; Akita, O. *Aspergillus oryzae* strains with a large deletion of the aflatoxin

biosynthetic homologous gene cluster differentiated by chromosomal breakage. *Appl Microbiol Biotechnol* **2006**, *72*, 339-45.

14. Chang, P. K.; Matsushima, K.; Takahashi, T.; Yu, J.; Abe, K.; Bhatnagar, D.; Yuan, G. F.; Koyama, Y.; Cleveland, T. E. Understanding nonaflatoxigenicity of *Aspergillus sojae*: a windfall of aflatoxin biosynthesis research. *Appl Microbiol Biotechnol* **2007**, *76*, 977-84.
15. Carbone, I.; Ramirez-Prado, J. H.; Jakobek, J. L.; Horn, B. W. Gene duplication, modularity and adaptation in the evolution of the aflatoxin gene cluster. *BMC Evol Biol* **2007**, *7*, 111.
16. Chang, P.-K.; Cary, J. W.; Bhatnagar, D.; Cleveland, T. E.; Bennett, J. W.; Linz, J. E.; Woloshuk, C. P.; Payne, G. A. Cloning of the *Aspergillus parasiticus apa-2* gene associated with the regulation of aflatoxin biosynthesis. *Appl. Environ. Microbiol.* **1993**, *59*, 3273-3279.
17. Woloshuk, C. P.; Foutz, K. R.; Brewer, J. F.; Bhatnagar, D.; Cleveland, T. E.; Payne, G. A. Molecular characterization of *aflR,* a regulatory locus for aflatoxin biosynthesis. *Appl. Environ. Microbiol.* **1994**, *60*, 2408-2414.
18. Price, M. S.; Yu, J.; Nierman, W. C.; Kim, H. S.; Pritchard, B.; Jacobus, C. A.; Bhatnagar, D.; Cleveland, T. E.; Payne, G. A. The aflatoxin pathway regulator AflR induces gene transcription inside and outside of the aflatoxin biosynthetic cluster. *FEMS Microbiol. Lett.* **2006**, *255*, 275-279.
19. Chang, P. K.; Yu, J.; Bhatnagar, D.; Cleveland, T. E. Repressor-AFLR interaction modulates aflatoxin biosynthesis in *Aspergillus parasiticus. Mycopathologia* **1999**, *147*, 105-112.
20. Ehrlich, K. C.; Cary, J. W.; Montalbano, B. G. Characterization of the promoter for the gene encoding the aflatoxin biosynthetic pathway regulatory protein AFLR. *Biochim. Biophys. Acta.* **1999**, *1444*, 412-417.
21. Chang, P. K.; Yu, J.; Bhatnagar, D.; Cleveland, T. E. Characterization of the *Aspergillus parasiticus* major nitrogen regulatory gene, *areA*. *Biochim. Biophys. Acta.* **2000**, *1491*, 263-266.
22. Chang, P. K.; Bennett, J. W.; Cotty, P. J. Association of aflatoxin biosynthesis and sclerotial development in *Aspergillus parasiticus. Mycopathologia* **2002**, *153*, 41-48.
23. Meyers, D. M.; Obrian, G.; Du, W. L.; Bhatnagar, D.; Payne, G. A. Characterization of *aflJ*, a gene required for conversion of pathway intermediates to aflatoxin. *Appl. Environ. Microbiol.* **1998**, *64*, 3713-3717.
24. Chang, P. K. The *Aspergillus parasiticus* protein AFLJ interacts with the aflatoxin pathway-specific regulator AFLR. *Mol. Genet. Genomics* **2003**, *268*, 711-719.
25. Du, W.; Obrian, G. R.; Payne, G. A. Function and regulation of *aflJ* in the accumulation of aflatoxin early pathway intermediate in *Aspergillus flavus. Food. Addit. Contam.* **2007**, *24*, 1043-1050.
26. Kale, S. P.; Cary, J. W.; Bhatnagar, D.; Bennett, J. W. Characterization of experimentally induced, nonaflatoxigenic variant strains of *Aspergillus parasiticus*. *Appl. Environ. Microbiol.* **1996**, *62*, 3399-3404.
27. Hicks, J. K.; Yu, J. H.; Keller, N. P.; Adams, T. H. *Aspergillus* sporulation and mycotoxin production both require inactivation of the FadA G alpha protein-dependent signaling pathway. *EMBO J.* **1997**, *16*, 4916-4923.

28. Yu, J. H.; Keller, N. Regulation of secondary metabolism in filamentous fungi. *Annu. Rev. Phytopathol.* **2005**, *43*, 437-458.
29. Calvo, A. M.; Wilson, R. A.; Bok, J. W.; Keller, N. P. Relationship between secondary metabolism and fungal development. *Microbiol. Mol. Biol. Rev.* **2002**, *66*, 447-459.
30. Brodhagen, M.; Keller, N. P. Signalling pathways connecting mycotoxin production and sporulation. *Mol. Plant Pathol.* **2006**, *7*, 285-301.
31. Paoletti, M.; Seymour, F. A.; Alcocer, M. J.; Kaur, N.; Calvo, A. M.; Archer, D. B.; Dyer, P. S. Mating type and the genetic basis of self-fertility in the model fungus *Aspergillus nidulans*. *Curr Biol* **2007**, *17*, 1384-9.
32. Atoui, A.; Bao, D.; Kaur, N.; Grayburn, W. S.; Calvo, A. M. *Aspergillus nidulans* Natural Product Biosynthesis is Regulated by MpkB, a Putative Pheromone-Response MAP-Kinase. *Appl. Environ. Microbiol.* **2008**, AEM.02842-07.
33. Yager, L. N. Early developmental events during asexual and sexual sporulation in *Aspergillus nidulans*. *Biotechnology* **1992**, *23*, 19-41.
34. Kato, N.; Brooks, W.; Calvo, A. M. The expression of sterigmatocystin and penicillin genes in *Aspergillus nidulans* is controlled by *veA*, a gene required for sexual development. *Eukaryot. Cell* **2003**, *2*, 1178-1186.
35. Calvo, A. M.; Bok, J.-W.; Brooks, W.; Keller, N. P. VeA is required for toxin and sclerotial production in *Aspergillus parasiticus*. *Appl. Environ. Microbiol.* **2004**, *70*, 4733-4739.
36. Duran, R.; Cary, J.; Calvo, A. Production of cyclopiazonic acid, aflatrem, and aflatoxin by *Aspergillus flavus* is regulated by *veA,* a gene necessary for sclerotial formation. *Appl. Microbiol. Biotechnol.* **2007**, *73*, 1158-1168.
37. Cary, J.; Obrian, G.; Nielsen, D.; Nierman, W.; Harris-Coward, P.; Yu, J.; Bhatnagar, D.; Cleveland, T.; Payne, G.; Calvo, A. Elucidation of *veA* -dependent genes associated with aflatoxin and sclerotial production in *Aspergillus flavus* by functional genomics. *Appl. Microbiol. Biotechnol.* **2007**, *76*, 1107-1118.
38. Calvo, A. M. The VeA regulatory system and its role in morphological and chemical development in fungi. *Fungal Genet Biol* **2008**, *45*, 1053-61.
39. Geiser, D. M.; Timberlake, W. E.; Arnold, M. L. Loss of meiosis in Aspergillus. *Mol Biol Evol* **1996**, *13*, 809-17.
40. Squire, R. A. Ranking animal carcinogens: a proposed regulatory approach. *Science* **1981**, *214*, 877-880.
41. Wu, F. In *Proceedings of the 2006 annual multi-crop aflatoxin/fumonisin elimination & fungal genomics workshops* Ft. Worth, TX, Oct 16-18 2006, p 49.
42. Stinnett, S. M.; Espeso, E. A.; Cobeno, L.; Araujo-Bazan, L.; Calvo, A. M. *Aspergillus nidulans* VeA subcellular localization is dependent on the importin alpha carrier and on light. *Mol. Microbiol.* **2007**, *63*, 242-255.
43. Blumenstein, A.; Vienken, K.; Tasler, R.; Purschwitz, J.; Veith, D.; Frankenberg-Dinkel, N.; Fischer, R. The *Aspergillus nidulans* phytochrome FphA represses sexual development in red light. *Curr. Biol.* **2005**, *15*, 1833-1838.
44. Purschwitz, J.; Muller, S.; Kastner, C.; Schoser, M.; Haas, H.; Espeso, E. A.; Atoui, A.; Calvo, A. M.; Fischer, R. Functional and physical interaction

of blue- and red-light sensors in *Aspergillus nidulans*. *Curr. Biol.* **2008**, *18*, 255-259.

45. Bayram, O.; Krappmann, S.; Seiler, S.; Vogt, N.; Braus, G. H. *Neurospora crassa ve-1* affects asexual conidiation. *Fungal Genet. Biol.* **2008**, *45*, 127-138.
46. Bayram, O.; Krappmann, S.; Ni, M.; Bok, J. W.; Helmstaedt, K.; Valerius, O.; Braus-Stromeyer, S.; Kwon, N. J.; Keller, N. P.; Yu, J. H.; Braus, G. H. VelB/VeA/LaeA complex coordinates light signal with fungal development and secondary metabolism. *Science* **2008**, *320*, 1504-6.
47. Shieh, M. T.; Brown, R. L.; Whitehead, M. P.; Cary, J. W.; Cotty, P. J.; Cleveland, T. E.; Dean, R. A. Molecular genetic evidence for the involvement of a specific polygalacturonase, P2c, in the invasion and spread of *Aspergillus flavus* in cotton bolls. *Appl Environ Microbiol* **1997**, *63*, 3548-52.
48. Kale, S. P.; Milde, L.; Trapp, M. K.; Frisvad, J. C.; Keller, N. P.; Bok, J. W. Requirement of LaeA for secondary metabolism and sclerotial production in *Aspergillus flavus*. *Fungal Genet Biol* **2008**.
49. Hoffmeister, D.; Keller, N. P. Natural products of filamentous fungi: enzymes, genes, and their regulation. *Nat. Prod. Rep.* **2007**, *24*, 393-416.
50. Perrin, R. M.; Fedorova, N. D.; Bok, J. W.; Cramer, R. A.; Wortman, J. R.; Kim, H. S.; Nierman, W. C.; Keller, N. P. Transcriptional regulation of chemical diversity in *Aspergillus fumigatus* by LaeA. *PLoS Pathol.* **2007**, *3*, e50.
51. Bok, J. W.; Balajee, S. A.; Marr, K. A.; Andes, D.; Nielsen, K. F.; Frisvad, J. C.; Keller, N. P. LaeA, a regulator of morphogenetic fungal virulence factors. *Eukaryot. Cell* **2005**, *4*, 1574-1582.
52. Rohlfs, M.; Albert, M.; Keller, N. P.; Kempken, F. Secondary chemicals protect mould from fungivory. *Biol Lett* **2007**, *3*, 523-5.
53. Bok, J.-W.; Keller, N. P. LaeA, a regulator of secondary metabolism in *Aspergillus* spp. *Eukaryot. Cell* **2004**, *3*, 527-535.
54. Bok, J. W.; Hoffmeister, D.; Maggio-Hall, L. A.; Murillo, R.; Glasner, J. D.; Keller, N. P. Genomic mining for *Aspergillus* natural products. *Chem. Biol.* **2006b**, *13*, 31-37.
55. Bok, J. W.; Noordermeer, D.; Kale, S. P.; Keller, N. P. Secondary metabolic gene cluster silencing in *Aspergillus nidulans*. *Mol. Microbiol.* **2006c**, *61*, 1636-1645.
56. Keller, N. P.; Turner, G.; Bennett, J. W. Fungal secondary metabolism - from biochemistry to genomics. *Nat Rev Microbiol* **2005**, *3*, 937-47.
57. Roze, L. V.; Miller, M. J.; Rarick, M.; Mahanti, N.; Linz, J. E. A novel cAMP-response element, CRE1, modulates expression of *nor-1* in *Aspergillus parasiticus*. *J. Biol. Chem.* **2004b**, *279*, 27428-27439.
58. Shahbazian, M. D.; Grunstein, M. Functions of site-specific histone acetylation and deacetylation. *Annu. Rev. Biochem.* **2007**, *76*, 75-100.
59. Spencer, V. A.; Davie, J. R. Role of covalent modifications of histones in regulating gene expression. *Gene* **1999**, *240*, 1-12.
60. Daniel, P. B.; Walker, W. H.; Habener, J. F. Cyclic AMP signaling and gene regulation. *Annu. Rev. Nutr.* **1998**, *18*, 353-383.

61. Roze, L. V.; Arthur, A. E.; Hong, S. Y.; Chanda, A.; Linz, J. E. The initiation and pattern of spread of histone H4 acetylation parallel the order of transcriptional activation of genes in the aflatoxin cluster. *Mol. Microbiol.* **2007**, *66*, 713-726.
62. Liang, S.-H.; Wu, T.-S.; Lee, R.; Chu, F. S.; Linz, J. E. Analysis of mechanisms regulating expression of the *ver-1* gene involved in aflatoxin biosynthesis. *Appl. Environ. Microbiol.* **1997**, *63*, 1058-1065.
63. Shwab, E. K.; Keller, N. P. Regulation of secondary metabolite production in filamentous ascomycetes. *Mycol Res* **2008**, *112*, 225-30.
64. Niehaus, W. G. J.; Jiang , W. P. Nitrate induces enzymes of the mannitol cycle and suppresses versicolorin synthesis in *Aspergillus parasiticus. Mycopathologia* **1989**, *107*, 131-137.
65. Feng, G. H.; Leonard, T. J. Culture conditions control expression of the genes for aflatoxin and sterigmatocystin biosynthesis in *Aspergillus parasiticus and A. nidulans. Appl. Environ. Microbiol.* **1998**, *64*, 2275-2277.
66. Marzluf, G. A. Genetic regulation of nitrogen metabolism in the fungi. *Microbiol. Molec. Biol. Rev.* **1997**, *61*, 17-32.
67. Chang, P. K.; Ehrlich, K. C.; Yu, J.; Bhatnagar, D.; Cleveland, T. E. Increased expression of *Aspergillus parasiticus aflR*, encoding a sequence-specific DNA-binding protein, relieves nitrate inhibition of aflatoxin biosynthesis. *Appl. Environ. Microbiol.* **1995**, *61*, 2372-2377.
68. Ehrlich, K. C.; Cotty, P. J. Variability in nitrogen regulation of aflatoxin production by *Aspergillus flavus* strains. *Appl. Microbiol. Biotechnol.* **2002**, *60*, 174-178.
69. Bennett, J. W.; Horowitz, P. C.; Lee, L. S. Production of sclerotia by aflatoxigenic and nonaflatoxigenic strains of *Aspergillus flavus* and *A. parasiticus*. *Mycologia* **1979**, *71*, 415-422.
70. Buchanan, R. L.; Lewis, D. F. Regulation of aflatoxin biosynthesis: effect of glucose on activities of various glycolytic enzymes. *Appl. Environ. Microbiol.* **1984**, *48*, 306-310.
71. Woloshuk, C. P.; Cavaletto, J. R.; Cleveland, T. E. Inducers of Aflatoxin Biosynthesis from Colonized Maize Kernels Are Generated by an Amylase Activity from *Aspergillus flavus*. *Phytopathology* **1997**, *87*, 164-169.
72. Chang, P. K.; Hua, S. S. Molasses supplementation promotes conidiation but suppresses aflatoxin production by small sclerotial *Aspergillus flavus. Lett Appl Microbiol* **2007**, *44*, 131-7.
73. Yu, J.; Chang, P.; Bhatnagar, D.; Cleveland, T. E. Cloning of a sugar utilization gene cluster in *Aspergillus parasiticus*. *Biochim Biophys Acta* **2000**, *1493*, 211-4.
74. Keller, N. P.; Nesbitt, C.; Sarr, B.; Phillips, T. D.; Burow, G. B. pH regulation of sterigmatocystin and aflatoxin biosynthesis in *Aspergillus* spp. *Phytopathol.* **1997**, *87*, 643-648.
75. Ehrlich, K. C.; Montalbano, B. G.; Cotty, P. J. Divergent regulation of aflatoxin production at acidic pH by two *Aspergillus* strains. *Mycopathologia* **2005**, *159*, 579-581.
76. Tilburn, J.; Sarkar, S.; Widdick, D. A.; Espeso, E. A.; Orejas, M.; Mungroo, J.; Penalva, M. A.; Arst Jr, H. N. The *Aspergillus* PacC zinc finger

transcription factor mediates regulation of both acid- and alkaline expressed genes by ambient pH. *EMBO. J.* **1995**, *14*, 779-790.

77. Ehrlich, K. C.; Montalbano, B. G.; Cary, J. W.; Cotty, P. J. Promoter elements in the aflatoxin pathway polyketide synthase gene. *Biochim. Biophys. Acta* **2002**, *1576*, 171-175.
78. Cotty, P. Aflatoxin and sclerotial production by *Aspergillus flavus*: influence of pH. *Phytopathology* **1988**, *78*, 1250-1253.
79. Holmes, R. A.; Boston, R. S.; Payne, G. A. Diverse inhibitors of aflatoxin biosynthesis. *Appl Microbiol Biotechnol* **2008**, *78*, 559-72.
80. Zeringue, H. J., Jr.; McCormick, S. P. Aflatoxin production in cultures of *Aspergillus flavus* incubated in atmospheres containing selected cotton leaf-derived volatiles. *Toxicon* **1990**, *28*, 445-448.
81. Greene-McDowelle, D. M.; Ingber, B.; Wright, M. S.; Zeringue, H. J., Jr.; Bhatnagar, D.; Cleveland, T. E. The effects of selected cotton-leaf volatiles on growth, development and aflatoxin production of *Aspergillus parasiticus*. *Toxicon* **1999**, *37*, 883-893.
82. Wright, M. S.; Greene-McDowelle, D. M.; Zeringue, H. J.; Bhatnagar, D.; Cleveland, T. E. Effects of volatile aldehydes from *Aspergillus*-resistant varieties of corn on *Aspergillus parasiticus* growth and aflatoxin biosynthesis. *Toxicon* **2000**, *38*, 1215-1223.
83. Norton, R. A. Inhibition of aflatoxin B(1) biosynthesis in *Aspergillus flavus* by anthocyanidins and related flavonoids. *J. Agric. Food Chem.* **1999**, *47*, 1230-1235.
84. Bhatnagar, D.; McCormick, S. P. The inhibitory effect of neem leaf extracts on aflatoxin synthesis in *Aspergillus parasiticus*. *JAOCS* **1988**, *65*, 1166-1167.
85. Goodrich-Tanrikulu, M.; Mahoney, N. E.; Rodriguez, S. B. The plant growth regulator methyl jasmonate inhibits aflatoxin production by Aspergillus flavus. *Microbiology* **1995**, *141*, 2831-2837.
86. Feussner, I.; Wasternack, C. The lipoxygenase pathway. *Annu. Rev. Plant Biol.* **2002**, *53*, 275-297.
87. Roze, L. V.; Calvo, A. M.; Gunterus, A.; Beaudry, R.; Kall, M.; Linz, J. E. Ethylene modulates development and toxin biosynthesis in *Aspergillus* possibly via an ethylene sensor-mediated signaling pathway. *J. Food Prot.* **2004a**, *67*, 438-447.
88. Gunterus, A.; Roze, L. V.; Beaudry, R.; Linz, J. E. Ethylene inhibits aflatoxin biosynthesis in *Aspergillus parasiticus* grown on peanuts. *Food Microbiol.* **2007**, *24*, 658-663.
89. Roze, L. V.; Beaudry, R. M.; Arthur, A. E.; Calvo, A. M.; Linz, J. E. Aspergillus volatiles regulate aflatoxin synthesis and asexual sporulation in *Aspergillus parasiticus*. *Appl Environ Microbiol* **2007**, *73*, 7268-76.
90. Molyneux, R. J.; Mahoney, N.; Kim, J. H.; Campbell, B. C. Mycotoxins in edible tree nuts. *Int J Food Microbiol* **2007**, *119*, 72-8.
91. Oh, K. B.; Miyazawa, H.; Naito, T.; Matsuoka, H. Purification and characterization of an autoregulatory substance capable of regulating the morphological transition in *Candida albicans*. *Proc. Natl. Acad. Sci. U S A* **2001**, *98*, 4664-4668.

92. Champe, S. P.; Rao, P.; Chang, A. An endogenous inducer of sexual development in *Aspergillus nidulans*. *J. Gen. Microbiol.* **1987**, *133*, 1383-1387.
93. Calvo, A. M.; Hinze, L. L.; Gardner, H. W.; Keller, N. P. Sporogenic effect of polyunsaturated fatty acids on development of *Aspergillus* spp. *Appl. Environ. Microbiol.* **1999**, *65*, 3668-3673.
94. Burow, G. B.; Nesbitt, J. D.; Keller, N. P. Seed lipoxygenase products modulate *Aspergillus* mycotoxin synthesis. *Mol. Plant Microbe Int.* **1997**, *10*, 380-387.
95. Wilson, R. A.; Gardner, H. W.; Keller, N. P. Cultivar-dependent expression of a maize lipoxygenase responsive to seed infesting fungi. *Mol. Plant Microbe Interact.* **2001**, *14*, 980-987.
96. Tsitsigiannis, D. I.; Kunze, S.; Willis, D. K.; Feussner, I.; Keller, N. P. *Aspergillus* infection inhibits the expression of peanut 13S-HPODE-forming seed lipoxygenases. *Mol. Plant Microbe Interact.* **2005**, *18*, 1081-1089.
97. Brodhagen, M.; Tsitsigiannis, D. I.; Hornung, E.; Goebel, C.; Feussner, I.; Keller, N. P. Reciprocal oxylipin-mediated cross-talk in the *Aspergillus*-seed pathosystem. *Mol. Microbiol.* **2008**, *67*, 378-391.
98. Gao, X.; Shim, W. B.; Gobel, C.; Kunze, S.; Feussner, I.; Meeley, R.; Balint-Kurti, P.; Kolomiets, M. Disruption of a maize 9-lipoxygenase results in increased resistance to fungal pathogens and reduced levels of contamination with mycotoxin fumonisin. *Mol Plant Microbe Interact* **2007**, *20*, 922-33.
99. *Aspergillus flavus* Genome Database http://www.aspergillusflavus.org/
100. Machida, M.; Asai, K.; Sano, M.; Tanaka, T.; Kumagai, T.; Terai, G.; Kusumoto, K.; Arima, T.; Akita, O.; Kashiwagi, Y.; Abe, K.; Gomi, K.; Horiuchi, H.; Kitamoto, K.; Kobayashi, T.; Takeuchi, M.; Denning, D. W.; Galagan, J. E.; Nierman, W. C.; Yu, J.; Archer, D. B.; Bennett, J. W.; Bhatnagar, D.; Cleveland, T. E.; Fedorova, N. D.; Gotoh, O.; Horikawa, H.; Hosoyama, A.; Ichinomiya, M.; Igarashi, R.; Iwashita, K.; Juvvadi, P. R.; Kato, M.; Kato, Y.; Kin, T.; Kokubun, A.; Maeda, H.; Maeyama, N.; Maruyama, J.; Nagasaki, H.; Nakajima, T.; Oda, K.; Okada, K.; Paulsen, I.; Sakamoto, K.; Sawano, T.; Takahashi, M.; Takase, K.; Terabayashi, Y.; Wortman, J. R.; Yamada, O.; Yamagata, Y.; Anazawa, H.; Hata, Y.; Koide, Y.; Komori, T.; Koyama, Y.; Minetoki, T.; Suharnan, S.; Tanaka, A.; Isono, K.; Kuhara, S.; Ogasawara, N.; Kikuchi, H. Genome sequencing and analysis of Aspergillus oryzae. *Nature* **2005**, *438*, 1157-61.
101. *Aspergillus nidulans* Database- Broad Institute http://www.broad.mit.edu/annotation/genome/aspergillus_nidulans/Home.html
102. *Aspergillus niger v1.0* http://genome.jgi-psf.org/Aspni1/Aspni1.home.html
103. Yu, J.; Payne, G. A.; Nierman, W. C.; Machida, M.; Bennett, J. W.; Campell, B. C.; Robens, J. F.; Bhatnagar, D.; Dean, R. A.; Cleveland, T. E. *Aspergillus flavus* genomics as a tool for studying the mechanism of aflatoxin formation. *Food Addit Contam* **2008**, 1-6.
104. Yu, J.; Whitelaw, C. A.; Nierman, W. C.; Bhatnagar, D.; Cleveland, T. E. *Aspergillus flavus* expressed sequence tags for identification of genes with

putative roles in aflatoxin contamination of crops. *FEMS Microbiol Lett* **2004**, *237*, 333-40.

105. O'Brian, G. R.; Georgianna, D. R.; Wilkinson, J. R.; Yu, J.; Abbas, H. K.; Bhatnagar, D.; Cleveland, T. E.; Nierman, W.; Payne, G. A. The effect of elevated temperature on gene transcription and aflatoxin biosynthesis. *Mycologia* **2007**, *99*, 232-9.
106. Price, M. S.; Conners, S. B.; Tachdjian, S.; Kelly, R. M.; Payne, G. A. Aflatoxin conducive and non-conducive growth conditions reveal new gene associations with aflatoxin production. *Fungal Genet Biol* **2005**, *42*, 506-18.
107. Wilkinson, J. R.; Yu, J.; Bland, J. M.; Nierman, W. C.; Bhatnagar, D.; Cleveland, T. E. Amino acid supplementation reveals differential regulation of aflatoxin biosynthesis in *Aspergillus flavus* NRRL 3357 and *Aspergillus parasiticus* SRRC 143. *Appl Microbiol Biotechnol* **2007**, *74*, 1308-19.
108. Guo, B.; Chen, X.; Dang, P.; Scully, B. T.; Liang, X.; Holbrook, C. C.; Yu, J.; Culbreath, A. K. Peanut gene expression profiling in developing seeds at different reproduction stages during *Aspergillus parasiticus* infection. *BMC Dev Biol* **2008**, *8*, 12.

Analytical Approaches

Chapter 14

Sampling Plans of Mycotoxins in Foods and Dietary Supplements (Ginger Capsules)

Mary W. Trucksess[1] and Thomas B. Whitaker[2]

[1]Food and Drug Administration, 5100 Paint Branch Parkway, College Park, MD 20740, USA
[2]US Department of Agriculture, Agricultural Research Service (Retired), Box 7625, NC State University, Raleigh, Raleigh, NC, 27695

Detection and quantitation of mycotoxins in foods and feeds are important to human and animal health. Regulatory decisions for edible quality of imported and domestic foods depend on the classification of the bulk lot as acceptable or unacceptable. The classification is made by comparing the measured mycotoxin concentration in a test sample taken from the bulk lot to a legal limit. The assumption is that the lot concentration is the same as the test sample concentration. If the test sample concentration does not represent the lot concentration then the lot may be misclassified. Some good lots may be rejected by the sampling plan (seller's risk) and some bad lots may be accepted by the sampling plan (buyer's risk). The results will be either economic loss or adverse health impacts. Because of the large variability among sample test results taken from the same lot, it is difficult to obtain a precise and accurate estimate of the true contaminant concentration of a bulk lot when using small test samples. Proper sampling plans must be developed according to "suitable for purpose" principle. Some of the sampling plans for grains and nuts will be covered. Recent research on sampling and analytical variability associated with determination of aflatoxins and ochratoxin A in dietary supplement ginger capsules is presented to illustrate the development of sampling plans.

Introduction

Accurate estimates of mycotoxin concentration in food are important for food safety and wholesomeness. Since the discovery of aflatoxins in the early 1960s, considerable effort has been focused on test procedures for mycotoxin in foods. All test procedures contain three critical elements: sampling, sample preparation, and analysis. Sampling is the most difficult step in mycotoxin determination. It is known that mycotoxins are heterogeneously distributed in agricultural commodities. There have been reports of finding aflatoxin concentrations in excess 1,000,000 ng/g for individual peanut kernels, 5,000,000 ng/g for a cottonseed and over 400,000 ng/g in a corn kernel. Therefore, sampling variability is usually the largest source of variability associated with a mycotoxin test procedure. Sample preparation variability is lower than that of the sampling and depends primarily on the size of the test portion and the type of mill used to comminute the kernels in the sample. The use of a validated performance based analytical method keeps the analytical variability at a minimum.

A sampling plan is defined by the test procedure used to quantify the mycotoxin and an accept-reject limit (1). The accept-reject limit is often equal to a country's regulatory limit, which varies greatly from one country to another. Subsequently, through the years many sampling plans have been developed for mycotoxins, mainly for aflatoxins in corn and nuts. The control of the occurrence of mycotoxins in foods and feeds requires effective surveillance and quality control procedures which involve a sequence of sampling, sample preparation and analysis steps (2). The certainty of an estimate of the true mycotoxin concentration in a lot depends on the variability associated with each step of the mycotoxin test procedure. Each step contributes to the total uncertainty of the test procedure. Using variance as a statistical measure of uncertainty, the variance associated with each step of a test procedure used to detect various mycotoxins in various commodities has been investigated (3,4,5,6). For example, the sampling, sample preparation, and analytical steps account for 85, 10, and 5% of the total variability (as measured by the variance) associated with the aflatoxin test procedure used to measure aflatoxin in shelled peanuts when using a 10 kg sample, USDA hammer mill, 250 g test portion, and HPLC analytical methods (4).

Using the variance and distribution among sample test results, methods have been developed to predict the performance of mycotoxin sampling plans. For examples, evaluation methods have been developed for sampling plans for aflatoxin in farmers' stock peanuts (7), fumonisin in shelled corn (8), and ochratoxin in green coffee (9). There are very few sampling studies for mycotoxins in consumer products. One study used different ways of sampling figs for aflatoxins and estimated the confidence that can be associated with each of these (10). Another study assessed appropriate statistical models for use in evaluating retail sampling plans for the determination of mycotoxins in food (11). A compound gamma model was found to be a suitable fit. A simulation model based on the compound gamma model produced fit-for-purpose results for the measurement of OTA in retail dried fruit samples but was not suitable for the measurement of other toxins in other food products. Another example is a study that evaluated sampling plans used in the United States (US), United Kingdom (UK) and The Netherlands to test raw shelled peanuts for aflatoxin in the 1990s (12). The acceptance limit for US, UK, and the Dutch were 15 ng total aflatoxin/g, 10 ng total aflatoxin/g, and 3 ng aflatoxin B_1/g, respectively. The U.S. plan used 3 sampling units, 21.8 kg each; the UK plan used a single sampling unit of 10 kg; and the Dutch plan used 4 sampling units, 7.5 kg each. The sampling variance was lowest for the U.S. plan and highest for the Dutch plan. The sample preparation variance was lower for the Dutch and UK plans than for the U.S. plan, because of the mill type used to comminute the kernels in the sample. The U.S. plan accepted the greatest number of lots with the highest aflatoxin concentration and the Dutch plan rejected the greatest number of lots with the lowest aflatoxin concentration. The variations of sampling plans for aflatoxins have resulted in many international trade issues.

US Food and Drug Administration Sampling Plans for Aflatoxins

The maximum limit of aflatoxins in foods in the United States is 20 ng/g total aflatoxins and in milk is 0.5 ng/g aflatoxin M_1 (13). One analysis is performed per sample. If the aflatoxin test result is less than or equal to 20 ng/g total aflatoxin, then the lot is acceptable. If the test result is greater than 20 ng/g then a second or check analysis is performed on the same sample. If the check test result is also greater than 20 ng/g, then the lot is detained for further consideration. The identities of aflatoxins B_1 and G_1 must be confirmed. If the check test result is less than 20 ng/g, then further testing is required. Statistical evaluation of all results would be necessary for regulatory decision. Analytical methods used are based on performance characteristics. In general, only AOAC® Official Analytical Methods℠ are used unless the methods are not applicable for the products. Sample sizes vary depending on the products, as shown on Table 1 (14).

Table 1. Product sample sizes used by the United States Food and Drug Administration

Product	Description	Package Type	Lot Size	Number of units	Unit Size	Sample (lbs)
Peanut Butter	Smooth	Consumer Bulk		24	0.5	12
Peanuts	Crunch butter, raw, roasted, ground topping	Consumer & bulk		48	1	48
Tree nuts	Inshell, shelled, slices or flour paste	Consumer & bulk		10 50 12	1 1 1	10 50 12
Brazil nuts	Inshell in Import status	Bulk	<200 bags 201-800 801-2,000	20 40 60	1 1 1	
Pistachio nuts	Inshell in import status	Bulk	75,000 lbs <75,000	20% of Units		50 25
Corn	Shelled, meal, flour, grits	Consumer & bulk		10	1	10
Cottonseed		Bulk		15	4	60
Oilseed meals	Peanut, cottonseed	Bulk		20	1	20
Edible seeds	Pumpkin, melon, seasame,etc.	Bulk		50	1	50
Ginger root	Dried, whole ground	Consumer & bulk	"n" units	?? 10	10 x .06	15 10
Milk	Whole, low fat, skim	Bulk consumer		10	1	10 10
Small grains	Wheat, barley sorghum, etc.	Bulk		10	1	10
Dried fruit	Figs	Consumer & bulk		50		50
Mixtures	-commodity particles large -commodity particles finely ground	Consumer & bulk		50 10		50 10

Note: from reference (14)

European Commission sampling plan for aflatoxins

The European Commission (EC) food safety legislation specifies the maximum levels for both B_1 and total aflatoxins and sampling plan designs for various food products (15). EC regulations are directly applicable in the UK and other members of the European Union. They give further details of maximum aflatoxin levels set for individual commodities (Table 2). Both the total aflatoxin and aflatoxin B_1 levels must be at or below the maximum levels. The sample sizes are specified on Table 3. Sampling plans on Tables 2 and 3 are for bulk lots and are not applicable to sampling

Table 2. Maximum aflatoxin levels in foods set by European Commission regulation

Products	Maximum Limit ng/g	
	Aflatoxin B_1	Total aflatoxins
Groundnuts to be subjected to sorting or other physical treatment before consumption or use as an ingredient	8.0	15.0
Maize, nuts and dried fruit to be subjected to sorting or other physical treatment before consumption or use as an ingredient; Spices: chilies, chili powder, cayenne, paprika, white and black pepper, nutmeg, ginger, turmeric	5.0	10.0
Cereal and cereal products; groundnuts, nuts, and dried fruit and processed products thereof, intended for direct consumption or use as an ingredient: except those commodities list below	2.0	4.0
Processed cereal-based foods and baby foods for infants and young children, dietary foods for medicinal purposes intended specifically for infants	0.1	

Note: information from reference (15)

food at the retail stage. In addition, the sampling plans provide guidance for official methods of analysis and interpretation of analytical results. The EC does not specify which analytical methods should be used, but requires an analytical method to meet minimum performance standards such as repeatability and extraction efficiency. All analytical results have variability or measurement uncertainty. In the UK, the analytical result is reported as "x ± 2u", where x is the analytical measurement and u is the standard deviation associated with the analytical measurement uncertainty. The expanded measurement uncertainty, 2u, gives approximately a 95% confidence level. For example, if the contamination level is 100 ng/g then 2u is 44% and the range of acceptable concentrations is 56 – 144 ng/g.

Table 3. Sampling procedures set by European Commission regulation

Commodity	Lot weight	No. of Incremental Samples	Aggregate Sample wt. (kg)
Cereal and cereal products	<50 kg	3	1
	> 50 kg ≤ 1 tonne	5	1
	> 500 kg ≤ 3tonnes	10	1
	> 1 tonne ≤ 3 tonnes	20	2
	> 3 tonnes ≤ 10 tonnes	40	4
	> 10 ≤ 20 tonnes	60	6
	> 20 tonnes ≤ 50 tonnes	100	10
Dried Fruit (dried figs not included)	≤ 100 kg	10	1
	>100 kg ≤ 200 kg	15	1.5
	> 200 kg ≤ 500 kg	20	2
	> 500 kg ≤ 1000 kg	30	3
	> 1 tonne ≤ 2 tonnes	40	4
	> 2 tonnes ≤ 5 tonnes	60	6
	> 5 tonnes ≤ 10 tonnes	80	8
	> 10 tonnes	100	10
Dried figs, groundnuts & nuts	≤ 100 kg	10	3
	>100 kg ≤ 200 kg	15	4.5
	> 200 kg ≤ 500 kg	20	6
	> 500 kg ≤ 1000 kg	30	9
	> 1 tonne ≤ 2 tonnes	40	12
	> 2 tonnes ≤ 5 tonnes	60	18
	> 5 tonnes ≤ 10 tonnes	80	24
	> 10 tonnes	100	30

Note: information from reference (15)

Codex Alimentarius Commission Sampling Plan for Aflatoxins in Tree Nuts

Since 2003, approximately 100 countries have specified regulations or detailed guidelines for mycotoxins in foods (16). Seventy-six countries have legislated limits on aflatoxins ranging from zero to 35 ng/g. For this reason, the Codex Committee on Contaminants in Foods (CCCF) began work in 2004 to harmonize aflatoxin limits and sampling plans for aflatoxins in treenuts, specifically almonds, hazelnuts, pistachios, and Brazil nuts. Sampling plans for almonds and hazelnuts have also been developed (17,18,19). In June 2008 at the 31st Codex Alimentarius Commission (CAC) conference, CAC adopted CCCF aflatoxin sampling plans for aflatoxin contamination in ready-to-eat treenuts and treenuts destined for further processing: almonds, hazelnuts and pistachios (20). Table 4 gives the elements and criteria of each element of the sampling plans. These two plans have been designed for enforcement and controls concerning total aflatoxin in bulk consignments (lots) of treenuts traded in the export market.

Table 4. Codex Alimentarius Commission sample plan for treenuts: almonds, hazelnuts, and pistachio nuts.

Element	Destined for further processing Criteria	Destined for ready-to-eat Criteria
Maximum linit	15 ng/g total aflatoxins	10 ng/g total aflatoxins
Analysis	1	2
Sample size	20 kg, almonds (shelled) hazelnuts (shelled) pistachio nuts (inshell, equivalent to about 10 kg shelled nuts)	10 kg, almonds (shelled) hazelnuts (shelled) pistachio nuts (inshell, equivalent to about 10 kg shelled nuts)
Sample preparation	Dry grind with vertical cutter mixer type mill 50 g test portion	Dry grind with vertical cutter mixer type mill 50 g test portion
Analytical method	Performance based blanks: negligible recovery: 1-15 ng/g, 70-110% >15 ng/g, 80-110% RSD_R: 1-120 ng/g, 22% RSD_r 1-120 ng/g, 0.66 x RSD_R	Performance based blanks: negligible recovery: 1-15 ng/g, 70-110% >15 ng/g, 80-110% RSD_R: 1-120 ng/g, 22% RSD_r 1-120 ng/g, 0.66 x RSD_R
Decision rule	If aflatoxin test result is less than or equal to 15 ng/g total aflatoxins, then accept the lot. Otherwise, reject the lot.	If aflatoxin test result is less than or equal to 10 ng/g total aflatoxins for both samples, then accept the lot. Otherwise, reject the lot.

Note: information form reference (20)

The Recent Join Research Projects On Sampling For Mycotoxins In Botanical Roots At The USFDA and USDA

A collaborative project between USFDA and USDA, ARS on sampling for aflatoxin (AF) and ochratoxin A (OTA) in botanical dietary supplements such as ginger capsules was conducted. The safety of botanicals is of primary concern to the consumers of medicinal plants and dietary supplements. In the United States, the FDA generally recognizes ginger as safe. It is not approved for the treatment or cure of any disease and is consumed as food or as dietary supplements. Powdered dried ginger root is marketed in capsules and sold in herbal supplement stores. AF and OTA are frequently found in powdered ginger (21).

The ginger capsule study

The study was designed to (a) determine if the dietary supplement ginger sold as a powder in capsule form and packaged in individual bottles is contaminated with AF and OTA and (b) determine both the within-bottle variability among 5 g test portion results taken from individual bottles of powdered ginger and the bottle-to-bottle variability among AF and OTA test portion results. The bottle-to-bottle variance can be used to make recommendations for the number of incremental samples (bottles) that need to be selected from a lot and pooled to form a bulk sample to overcome any lack of homogeneity among contaminated particles in the lot. The within- bottle variance can be used to recommend the size of the test portion (or number of test portions) to be selected from a bulk sample to obtain an accurate estimate of the true mycotoxin concentration in the lot.

Material and sampling design

Twenty bottles of ginger (*Zingiber officinale*), each containing 60 capsules (625 mg ginger per capsule, a total 37.5 g of powdered ginger) were purchased from a retail store in Philadelphia, PA at three separate six month intervals (total 60 bottles). It was assumed that the three groups of 20 bottles each came from three separate lots. However, this could not be confirmed since no identification codes were found on the bottles. For a given group of 20 bottles, all 60 capsules in a bottle were opened and the powdered ginger was combined for a total of 37.5 g of powdered ginger per bottle. The 37.5 g of ginger from each bottle was tumbled for 4 hours to thoroughly mix the ginger. Initially, two 5 g test portions were removed from each of the 20 bottles in lot 1. It was later decided to remove four 5 g test portions from each bottle for lots 2 and 3. The 5 g test portions were identified by lot (1, 2, or 3), bottle (1 to 20), and sample number (1 to 4). For lots 1 and 2, one aflatoxin and one OTA measurement was made per 5 g test portion. However, for lot 3, two aflatoxin and two OTA measurements were made for sample 1 (one AF and one OTA measurement for samples 2, 3, and 4) from each of the 20 bottles to get an estimate of analytical variability. The nested design is represented in Figure 1.

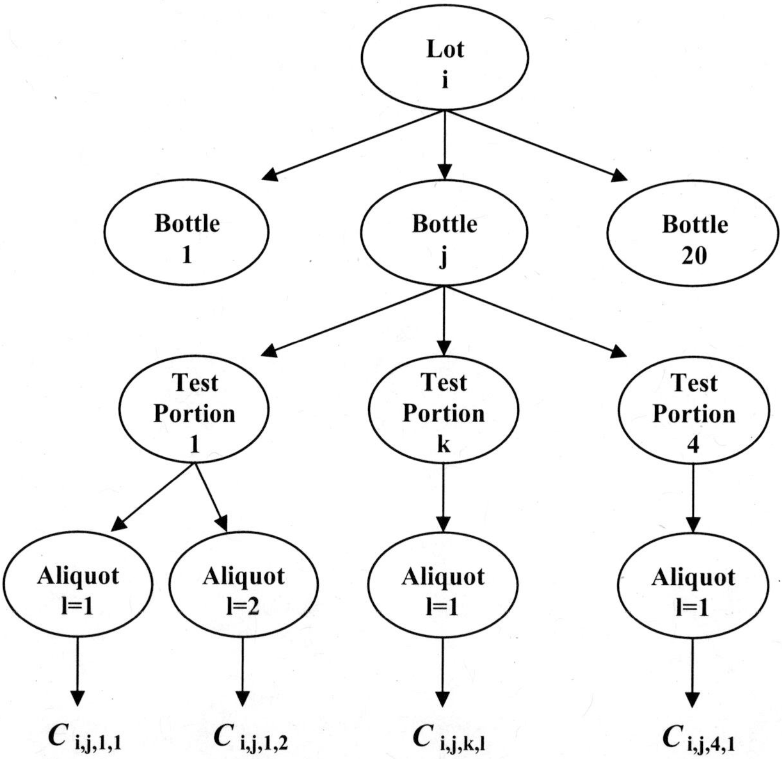

Figure 1. Nested design used to partition the total variance associated with measuring AF and OTA concentration (C) in powdered ginger for lot 3. Lot 1 had a similar design except number of test portions (k) was equal to 2 and number of aliquots quantified (l) was 1 for all test portions. Lot 2 was similar in design except the number of aliquots quantified was 1 for all test portions.

Chemical Analysis of Aflatoxins and Ochratoxin A

A published method was used to simultaneously extract and purify the AF and OTA in the powered ginger (21). The accuracy, repeatability and reproducibility characteristics of the method has been validated by an international collaborative and has been adopted as AOAC® Official Method℠ 2008.02 (22). The entire test portion was extracted with methanol and 0.5% aqueous sodium hydrogen carbonate solution (70+30, v/v). The extract was centrifuged, diluted with phosphate buffer (PB), filtered and applied to an immunoaffinity column containing antibodies specific for AF and OTA. After washing, the toxins were eluted from the column with methanol, and determined and quantified by high performance liquid chromatography with fluorescence detector. AF was separated and determined by reversed phase liquid chromatography (RPLC) using a Waters 4.6 x 150 mm column, a post-column photochemical derivatization cell and fluorescence detection at excitation wavelength 362 nm and emission wavelength 440 nm. The mobile phase was methanol – acetonitrile – water (25 + 15 + 60, v/v) and flow rate was 0.8 mL/min. OTA was separated and determined by RPLC using a Beckman 4.6 x 250 mm, 5 µm, C-18 column and fluorescence detection at excitation wavelength 333 nm and emission wavelength 460 nm. The mobile phase was acetonitrile – water – acetic acid (47 + 53 + 1, v/v) and the flow rate was 1 mL/min.

Measurement of variability (variance)

From the nested design (Figure 1), the total variance (S^2_t) among all sample test results per lot is the sum of the bottle-to-bottle variance (S^2_{btb}), within- bottle variance (S^2_{wib}), and analytical variance (S^2_a).

$$s^2_t = s^2_{btb} + s^2_{wib} + s^2_a \qquad (1)$$

For lots 1 and 2, the experimental design didn't allow for a direct measurement of the analytical variance. Instead, the bottle-to-bottle (s^2_{btb}) and the combined within-bottle and analytical variance (s^2_{wiba}) were measured.

$$s^2_t = s^2_{btb} + s^2_{wiba} \qquad (2)$$

where

$$s^2_{wiba} = s^2_{wib} + s^2_a \qquad (3)$$

For lot 3, the experimental design allowed the combined within-bottle plus analytical variances to be separated into within-bottle variance and analytical variance.

The variance components in Equations 1 and 2 were determined using Proc Nested in the Statistical Analysis System (SAS). For lots 1 and 2, estimates of s^2_t, s^2_{btb}, and s^2_{wiba} were made. For lot 3, estimates of s^2_t, s^2_{btb}, s^2_{wib}, and s^2_a were made.

Results of study

Variances Associated with Measurements of AF. The total variance associated with measuring AF in a 5 g test portion by RPLC methods was partitioned into bottle-to-bottle and combined within-bottle plus analytical variances for lots 1, 2, and 3 (Table 5). The total, bottle-to-bottle, and combined within-bottle plus analytical variances averaged across the three lots was 1.544, 0.354, and 1.190. If the analytical variance from lot 3 is subtracted from the average combined within-bottle plus analytical variance (Equation 2), the within-bottle variance is 0.662 (assume the analytical variance for lots 1 and 2 is the same as that for lot 3). The total variance (Equation 1) is equal to the sum of the bottle-to-bottle, within-bottle, and analytical variances or 1.544 (0.354 +

0.662 + 0.528). The bottle-to-bottle, within bottle, and analytical variances account for 22.9, 42.9, and 34.2 percent of the total variance, respectively.

Table 5. Total, bottle-to-bottle, within bottle, and analytical variances associated when sampling ginger for aflatoxins

Lot	No. analyzed	Ave. AF ng/g	Median AF ng/g	Variance Components (proc mixed)			
				Total	Bottle to Bottle	Within Bottle & Analytical	Analytical
1	39	6.54	6.36	0.617	0.052	0.565	
2	80	8.26	7.96	2.966	1.010	1.956	
3	75	7.22	7.25	1.048	0.000	1.048	0.528
Ave		7.34	7.19	1.544	0.354	1.190	0.528

The analytical variance (0.528) associated with measuring AF in one aliquot by RPLC can be used to predict the analytical variance for any number of aliquots, na, quantified for AF. The analytical variance for any number of aliquots is

$$s^2_a = (1/na)\ 0.528 \qquad (4)$$

The within-bottle variance (0.662) associated with measuring AF in a 5 g test portion taken from an aggregate sample can be used to predict the within-bottle variance for any given test portion size, tns, in grams.

$$s^2_{wib} = (5/tns)\ 0.662 \qquad (5)$$

The bottle-to-bottle variance (S^2_{btb}) for any given aggregate sample size, bns, in number of bottles can be estimated from 0.354 for a single bottle of ginger.

$$S^2_{btb} = (1/bns)\ 0.354 \qquad (6)$$

The total variance associated with pooling ginger powder taken from capsules from bns bottles to form an aggregate sample, taking a test portion of tns grams from the aggregate sample, and quantifying the AF by RPLC in any number of aliquots, na, can be determined by summing Equations 4, 5, and 6 (as shown in Equation 1).

$$S^2_t = (1/bns)\ 0.354 + (5/tns)\ 0.662 + (1/na)\ 0.528 \qquad (7)$$

Increasing test portion size and number of aliquots quantified reduces variability of the overall AFL test procedure. However, it appears that the most cost effective method of reducing the overall uncertainty of the AFL test procedure is to increase test portion size.

Variances Associated with Measurements of OTA. Since OTA was measured in the same 5 g test portion as AF, the same statistical analysis was applied to OTA sample test results as described above for AF. The total variance associated with measuring OTA in a 5 g test portion was partitioned into bottle-to-bottle and combined within-bottle plus analytical variances for lots 1, 2, and 3 (Table 6). The total, bottle-to-bottle, and combined within-bottle plus analytical variances averaged across all three lots was 0.228, 0.015, and 0.213, respectively. If the analytical variance from lot 3 (0.020) is subtracted from the average combined within-bottle plus analytical variance (0.213) the within-bottle variance is 0.193 (assume the analytical variance for lots 1 and 2 are the same as lot 3). The total variance (Equation 1) is equal to the sum of bottle-to-bottle, within-bottle, and analytical variances or 0.228 (0.015 + 0.193 + 0.020). The bottle-to-bottle, within-bottle, and analytical variances accounts for 6.6, 84.6, and 8.8 percent of the total variance, respectively. The above variances are specific to using RPLC methods to measure OTA in powdered ginger with 5 g test portions taken from a lot at 1.93 ug/g OTA.

Table 6. Total, bottle-to-bottle, within bottle, and analytical variances associated when sampling ginger for ochratoxin A

Lot	No. analyzed	Ave. OTA ng/g	Median OTA ng/g	Variance Components (proc mixed)			
				Total	Bottle to Bottle	Within Bottle & Analytical	Analytical
1	40	1.96	1.86	0.161	0.000	0.161	
2	78	2.19	2.15	0.166	0.003	0.163	
3	75	1.65	1.45	0.355	0.041	0.314	0.020
Ave		1.93	1.82	0.228	0.015	0.213	0.020

The total variance associated with pooling ginger powder from bns bottles to form an aggregate sample, taking a test portion of tns grams from the aggregate sample, and quantifying the OTA in any number of aliquots by the RPLC method can be determined:

$$s^2_t = (1/bns)\ 0.015 + (5/tns)\ 0.193 + (1/na)\ 0.020 \qquad (8)$$

Since within-bottle variance is the largest source of variability (accounts for 84.6% of the total variability), increasing test portion size should be the first consideration to reduce the total variance associated with the OTA test procedure.

Conclusion of ginger capsule study

When measuring AF and OTA, emphasis on error (variability) reduction should focus on using larger or more test portion sizes since the within-bottle variance associated with 5 g test sample accounted for largest percent of the total variance. The bottle-to-bottle variance (reflecting non-uniform distribution of contaminated ginger particles throughout the lot) was the smallest percent of the total variance and can be minimized by pooling the contents of a relatively few bottles.

Summary

More than 100 countries have established maximum levels and sampling plans for various mycotoxins in food and feed. Because of the uncertainty associated with the mycotoxin test procedure, the true mycotoxin concentration in a bulk lot cannot be determined with 100% certainty by measuring the concentration in samples taken from the lot. By measuring the variability and distribution among sample test results, researchers have developed methods to predict the chances of accepting bad lots and rejecting good lots. These methods have been used by commodity industries, regulatory agencies, and international organizations to design mycotoxin sampling plans that minimize safety risks within available resources. FAO/WHO, working through the Codex Committee process, is attempting to improve both consumer safety and export trade by harmonizing mycotoxin limits and sampling plans used in the export market. These efforts by Codex have led to the harmonization of aflatoxin limits and sampling plans for peanuts, almonds, pistachios, and hazelnuts. The recent USFDA and USDA joint research project on sampling for mycotoxins in powdered ginger capsules marketed in bottles sold as capsules found that the within-bottle variance was the largest source of variability and that increasing test portion size should be the first consideration to reduce sampling error.

References

1. Whitaker, T. B.; Johansson, A. S. Sampling uncertainties for the detection of chemical agents in complex food matrices. *J. Food Prot.* **2005**, *68*, 1306-1313.
2. Coker, R. D.; Nagler, M. J.; Blunden, G.; Sharkey, A. J.; Defize, P. R.; Derksen, G. B.; Whitaker, T. B. Design of sampling plans for mycotoxins in foods and feeds. *Nat. Toxins* **1995**, *3*, 257-262.
3. Whitaker, T. B., Dickens, J. W., and Monroe, R. J. Variability of aflatoxin test results. *J. AOAC, Int.* **1974,** *51*, 214-218.
4. Whitaker, T. B., Dowell, F. E., Hagler, W. M., Jr., Giesbrecht, F. G., and Wu, J. Variability Associated with sampling, sample preparation, and chemical testing of farmers' stock peanuts. *J. AOAC Int.* **1994**, *77*, 107-116.
5. Johansson, A. S., T. B. Whitaker, W. M. Hagler, Jr., F. G. Giesbrecht, J. H. Young, and D. T. Bowman. Testing shelled corn for aflatoxin, Part I: Estimation of variance components. *J. AOAC Int.* **2000,** *83*, 1264-1269.

6. Whitaker, T. B., Hagler, Jr., W. M., Giesbrecht, F. G., and Johansson, A. S. Sampling, sample preparation, and analytical variability associated with testing wheat for deoxynivalenol. *J. AOAC Int.* **2000,** *83*, 1285-1292.
7. Whitaker, T. B.; Wu, J.; Dowell, F. E.; Hagler, W. M., Jr.; Giesbrecht, F. G. Performance of sampling plans to determine aflatoxin in farmers' stock peanut lots by measuring aflatoxin in high-risk-grade components. *J. AOAC Int.* **1994**, *77*, 1672-1680.
8. Whitaker, T. B.; Trucksess, M. W.; Johansson, A. S.; Giesbrecht, F. G.; Hagler, W. M., Jr.; Bowman, D. T. Variability associated with testing corn for fumonisin. *J. AOAC Int.* **1998**, *81*, 1162-1168.
9. Vargas, E. A.; Whitaker, T. B.; Santos, E. A.; Slate, A. B.; Lima, F. B.; Franca, C. A. Testing green coffee for ochratoxin a, part iii: performance of ochratoxin a sampling plan. *J. Food Addit. Contam.* **2006**, *23*, 62-72.
10. Sharman, M.; Macdonald, S.; Sharkey, A. J.; Gilbert, J. Sampling bulk consignments of dried figs for aflatoxin analysis. *J. Food Addit. Contam.* **1994**, *11*, 17-23.
11. MacArthur, R.; MacDonald, S.; Brereton, P.; Murray, A. Statistical modelling as an aid to the design of retail sampling plans for mycotoxins in food. J. *Food Addit. Contam.* **2006**, *23*, 84-92.
12. Whitaker, T. B.; Springer, J.; Defize, P. R.; deKoe, W.J.; Coker, R. Evaluation of sampling plans used in the United States, United kingdom, and The Netherlands to test raw shelled peanuts for aflatoxin. *J. AOAC Int.* **1995**, *78*, 1010-1018.
13. Food and Drug Administration (FDA), Action levels for added poisonous or deleterious substances in human food and animal feed. 2000. http://www.cfsan.fda.gov/~lrd/fdaact.html. accessed 13 November 2007.
14. FDA-Office of Regulatory Affairs, Investigations Operations Mannual 2009, Chapter 4, Sample schedule 6, pg. 189.
15. Food Standard Agency (2007) Sampling advice: mycotoxins in foodstuffs. 19 July 2007. pg. 1-24.
16. van Egmond, H. P.; Schothorst, R. C.; Jonker, M. A. Regulations relating to mycotoxins in food: perspectives in a global and European context. *J. Anal. Bioanal. Chem.* **2007**, *389*, 147-157.
17. Whitaker, T. B.; Slate, A. B.; Jacobs, M.; Hurley, J. M.; Adams, J. G.; Giesbrecht, F. G. Sampling almonds for aflatoxin, part I: estimation of uncertainty associated with sampling, sample preparation, and analysis. *J. AOAC Int.* **2006**, *89*, 1027-1034.
18. Whitaker, T. B.; Slate, A. B.; Hurley, J. M.; Giesbrecht, F. G. Sampling almonds for aflatoxin, part II: estimating risks associated with various sampling plan designs. *J. AOAC Int.* **2007**, *90*, 778-785.
19. Ozay, G.; Seyhan, F.; Yilmaz, A.; Whitaker, T. B.; Slate, A. B.; Giesbrecht, F. G. Sampling hazelnuts for aflatoxin: effect of sample size and accept/reject limit on reducing the risk of misclassifying lots. *J. AOAC Int.* **2007**, *90*, 1028-1035.

20. Codex Alimentarius Commission (CAC), Alinorm 08/31/41 Appendix IX. 31st CAC session, Geneva, Switzerland, 30 June – 4 July, 2008.
21. Trucksess, M. W.; Weaver, C. M.; Oles, C. J.; Rump, L.V.; White, K. D.; Betz, J. M.; Rader, J. I. The use of multi-toxin immunoaffinity columns for determination of aflatoxins and ochratoxin A in ginseng and ginger. *J. AOAC Int.* **2007**, *90*, 1042-1049.
22. Trucksess, M. W.; Weaver, C. M.; Oles, C. J.; Fry, F. S. Jr.: Noonan, G. O.; Betz, J. M.; Rader, J. I. Determination of aflatoxins B_1, B_2, G_1, and G_2 and ochratoxin A in ginseng and ginger by multitoxin immunoaffinity column cleanup and liquid chromatographic quantitation: collaborative study *J. AOAC Int.* **2008**, *91*, 511-523.

Chapter 15

Extraction Efficiency Studies for Mycotoxins in Naturally Contaminated Commodities

Ryan J. Malone

Trilogy Analytical Laboratory, Inc. Washington, MO 63090

An evaluation was conducted to compare the extraction efficiency of various mycotoxins from naturally contaminated commodities. The aflatoxin extraction solvents compared were acetonitrile/water (84/16), acetonitrile/water (90/10), methanol/water (80/20), methanol/water (70/30), and chloroform. The commodities evaluated for aflatoxins were cottonseed meal, corn gluten meal, corn gluten feed, and dried distillers grain. The ochratoxin A extraction solvents were methanol/water (70/30), methanol/water (70/30) with sodium bicarbonate, and acetonitrile/water (60/40) evaluated using wheat and corn. In addition, the extraction efficiency of methanol and ethanol at different ratios was evaluated for ochratoxin A in wheat, aflatoxins, zearalenone, and fumonisins in corn. Each naturally contaminated matrix was finely ground, well-homogenized, and extracted using the various solvents and analyzed by HPLC using AOAC methodology. This comparison of extraction solvents demonstrates the importance of the evaluation of naturally contaminated matrices when validating new methodology.

Introduction

Mycotoxins, secondary metabolites of various fungi, can be found in a variety of different food and feed matrices. These matrices can vary from simple commodities including corn, wheat, barley, and other grains to complex feeds, grain-processing byproducts, and silages (*1,2,3*). The identification and quantification of mycotoxins in these matrices typically consist of several basic steps including extraction, extract purification, and analysis of the purified extract. There are wide varieties of solvents used for the extraction of mycotoxins. Solvent system selection varies depending on the mycotoxin, matrix, compatibility with the detection method, and health and environmental hazards. The most efficient solvents will obtain a complete extract of the mycotoxin from the matrix. In addition to extraction solvent selection, other extraction parameters include the amount of sample to extraction solvent ratio, duration and speed of extraction, and the type of extraction method (for example, blending versus shaking).

This extraction efficiency evaluation was comprised of three separate studies. The first study compared the efficiency of various solvents on the extraction of aflatoxins in complex matrices including cottonseed meal, corn gluten meal, corn gluten feed, and dried distillers grain produced from corn. Cottonseed meal is the byproduct remaining after cotton is ginned, the seeds crushed, and the oils extracted. Corn gluten meal is the dried residue produced from the wet milling of corn after the starch, germ, and bran are removed; corn gluten feed is the residue after the starch, gluten, and germ has been removed from the same process. Dried distillers grain is the cereal byproduct of the distillation process after the ethanol is removed and the resultant grain fraction is dried. The extraction solvents selected were acetonitrile/water (84/16), acetonitrile/water (90/10), methanol/water (80/20), methanol/water (70/30), and chloroform. These solvents selected for the evaluation are used in chromatography and test kit methodologies for the extraction of aflatoxins (*4, 5, 6, 7*). Methanol/water extraction solvents are widely used in aflatoxin test kits due to their compatibility with these analytical techniques.

The second study compared efficiency of three solvents on the extraction of ochratoxin A in wheat and corn. Extraction solvents selected were methanol/water (70/30), methanol/water (70/30) with sodium bicarbonate, and acetonitrile/water (60/40). These particular solvents are used in chromatography and test kit methodologies for the extraction of ochratoxin A (*8,9*). In addition, extraction time and extraction method (blending versus shaking) were evaluated as variables in the extraction process.

The third study compared the efficiency of various concentrations of methanol and ethanol for the extraction of ochratoxin A in wheat and aflatoxins, fumonisins, and zearalenone in corn. Various concentrations of methanol/water and ethanol/water ranging from 50% to 80% were used for this evaluation. Comparable extraction efficiencies between the two solvents would demonstrate the feasibility of replacing the ethanol with methanol as a mycotoxin extraction solution. Ethanol may be a favorable replacement for methanol because it is less toxic to humans, the environment, and it has a lower procurement cost than methanol.

Mycotoxin reference materials, prepared by Trilogy Analytical Laboratory, were used for the evaluation of the extraction solvent efficiency for each study. A reference material is a naturally contaminated homogeneous product that has been verified to contain a specific concentration of mycotoxin. These reference materials can be utilized for a number of different applications including technician training, technician certification, proficiency samples, quality assurance, quality documentation, and method validation. Reference materials are especially suitable for method validation since they are a homogeneous naturally contaminated materials. These materials represent a natural product that is more difficult to extract and more realistic than utilizing spiked samples to evaluate mycotoxin extraction efficiency.

Preparation of Mycotoxin Reference Materials

Mycotoxin reference materials were prepared prior to this study. Naturally contaminated commodities for each mycotoxin evaluated were sourced. Each sample was finely ground to allow 90% of the sample to pass through a 20 mesh screen. The ground material then was homogenized in a rotary mixer for a minimum of 12 hours per sample. Sample homogeneity was verified by analyzing the samples using AOAC HPLC methodology (*10, 11, 12, 13*) for a minimum of 10 replications on three separate days. Each sample was homogenized to obtain a variability of less than 10% for each mycotoxin.

Procedure

Trial 1: Aflatoxin Extraction Efficiency

An extraction efficiency study of aflatoxins in naturally contaminated commodities with various concentrations of acetonitrile/water, methanol/water, and chloroform was performed. Commodities evaluated included cottonseed meal, corn gluten meal, dried distillers grain, and corn gluten feed. These naturally contaminated commodities contained only aflatoxins B_1 and B_2. The extraction solvent systems evaluated were acetonitrile/water (84/16), acetonitrile/water (90/10), methanol/water (80/20), methanol/water (70/30), and 100% chloroform. These systems were chosen due to their utilization in chromatographic and test kit methods for the analysis of aflatoxins.

Each of the reference material commodities were extracted in replicates of 10 using 25-gram samples and 100 ml of each extraction solvent. All the samples were shaken on an Eberbach reciprocating shaker at 180 rotations per minute (rpm) for one hour and then filtered. The procedure used to evaluate the aflatoxin extraction efficiency of these various solvents was AOAC #994.08 (*10*) with some necessary modifications depending on the extraction solvent used. For the extraction solvents, acetonitrile/water (84/16) and acetonitrile/water (90/10), the samples extracts were purified through a solid phase extraction (SPE) column (Trilogy M160). To optimize the extract purification and aflatoxin recovery, the methanol/water extracts were first diluted 1:1 with acetonitrile prior to the SPE column cleanup procedure. Chloroform extracts were modified by evaporating a portion to dryness and reconstituting with acetonitrile/water (84/16). Fortified samples were run for each matrix and solvent system to validate the purification procedure. Each purified extract was then diluted with water per the AOAC #994.08 method and analyzed by HPLC. The pre-column derivatization method using trifloroacetic acid (TFA) was substituted with the post-column bromination derivatization reaction using a KOBRA cell.

Trial 2: Ochratoxin A Extraction Efficiency

The extraction efficiency of various solvents on ochratoxin A in naturally contaminated wheat and corn at a low and high contamination level was tested. Extraction solvents evaluated were acetonitrile/water (60/40), methanol/water (70/30), and methanol/water (70/30) with sodium bicarbonate utilizing certified reference materials. In addition to evaluating the extraction solvents, three different extraction methods were also evaluated including a three minute blend at high speed, a one-hour shake, and 16-hour shake on a reciprocating shaker.

Each of the reference material commodities were weighted in triplicate using 25-gram samples and 100 ml of each extraction solvent. The samples were then extracted using a three-minute blend on high speed, shaken for one and 16 hours on an Eberbach reciprocating shaker (180rpm), and filtered. All sample extracts were analyzed for ochratoxin A utilizing AOAC method #2000.03 (*12*). Sample extracts were diluted with phosphate-buffered saline, passed through an ochratoxin A immunoaffinity column, washed with water, and eluted with methanol. The methanol was then evaporated, reconstituted with mobile phase and analyzed by HPLC with fluorescence detection.

Trial 3: Comparison of Ethanol Versus Methanol for the Extraction of Mycotoxins

The efficiency of ethanol/water was compared to methanol/water for extraction of various mycotoxins from naturally contaminated reference materials. Mycotoxins included in the evaluation were ochratoxin A in wheat, aflatoxins B_1 and B_2 in corn, fumonisins B_1, B_2, and B_3 in corn, and zearalenone in corn. The ethanol/water and methanol/water solutions compared were 50%, 60%, 70%, and 80%.

Naturally contaminated ochratoxin A reference material in wheat and zearalenone in corn were extracted in triplicate. Aflatoxins and fumonisins in corn were extracted in replicates of five using the various solutions of methanol and ethanol. Wheat and corn, non-detect for the various mycotoxins, were spiked and extracted with each solvent solution to validate the procedure. All samples were extracted on a reciprocating shaker for one hour (180 rpm). The ochratoxin A extracts were analyzed by HPLC using AOAC method #2000.03 (*12*), aflatoxins using AOAC method #994.08 (*10*) with KOBRA cell derivatization, zearalenone using method by MacDonald (*13*), and fumonisins using AOAC method #995.15 with NDA derivative (*11*).

Results

Trial 1: Aflatoxin Extraction Efficiency

The extraction solvents evaluated showed varying results depending on the commodity tested. The average aflatoxins results, including aflatoxin B_1 and B_2, obtained on the cottonseed meal reference material using acetonitrile/water (84/16), acetonitrile/water (90/10), methanol/water (80/20), methanol/water (70/30), and chloroform, were 35.4, 16.9, 6.0, 5.6, and 33.6 ppb respectively. For comparison purposes, assuming acetonitrile/water (84/16) extracted 100% of the total aflatoxins from cottonseed meal, the extraction efficiencies using the acetonitrile/water (90/10), methanol/water (80/20), methanol/water (70/30), and chloroform, were 48%, 17%, 16%, and 95% respectively. The acetonitrile/water (84/16) and chloroform exhibited similar extraction results while the acetonitrile/water (90/10) and the two methanol/water solvents were significantly lower (Figure 1).

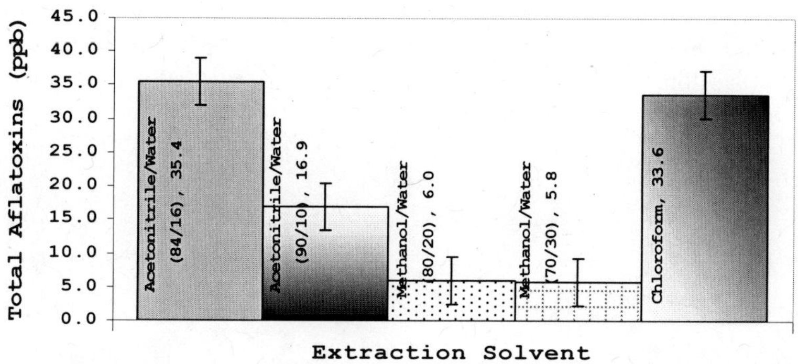

Figure 1. Extraction solvent comparison for aflatoxins in cottonseed meal.

Aflatoxin extraction efficiencies from dried distillers grain using five different extraction solutions were more similar than the cottonseed meal. The average results for the aflatoxins, including aflatoxin B_1 and B_2, extracted from this commodity using acetonitrile/water (84/16), acetonitrile/water (90/10), methanol/water (80/20), methanol/water (70/30), and chloroform, were 23.5, 22.2, 19.6, 18.1, and 19.3 ppb respectively. Assuming acetonitrile/water (84/16) extracted 100% of the aflatoxins from dried distillers grain, acetonitrile/water (90/10), methanol/water (80/20), methanol/water (70/30), and chloroform extracted 94%, 83%, 77%, and 82% of the aflatoxins respectively (Figure 2).

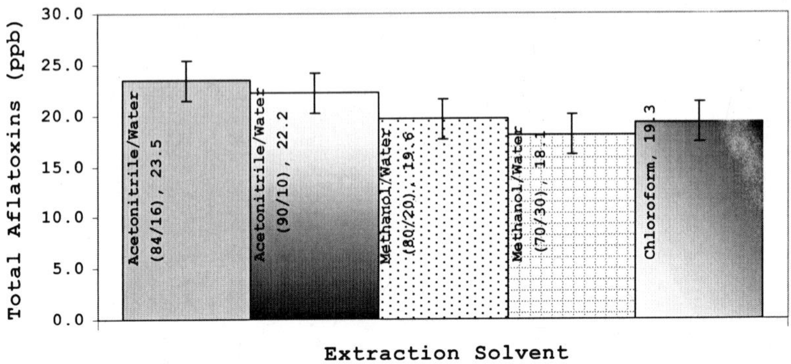

Figure 2. Extraction solvent comparison for aflatoxins in dried distillers grain.

Acetonitrile/water (84/16) obtained the best extraction efficiency for aflatoxins, including aflatoxin B_1 and B_2, from corn gluten meal. The average results for the aflatoxins extracted from this commodity using acetonitrile/water (84/16), acetonitrile/water (90/10), methanol/water (80/20), methanol/water (70/30), and chloroform, were 63.2, 44.1, 40.1, 33.8, and 54.0 ppb respectively. Assuming acetonitrile/water (84/16) extracted 100% of aflatoxins from corn gluten meal, acetonitrile/water (90/10), methanol/water (80/20), methanol/water (70/30), and chloroform, extracted 70%, 63%, 53%, and 85% of the aflatoxins respectively (Figure 3).

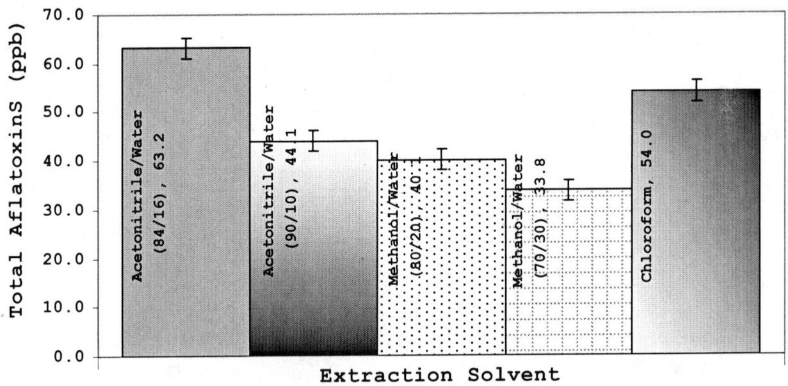

Figure 3. Extraction solvent comparison for aflatoxins in corn gluten meal.

Corn gluten feed was the final commodity evaluated with the solvent systems. Following the previous trend, acetonitrile/water (84/16) extracted aflatoxins, including aflatoxin B_1 and B_2, from corn gluten feed at the highest efficiency. The average results for aflatoxins extracted from this commodity using acetonitrile/water (84/16), acetonitrile/water (90/10), methanol/water (80/20), methanol/water (70/30), and chloroform were, 24.0, 18.3, 20.2, 16.6, and 20.0 ppb respectively. Methanol/water (70/30) showed the lowest efficiency of extraction at 69% compared to the acetonitrile/water (84/16) extraction solvent. The extraction efficiencies for acetonitrile/water (90/10), methanol/water (80/20), and chloroform were 76%, 84%, and 83% respectively (Figure 4).

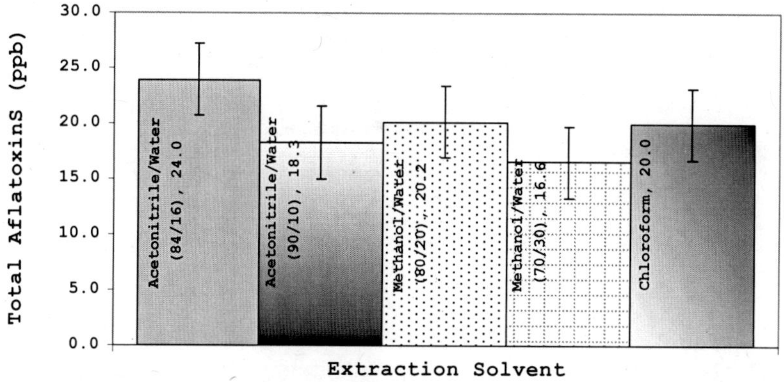

Figure 4. Extraction solvent comparison for aflatoxins in corn gluten feed.

Trial 2: Ochratoxin A Extraction Efficiency

The average ochratoxin A results obtained from the low level wheat for the methanol/water (70/30) extraction solvent ranged from 15.2 to 17.6 ppb for the three different extraction methods. Results using the methanol/water (70/30) with bicarbonate ranged from 17.5 to 21.4 ppb and the acetonitrile/water (60/40) results ranged from 20.7 to 21.9 ppb (Figure 5). For the high level wheat sample, the average ochratoxin A results obtained for the methanol/water (70/30), methanol/water (70/30) with bicarbonate, and acetonitrile/water 60/40) for the three extraction methods ranged from 84.9 to 93.7 ppb, 94.5 to 97.8 ppb, and 96.6 to 99.6 ppb respectively (Figure 6).

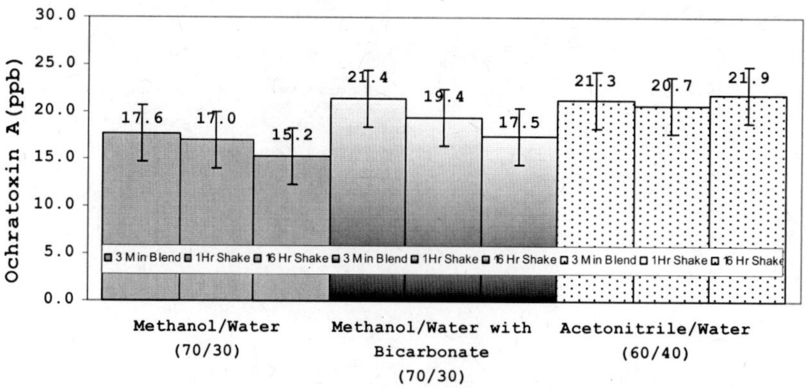

Figure 5. Low level ochratoxin A in wheat extraction solvent comparison.

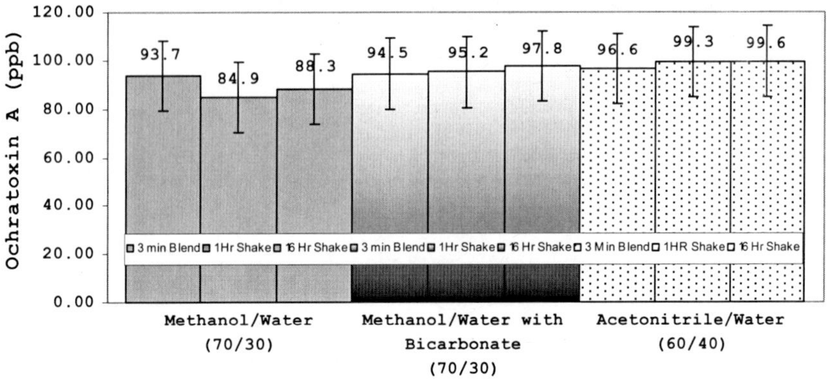

Figure 6. High level ochratoxin A in wheat extraction solvent comparison.

The average ochratoxin A results obtained from the low level corn for the methanol/water (70/30) extraction solvent ranged from 14.2 to 20.8 ppb for the three different extraction methods. Results using methanol/water (70/30) with bicarbonate ranged from 18.1 to 21.3 ppb and the acetonitrile/water (60/40) results ranged from 16.2 to 21.3 ppb (Figure 7). The average ochratoxin A results obtained for the high level corn for the methanol/water (70/30), methanol/water (70/30) with bicarbonate, and acetonitrile/water 60/40) for the three extraction methods ranged from 53.0 to 57.7 ppb, 55.5 to 55.9 ppb, and 61.0 to 72.9 ppb respectively (Figure 8).

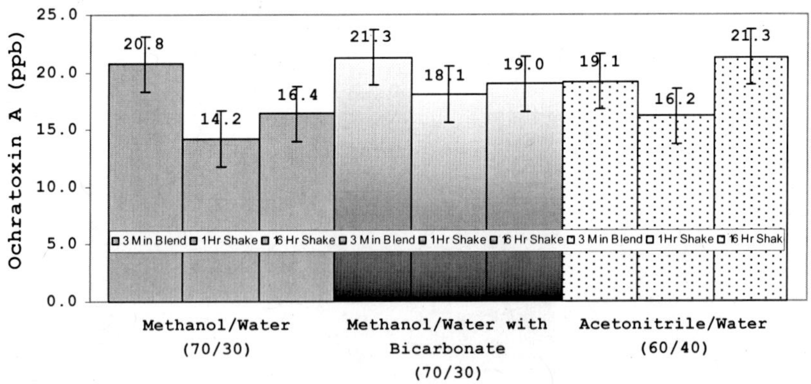

Figure 7. Low level ochratoxin A in corn extraction solvent comparison.

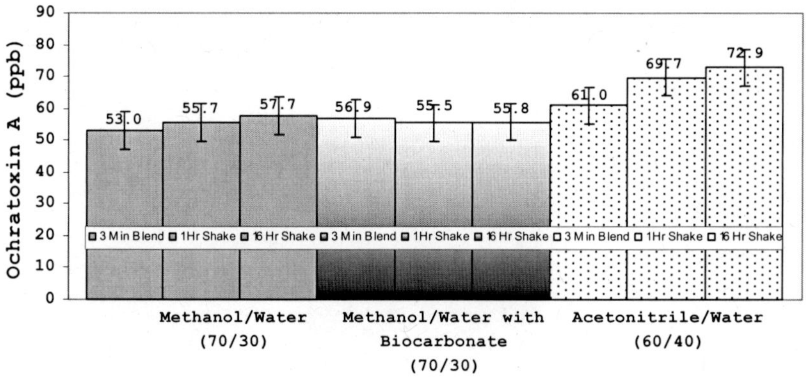

Figure 8. High level ochratoxin A in corn high extraction solvent comparison.

Trial 3: Comparison of Ethanol Versus Methanol for the Extraction of Mycotoxins

Extraction efficiency results of ochratoxin A in wheat were similar for all four solutions of methanol/water and ethanol/water. The averaged ochratoxin A results for methanol/water extraction solutions ranged from 7.5 to 9.0 ppb and the ethanol/water extraction solutions ranged from 7.2 to 9.6 ppb (Figure 9).

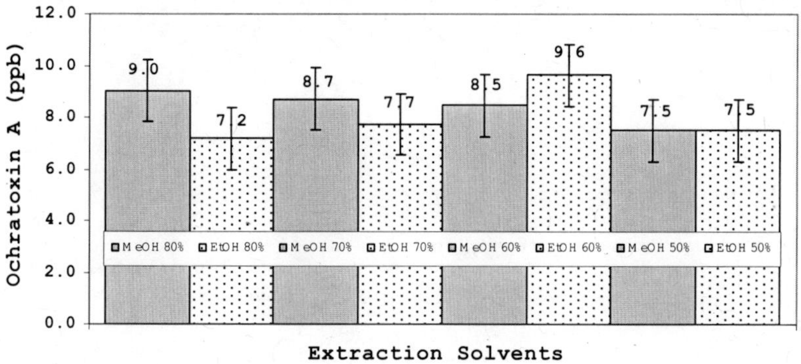

Figure 9. Comparison of ethanol to methanol for the extraction of ochratoxin A in wheat.

Results for the extraction efficiency of aflatoxins in corn demonstrated some significant differences between the methanol and ethanol solutions. Aflatoxin results for ethanol/water extraction solutions ranged from 16.5 to 17.9 ppb while the methanol/water sample extract results ranged from 10.8 to 15.7

ppb. The ethanol/water extract results were consistently higher than the methanol/water extract results. In addition, methanol/water extract results decreased significantly for the 60% and 50% solutions. These solutions exhibited results of 12.8 and 10.8 ppb respectively (Figure 10).

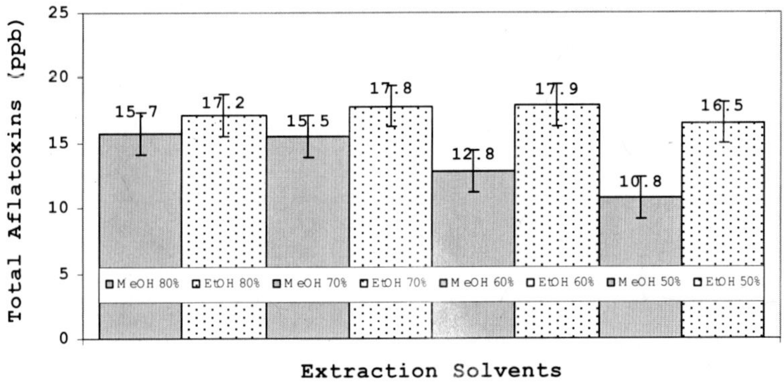

Figure 10. Comparison of ethanol to methanol for the extraction of total aflatoxins in corn.

There were also some significant differences in the extraction efficiencies of the ethanol and methanol solution for zearalenone in corn. Zearalenone results for ethanol/water extraction solutions ranged from 388.0 to 457.8 ppb while the methanol/water sample extract results ranged from 257.4 to 405.5 ppb. There was a sharp decrease in zearalenone results from 375.0 ppb for the 60% methanol/water extracts to 257.4 ppb for the 50% methanol/water extract (Figure 11).

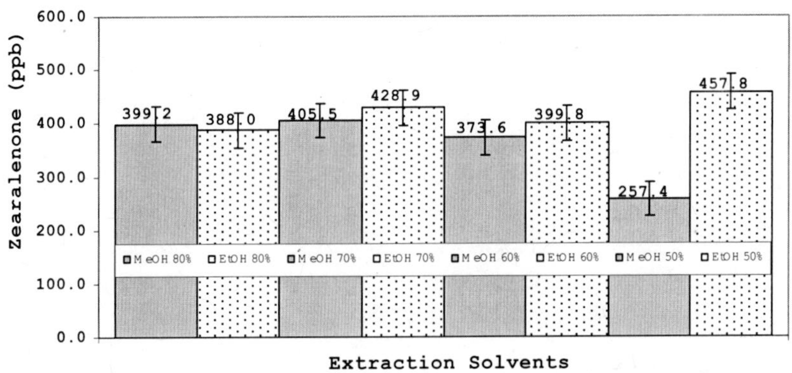

Figure 11. Comparison of ethanol to methanol for the extraction of zearalenone in corn.

Extraction efficiency results for fumonisins from corn was similar for all four solutions of methanol/water and ethanol/water. The averaged fumonisin results for the methanol/water extraction solutions ranged from 4.3 to 5.0 ppm and the ethanol/water extraction solutions ranged from 4.2 to 5.1 ppm (Figure 12).

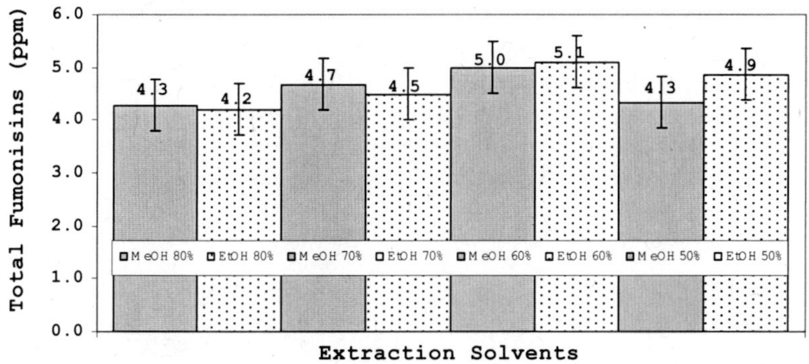

Figure 12. Comparison of ethanol to methanol for the extraction of tota fumonisins in corn.

Conclusions

Trial 1: Aflatoxin Extraction Efficiency

The efficiency of different organic solvents for the extraction of aflatoxins from naturally contaminated products can be highly variable and matrix dependent. Overall acetonitrile/water (84/16) was the most efficient solvent for extraction of aflatoxins from cottonseed meal, dried distillers grain, corn gluten meal, and corn gluten feed, followed by chloroform. The extraction efficiencies of the other solvents evaluated varied depending on sample matrix. The solvents: acetonitrile/water (90/10), methanol/water (80/20), and methanol/water (70/30), gave the lowest extraction efficiency compared to acetonitrile/water (84/16) for cotton seed meal with increasing efficiency for corn gluten meal, corn gluten feed, and dried distillers grain. This comparison of extraction solvents demonstrates the importance of the evaluation of naturally contaminated matrices when validating new methodology. Some extraction solvents commonly used for the analysis of aflatoxins should not be used on certain matrices due to their low extraction efficiencies.

Trial 2: Ochratoxin A Extraction Efficiency

The extraction solvents, methanol/water (70/30) with bicarbonate and acetonitrile/water (60/40), obtained similar extraction efficiencies for both the low and high ochratoxin A in wheat samples. There was a trend for the methanol/water (70/30) extraction solvent to be less efficient than the other two solutions. For both the low and high levels of ochratoxin A in wheat, the three-minute blend, one-hour shake, and 16-hour shake had similar extraction efficiencies for each solvent type.

For the ochratoxin A in corn evaluation, the three-minute blend, one-hour shake and 16-hour shake had similar extraction efficiencies for each solvent type with one exception. The one-hour shake extraction method using methanol/water (70/30) was significantly lower than the three-minute blend for the low level ochratoxin A in corn sample. The acetonitrile/water (60/40) solution overall was the most efficient in extracting the high level ochratoxin A from corn than the other two methanol/water solutions.

Trial 3: Comparison of Ethanol Versus Methanol for the Extraction of Mycotoxins

In this mycotoxin extraction evaluation, various concentrations of ethanol/water were compared to methanol/water for the extraction of ochratoxin A in wheat and aflatoxins, zearalenone, and fumonisins in corn. Each sample matrix was extracted with 50%, 60%, 70%, and 80% ethanol/water and methanol/water solutions. Results of this evaluation demonstrate that ethanol/water extraction solvents had the same or better extraction efficiency than methanol/water solutions. The ethanol/water solutions had similar efficiencies for each of the 50%-80% extraction solutions for all mycotoxins evaluated. For samples extracted for ochratoxin A and fumonisins the methanol/water solutions had similar efficiencies for each of the 50%-80% solutions. However, for samples extracted for aflatoxins and zearalenone, there was significantly lower extraction efficiency for the 50% and 60% methanol/water solutions for aflatoxin and 50% solution for zearalenone. For the matrices and mycotoxins evaluated, it appears that ethanol/water can be a replacement for methanol/water as an extraction solution.

Summary

Extraction solvent evaluation is a critical step in the total method validation for mycotoxin analysis. Not only is the solution selection important, but also the extraction procedure itself. The method of extraction (blending or shaking) and the length of extraction time are also variables that can affect extraction efficiencies.

Reference materials are valuable tools for the evaluation of mycotoxin extraction efficiencies. These naturally contaminated products are finely ground and mixed to provide a well-homogenized material. These products represent a natural mycotoxin contaminated matrix that is more difficult to extract than simple matrix spikes for the evaluation of extraction solvent efficiencies.

The efficiency of solutions for the extraction of mycotoxins can be matrix-dependant. One extraction solution may efficiently extract a particular mycotoxin from some naturally contaminated matrices but not others. It is critical that the extraction solution utilized in the analytical methodology be evaluated for each matrix of interest.

Acknowledgements

I would like to thank Bruce Malone for his help in editing and Danielle Gant for her assistance in creating the graphics for this publication.

References

1. Bullerman, L.B. Significance of mycotoxins to food safety and human health. *J. Food Prot.* 1979, 42, 65-86.
2. Bullerman, L.B. Public health significance of molds and mycotoxins in fermented dairy products. *J. Dairy Sci.* 1981, 63, 2439-2452.
3. Bullerman, L.B.; Tsai, W.J. Incidence and levels of *Fusarium moniforme*, *Fusarium proliferatum* and fumonisins in corn and corn-based foods and feeds. *J. Food Prot.* 1994, 57, 541-546.
4. Trucksess, M.W.; Albert, R.H.; Romer, T.R. Multifunctional column coupled with liquid chromatography for the determination of aflatoxins B1, B2, G1, G2 in corn, almonds, brazil nuts, peanuts, and pistachio nuts: collaborative study. *Journal of AOAC International*, 1994, 77, 1512-1520.
5. Malone, B.R.; Humphey, C.W.; Romer, T.R.; Richard, J.L. Determination of aflatoxins in grains and raw reanuts by a rapid procedure with fluorometric analysis. *Journal of AOAC International*, 83, 2000, 95-98.
6. *Official Methods of Analysis of AOAC International*, 16th ed.; Cunniff, P.Ed., AOAC International: Gaithersburg, MD, 1999; Method 991.31, Aflatoxins in corn, raw peanuts, and peanut butter, chapter 49, pp 20-21.
7. *Official Methods of Analysis of AOAC International*, 16th ed.; Cunniff, P.Ed., AOAC International: Gaithersburg, MD, 1999; Method 968.22, Aflatoxins in peanuts and peanut products, chapter 49, pp 8-10.
8. Trucksess, M.W.; Giler, J.;Young, K.; White, K.D.; Page, S.W. Determination and survey of ochratoxin A in wheat, barley and coffee – 1997, *Journal of AOAC International*, 82, 1999, 85-89.
9. Entwisle, A.C.; Williams, A.C.; Mann, P.J.; Slack, P.T.; Gilbert, J. Liquid chromatographic method with immunoaffinity column cleanup for determination of ochratoxin A in barley: collaborative study, J. *Journal of AOAC International*, 83, 2000, 1377-1383.
10. *Official Methods of Analysis of AOAC International*, 16th ed.; Cunniff, P.Ed., AOAC International: Gaithersburg, MD, 1999; Method 994.08, Aflatoxins in corn, almonds, brazil nuts, peanuts, and pistachio nuts, chapter 49, pp 24-24A.

11. *Official Methods of Analysis of AOAC International*, 16th ed.; Cunniff, P.Ed., AOAC International: Gaithersburg, MD, 1999; Method 995.15, Fumonisins B1, B2, and B3 in corn, chapter 49, pp 49-51
12. Entwisle, A.C.; Williams, A.C.; Mann, P.J.; Slack, P.T.; Gilbert, J. Liquid chromatographic method with immunoaffinity column cleanup for determination of ochratoxin A in barley: collaborative study, J. *Journal of AOAC International*, 83, 2000, 1377-1383
13. MacDonald, S.J.; Anderson, S.; Brereton, P.; Wood, R.; Damant, A. Determination of zearalenone in barley, maize and wheat flour, polenta, and maize-based baby food by immunoaffinity column cleanup with liquid chromatography: interlaboratory study, *Journal of AOAC International*, 88, 2005, 1733-1740.

Chapter 16

Fumonsins in materials related to ethanol production from corn

LC-MS analysis of fumonsins in ethanol manufacture by-products

Mark Busman and David F. Kendra

Mycotoxin Research Unit, Agricultural Research Service, United States Department of Agriculture, Peoria, IL 61604

Recently, an expansion in utilization of corn for production of fuel ethanol has occurred in the U.S. Production of ethanol from corn leaves a starch depleted residue from the grain. This by-product of ethanol production is commonly utilized as an ingredient in livestock feed. Mycotoxins are important contaminants in the by-products from production of ethanol from corn. In the northern part of the United States fumonisins are major mycotoxins in corn. Here we present results of development of mass spectrometric methods for the determination of fumonisins in matrices associated with ethanol production from corn.

There is a continuing effort to utilize agricultural products for the production of energy sources, particularly to replace petroleum based products utilized for motor fuels (*1*). Ethanol (*1*) and seed-oils (*2*) have been utilized to replace gasoline and diesel fuels, respectively. As processes for producing bio-fuels from agricultural products have developed, a need for the utilization or disposal of the by-products of these processes has also developed.

Recently, there has been an increase in the level of utilization of corn for the production of ethanol (*3*). This increase has resulted in a corresponding increase in production of the by-products from the ethanol production process (*4*).

Over the last thirty years large scale ethanol facilities have been constructed based on two general schemes. These two schemes are commonly called the wet-grind and the dry-grind processes (*5*). Despite the classification into the two schemes, there can be a variation from plant to plant in the details of the ethanol production process (*6*). The dry process can be summarized as follows: corn is ground, water and enzymes added to form a slurry, slurry is fermented utilizing an added yeast, ethanol is removed from the fermented liquid, solid and soluble residue is processed into a marketable by-product (*7*). The wet process differs in that the corn is soaked thoroughly in water prior to any mechanical treatment of the corn leading to the fermentation process (*5*). In either case, a starch-depleted corn residue is obtained as a by-product of the fermentation process. The corn residue may be supplemented by cellular debris from the population of yeast utilized in ethanol production (*8*).

The residue from ethanol process is processed into forms convenient for marketing. The characteristics of these by-products make them attractive for inclusion in livestock feeds for dairy (*9*) and beef (*10*) cattle, swine (*11*) and poultry (*12*). The common forms for the byproducts are distiller dried grains (DDG), distiller dried grains with solubles (DDGS) and wet distillers grains (*13*).

Mycotoxins are toxic secondary metabolites that are produced by fungi and can accumulate in grain in the field and during storage (*14*). Mycotoxins have long been recognized as harmful to humans and animals that consume contaminated grain. In the U.S. aflatoxins, ochratoxin, fumonisins, and deoxynivalenol have been noted as common contaminants in grains (*14*). Despite efforts to mitigate their levels and effects in harvested grains, mycotoxins remain a continuing problem for producers, processors and end users of grain and grain-derived products (*15*).

New U.S. ethanol facilities have been, in most cases, constructed for the utilization of corn (*16*). Corn purchased for use in the ethanol facilities can be contaminated with varying levels of mycotoxins (*17*). Management of ethanol production needs to be done with an awareness of the effects of mycotoxin contamination. While mycotoxins can have potential effects on yeast growth during fermentation and need to be considered in terms of inhalation exposure for ethanol plant workers, it is in the utilization of by-products that mycotoxins in corn could have their most significant impact (*18*).

It is generally thought that mycotoxins found in corn are concentrated three-fold in the by-products of the ethanol process (*17*). Production of ethanol from corn can concentrate mycotoxins because the mycotoxins remain in the corn residue and the residue undergoes a marked reduction in biomass as starch is converted to ethanol. This, combined with the fact that the fermentation process does not alter the structure or activity of mycotoxins can result in a three-fold increase in the levels of mycotoxins in the residue compared to the original corn (*19*).

Understanding of the effects of mycotoxins has progressed along with improvements in qualitiative and quantitative analysis of the toxins in grains(*20*). Both biological and chemical-based analytical methods for the determination of mycotoxins have been the developed as part of an effort to ensure the safety of grain products for human and animal consumption (*20*).

These methods for mycotoxin analysis from grains have long been adapted to monitor mycotoxin levels in materials related to production of fuel ethanol from grain (*19*). Similarly, investigators interested in safety of beverage ethanol products have utilized methods for the evaluation of fermentation processes, in terms of mycotoxin contamination, to ensure safety of alcoholic beverages (*21*).

Chemical analysis of mycotoxins has often utilized thin layer chromatography (TLC) (*22*) or high performance liquid chromatography (HPLC) (*23*) followed by an appropriate detection technique. In the last few decades atmospheric pressure ionization mass spectrometry has been demonstrated to be both a sensitive and selective detection technique for mycotoxins separated by liquid chromatography (*24*). Effective methods for quantitation of many important mycotoxins from a variety of grain matrices have been developed utilizing the combination of high performance liquid chromatography with mass spectrometry (LC-MS) (*25*).

TLC and LC coupled with fluorescence detection and GC-MS have been utilized by earlier workers to study the fate of mycotoxins in bench top and pilot scale experiments involving the production of ethanol from zearelanone (*26*), aflatoxin (*27*) and fumonisin (*19*) contaminated grains. Methods utilized for mycotoxin analysis in ethanol production matrices have often been modified from existing methods developed for analysis of mycotoxins from grains (*26*). Fuel grade ethanol manufacturers potentially require capability to monitor mycotoxin levels in purchased grain, process by-products, and intermediate process materials.

Here we report the application of liquid chromatography coupled to electrospray ionization mass spectrometry to the characterization of fumonisin contamination in by-products from ethanol production from corn. The method involves a convenient extraction of toxins, followed by direct analysis without use of sample clean-up.

Methods

Materials

Unless otherwise noted, all solvents were from Sigma Chemical Co. (St. Louis, MO). Water was from a Millipore water purification system. Purified fumonisins were obtained from cultures of *Fusarium verticillioides* as previously described (*28*). Samples of DDG and DDGS were obtained from a variety of ethanol production facilities in Illinois.

Sample preparation

Portions (~20 g) of DDG or DDGS were accurately weighed. Samples were extracted with 50 mL of 1:1 (v/v) water/acetonitrile for 2 hr with occasional swirling. Slurries of sample with solvent were centrifuged for 5 min at ~1500G.

One mL portions of the centrifugate were decanted by pipetted into sample vials for LC-MS analysis.

Fumonisin spike

To 20 g portions of DDGS, 200 uL aliquots of FB1, 2 and 3 were added during vigorous hand swirling. Spike solutions were formulated at concentrations to provide sample concentrations of 1, 10, 50 and 100 ppm. Except where noted, spiked samples were extracted with the same type of methodology as for other by-product samples.

LC-MS

Extracts were subjected to LC-MS analysis as previously described. Briefly, 10 µL were injected in to a gradient flow on a HPLC apparatus composed of a ThermoSpectraphysics AS4000 autosampler and a P4000 pump. Chromatographic separation of fumonisins was accomplished on a 150 x 3 mm MetaChem C18 column utilizing a 300 µL/min flow. The method used 35-85% gradient aq. methanol with 0.3% acetic acid over 25 min. The flow from the column was directed to the electrospray interface of a ThermoFinnigan LCQ-DECA ion trap mass spectrometer. The mass spectrometer was operated in full scan mode, scanning from 350-1000 m/z. Quantitation of fumonisins FB1(*m/z* 722), FB2(*m/z* 706) and FB3(*m/z* 706) was accomplished utilizing the ThermoFinnigan Xcalibur data system.

Results

Samples of DDG and DDGS were analyzed for fumonisin content. Figure 1 shows a chromatogram for FB1 from an analysis of a DDGS sample. The appearance of the chromatogram is similar to that seen for analyses of unprocessed corn samples utilizing a similar sample preparation and analysis method (data not shown).

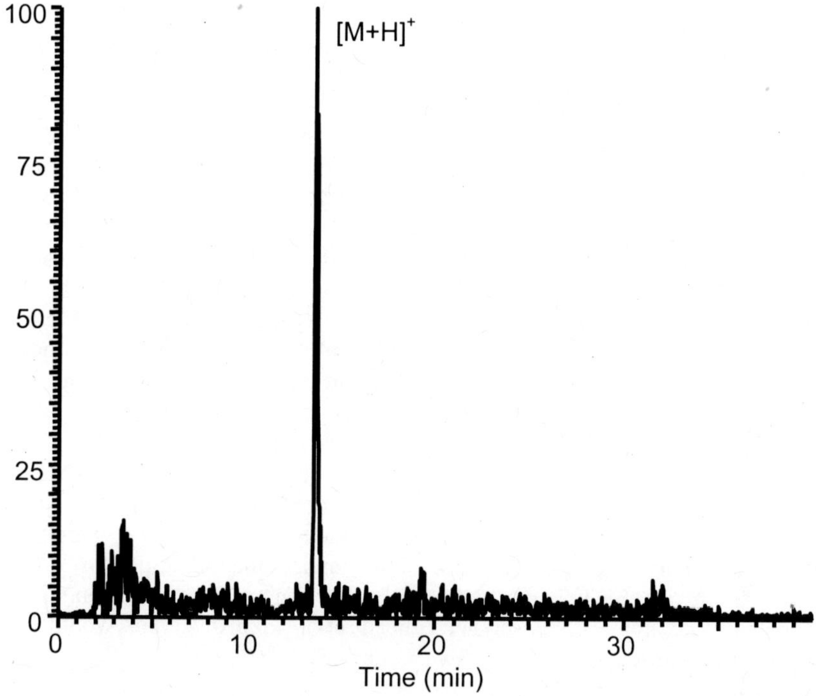

Figure 1. Example extracted ion chromatogram (722 m/z) for the detection of FB1 from a DDG sample.

Spike recovery experiments were conducted for fumonisin FB1. Table I shows results of a representative experiment evaluating FB1 levels in the DDGS samples. Fumonisins were extracted using several different solvent systems. Extractions were conducted utilizing water, methanol, acetonitrile, 1:1 methanol / water, 1:1 acetonitrile / water). Except for solvent used, identical treatments were used for each extraction. All spike recovery experiments were conducted in triplicate. The 1:1 methanol / water and 1:1 acetonitrile / water both gave ~65% efficiencies at lower FB1 concentrations. Water and acetonitrile gave much lower efficiencies.

Table I. Percent FB1 recovery from analysis of spiked DDGS samples with various extraction solvent systems.

FB1 Level(ppm)	*Water*	*Water /aceto-nitrile*	*Water /meth-anol*	*Aceto-nitrile*	*Meth-anol*
1	16	65	65	0	32
10	9	50	40	0	22
50	8	42	29	0	44
100	8	39	33	0	39

Several studies have suggested the advantages of introducing a mechanical procedure to deplete the corn of certain kernel components prior to fermentation (*29*). Because the yeasts commonly used in ethanol production utilize the starch components found primarily in the endosperm, efforts have been made to deplete the corn of low starch components prior to fermentation. A variety of schemes have been utilized to accomplish this prefractionation of the corn input (*30*). It is anticipated that the removed components will be directed to alternative uses. Such a depletion strategy will affect the ethanol production in several ways. First, removal of low starch components will reduce the amount of the grain that will be left undigested by the yeast (*30*). This will reduce the amount of by-product remaining at the completion of the fermentation. Second, the modification of the composition of input grain will change in the composition, including protein and lipid content, of the by-product (*30*).

Modification of the composition of corn that enters the fermentation process can influence the mycotoxin content of the resulting by-products. Studies have shown that mycotoxins are found at higher levels in certain sections of the kernel (*31*). Table II shows results of the analysis of a single sampling of prefractionation materials taken from a commercial ethanol facility. In this case, the levels of fumonisins were higher in bran and germ than in the endosperm. Potentially, depletion of the higher concentration toxin materials would likely result in a reduction of fumonisin levels in by-products.

Table II. Analysis of prefractionation components.

Sample ID	*[FB1](ppm)*
Corn bran	0.59
Corn germ	0.40
Corn endosperm	0.18

Discussion

Fumonisins commonly accumulate in U.S. corn (*32*). While the levels seen in corn are often below levels that would prevent use in livestock feeds, they are high enough that a three-fold concentration normally seen in by-products will result in fumonisin levels in the by-products at levels requiring some concern (*18*).

The results of DDGS fumonisin spike experiments here demonstrate the potential applicability of fumonisin methods developed for corn to by-products. However, it should be noted that the DDG and DDGS derived by-products are extremely variable from facility to facility. The by-products (DDG or DDGS) have been observed to vary widely in common nutritional characteristics (Protein: 25-30% typical, can range to >40%, Fiber: 10%, Fat: 10%, Amino acids, Lysine: 0.54-0.99%) (*33*). Variability of by-product composition is likely a consequence of variations in the details of operation procedures in the ethanol production process at each facility. Analysts are likely to require periodic evaluations of the performance of analytical methods.

Early work with chemical analysis of mycotoxins explored the possibility of chemical modification of the toxins by the yeasts during fermentation (*34*). The existing literature offers an incomplete picture of the effects of fermentation on mycotoxins (*18*). However, a goal for further work in the refinement of the ethanol production process could be the degradation of mycotoxins into less toxic forms. Certainly, chemical analysis, perhaps including LC-MS, of the products of degradation processes will be essential in understanding the effectiveness of reducing the toxicity of fermentation by-products.

The utilization of prefractionation processes prior to fermentation will modify the characteristics of the by-products of ethanol production. The evolving characteristics of the by-products will provide new challenges and opportunities for those looking to optimize utilization of the commodities consumed during ethanol production. The prefractionation process has the potential to improve efficiency of ethanol production and has the potential to yield more valuable by-products. However, prefractionation also has the potential to preferentially segregate mycotoxins into the certain fractions of the processed corn.

The products of the prefractionation processes will provide opportunities for new corn-derived products. Development of protocols for the evaluation of mycotoxins in these new products will be required. These new products may exhibit behavior unlike corn in the established analytical methods.

Conclusions

Ethanol production from corn in the U.S. has expanded rapidly in the last decade. This expansion has changed the utilization of the U.S. corn crop and has introduced a large quantity of a class of new, high value, commodities into the agricultural marketplace. These by-products of ethanol production from corn are nutritionally attractive ingredients for inclusion into livestock diets. The availability of robust tools for the evaluation of mycotoxin levels in these by-products is essential for their effective utilization.

As the ethanol from corn industry expands, the processes involved in ethanol production continue to evolve. The improvement in ethanol production is aimed at more efficient production of the ethanol and at the co-production of marketable by-products. The by-product commodities vary in composition depending upon the techniques used for operation of the ethanol production and, importantly, prefractionation procedures applied to the corn prior to the fermentation process. Those seeking to apply analytical methods for mycotoxins to the by-products must be aware of potential variability in this analytical matrix.

References

1. De La Torre Ugarte, D.G.; English, B.C.; Jensen, K. Sixty billion gallons by 2030: Economic and agricultural impacts of ethanol and biodiesel expansion. *Amer. J. Agr. Econ.* **2007**, *89*, 1290.
2. Knothe, G.; Dunn, R.O.; Bagby, M.O. Biodiesel: The Use of Vegetable Oils and Their Derivatives as Alternative Diesel Fuels, in *Fuels and Chemicals from Biomass*, edited by B.C. Saha and J. Woodword, ACS Symposium Series 666, American Chemical Society, Washington, DC, 1997, pp. 172–208.
3. Donner, S.D.; Kucharik, C.J. Corn-based ethanol production compromises goal of reducing nitrogen export by the Mississippi River. *Proc. Nat. Acad. Sci. U.S.A.,* **2008**, *105*, 4513.
4. Murthy, G.S.; Singh, V.; Johnston, D.B.; Rausch, K.D.; Tumbleson, M.E. Improvement in fermentation characteristics of degermed ground corn by lipid supplementation. *J. Ind. Microbiol. Biotechnol.* **2006**, *33*, 655.
5. Rausch, K.D.; Belyea, R.L.; Singh, V.; Tumbleson, M. E. "Corn Processing for Ethanol Production" in *Encyclopedia of Agricultural, Food, and Biological Engineering*, Heldman, D.R. ed. Taylor & Francis, 2007.
6. Belyea, R.; Eckhoff, S.; Wallig, M.; Tumbleson, M. Variability in the nutritional quality of distillers solubles. *Bioresour. Technol.*, **1998**, *66*, 207.
7. Murthy, G.S.; Singh, V.; Johnston, D.B.; Rausch, K.D.; Tumbleson, M.E. Evaluation and strategies to improve fermentation characteristics of modified dry-grind corn processes. *Cereal Chem.* **2006**, *83*, 455.
8. Kim, Y.; Mosier, N.S.; Hendrickson, R.; Ezeji, T.; Blaschek, H.; Dien, B.; Cotta, M.; Dale, B.; Ladisch. M.R. Composition of corn dry-grind ethanol by-products: DDGS, wet cake, and thin stillage. *Bioresour. Technol.*, **2008**, *99*, 5165.
9. Grings, E.E.; Roffler, R.E.; Deitelhoff, D.P. Responses of dairy cows to additions of distillers dried grains with solubles in alfalfa-based diets. *J. Dairy Sci.*, **1992**, *75*, 1946.
10. Klopfenstein, T.J.; Erickson, G.E.; Bremer, V.R. Board-invited review: Use of distillers by-products in the beef cattle feeding industry. *J. Anim. Sci*, **2008**, *86*, 1223.
11. Whitney, M.H.; Shurson, G.C.; Johnston, L.J.; Wulf, D.M.; Shanks, B.C. Growth performance and carcass characteristics of grower-finisher pigs fed high-quality corn distillers dried grain with solubles originating from a modern Midwestern ethanol plant. *J. Anim. Sci*, **2006**, *84*, 3356.
12. Wang, Z.; Cerrate, S.; Coto, C.; Yan, P.W. Utilization of distillers dried grains with solubles (DDGS) in broiler diets using a standardized nutrient matrix. *Int. J. Poult. Sci.*, **2007**, *6*, 470.
13. Rosentrater, K.A.; Lehman, R.M. Physical and chemical properties of corn distillers wet grains. *Applied Engineering in Agriculture*, **2008**, *24*, 57
14. Cleveland, T.E.; Dowd, P.F.; Desjardins, A.E.; Bhatnagar, D.; Cotty, P.J. United States Department of Agriculture - Agricultural Research Service research on pre-harvest prevention of mycotoxins and mycotoxigenic fungi in US crops. *Pest Manag. Sci.* **2003**, *59*, 629.

15. Munkvold, G.P. Cultural and Genetic Approaches to Managing Mycotoxins in Maize. *Annu. Rev. Phytopathol.* **2003**, *41*, 99.
16. Simpson, T.W.; Sharpley, A.N.; Howarth, R.W.; Paerl, H.W.; Mankin, K.R. The new gold rush: Fueling ethanol production while protecting water quality. *J. Environ. Qual.* **2008**, *37*, 318.
17. Bennett, G.A.; Richard, J.L. Fate of fumonisin B1 in naturally contaminated corn during ethanol fermentation. Influence of processing on Fusarium mycotoxins in contaminated grains. *Food Technol.*, **1996**, *50*, 235.
18. Wu, F.; Munkvold, G.P. Mycotoxins in ethanol co-products: Modeling economic impacts on the livestock industry and management strategies. *J. Agric. Food Chem.* **2008**, *56*, 3900.
19. Bothast, R.J.; Bennett, G.A.; Vancauwenberge, J.E.; Richard, J.L. Fate of fumonisin B1 in naturally contaminated corn during ethanol fermentation. *Appl. Environ. Microbiol.*, **1992**, *58*, 233.
20. Krska, R.; Schubert-Ullrich, P.; Molinelli, A.; Sulyok, M.; Macdonald, S.; Crews, C. Mycotoxin analysis: An update. *Food Add. Contam.*, **2008**, *25*, 152.
21. Shephard, G.S.; van der Westhuizen, L.; Gatyeni, P.M.; Somdyala, N.I.M.; Burger, H.-S.; Marasas, W.F.O. Fumonisin mycotoxins in traditional Xhosa maize beer in South Africa. *J. Agric. Food Chem.* **2005**, *53*, 9634.
22. Ross, P.F.; Rice, L.G.; Plattner, R.D.; Osweiler, G.D.; Wilson, T.M.; Owens, D.L.; Nelson, H.A.; Richard, J.L. Concentrations of fumonisin B1 in feeds associated with animal health problems. *Mycopathologia*, **1991**, *114*, 129.
23. Shephard, G.S.; Sydenham, E.W.; Thiel, P.G.; Gelderblom, W.A. Quantitative determination of fumonisins B1 and B2 by high-performance liquid chromatography with fluorescence detection. *Journal of Liquid Chromatography*, **1990**, *13*, 2077.
24. Wilkes, J.G.; Sutherland, J.B.; Churchwell, M.I.; Williams, A.J. Determination of fumonisins B1, B2, B3 and B4 by high-performance liquid chromatography with evaporative light-scattering detection. *J. Chromatogr. A* **1995**, *695*, 319.
25. Plattner, R.D. Detection of fumonisins produced in Fusarium moniliforme cultures by HPLC with electrospray MS and evaporative light scattering detectors. *Natural Toxins*, **1995**, *3*, 294.
26. Nofsinger, G.W.; Bothast, R.J. Ethanol production by Zymomonas mobilis and Saccharomyces uvarum on aflatoxin-contaminated and ammonia-detoxified corn. *Can. J. Microbiol.*, **1981**, *27*, 162.
27. Bennett, G.A.; Lagoda, A.A.; Shotwell, O.L.; Hesseltine, C.W. Utilization of zearalenone- contaminated corn for ethanol production. *J. Am. Oil Chem. Soc.*, **1981**, 974.
28. Poling, S.M.; Plattner, R.D. Rapid Purification of Fumonisins B3 and B4 with Solid Phase Extraction Columns. *J. Agric. Food Chem.* **1996**, *44*, 2792.
29. Bryan, T. Changing the game. *Ethanol Producer Magazine*, **2005**, *11*, 58.
30. Singh, V.; Johnston, D.B.; Naidu, K.; Rausch, K.D.; Belyea, R.L.; Tumbleson, M.E. Comparison of modified dry-grind corn processes for

fermentation characteristics and DDGS composition. *Cereal Chem.* **2005**, *82*, 187.

31. Katta, S.K.; Cagampang, A.E.; Jackson, L.S.; Bullerman, L.B. Distribution of fusarium molds and fumonisins in dry-milled corn fractions. *Cereal Chem.* **1997**, *74*, 858.
32. Nelson, P.E.; Plattner, Shackelford D.D.; Desjardins, A.E. Production of fumonisins by Fusarium moniliforme strains from various substrates and geographic areas. *Appl. Environ. Microbiol.*, **1991**, *57*, 2410.
33. Belyea, R.; Eckhoff, S.; Wallig, M.; Tumbleson, M. Variability in the nutritional quality of distillers solubles. *Biores. Technol.*, **1998**, *66*, 207.
34. Dam, R.; Tam, S.W.; Satterlee, L.D. Destruction of aflatoxins during fermentation and by-product isolation from artificially contaminated grains. *Cereal Chem.* **1977**, *54*, 705.

Chapter 17

LC/Ultraviolet/Fluorescence and LC/MS/MS Analyses of Multiple Mycotoxins in Spiked Beer Using Immunoaffinity Column Clean-Up

Darsa P. Siantar[1,*], Mark Benvenuti[2], Jim Krol[3], Joe Romano[2], Sarita Cardozo[1], Norma R. Hill[1], Abdul Mabud[4]

[1]Alcohol and Tobacco Tax and Trade Bureau (TTB), Walnut Creek, CA 94598
[2]Waters Corporation, Milford, MA 01757
[3]Currently at Applied Biosystems, Framingham, MA 01701
[4]Alcohol and Tobacco Tax and Trade Bureau (TTB), Beltsville, MD 20705

Ten different mycotoxins namely deoxynivalenol (DON), aflatoxins B_1, B_2, G_1, G_2, fumonisins B_1, B_2, B_3, zearalenone (ZON), and ochratoxin A (OTA) present in fortified beer matrices were extracted using mixed-bed ("Myco5in1") immunoaffinity columns and were detected using both the classical LC/Ultraviolet/Fluorescence with post-column derivatization and modern LC/MS/MS techniques. HPLC separations were performed using a reverse phase C_{18} column. Mass Spectroscopy measurements were done in the positive ESI mode. Multiple reaction monitoring (MRM) analyses were performed for each mycotoxin. The LC/u.v./Fluorescence method detection limit (MDL) ranged from 0.02 to 6 ppb and the LC/MS/MS MDL ranged from 0.5 to 2.5 ppb, respectively. The recoveries of multi-mycotoxin fortified triplicate beer samples using the "Myco5in1" column ranged from 73% to 102%. Both analytical methods achieved good separations and detections.

Introduction

Agricultural raw commodity foodstuffs such as grains, fruits and vegetables, are subject to microbiological contamination during their growth cycles, harvesting, storage and transport. Beers and wines use various cereal grains and fruits as the substrate for the fermentation process. These agricultural raw materials are also the prime substrate for fungi and mold colonization that could produce toxic secondary metabolites commonly known as mycotoxins. Since mycotoxins may contaminate various raw food ingredients and affect food safety, the analysis of these toxins in animal feeds, beverages and foodstuffs is critical. If not detected, suspected consequences may include liver cancer, kidney failure, intestinal distress, teratogenic and reproductive effects.

The Food and Drug Administration (FDA) uses guidance levels to provide guidance for industry in food or feed. An example of FDA established guidance level for deoxynivalenol (DON) is 1,000 ppb limit on finished wheat products for human consumption. An example of total fumonisins B_1+B_2+B_3 guidance level in human foods is 2,000 ppb limit for degermed dry milled corn products for human consumption. FDA current action level for total aflatoxins B_1+B_2+G_1+G_2 in human food is 20 ppb (*1*). FDA has not established any regulatory levels for ochratoxin A (OTA) or zearalenone (ZON) in food. Although there is no regulatory limit for ochratoxin A in the U.S., under the Food, Drug, and Cosmetic Act, any food containing any substance at a level that may be injurious to health is considered adulterated and may be subject to regulatory action. If ochratoxin A is found in a food product, FDA would determine on a case-by-case basis whether the level found in that product might pose a health hazard such that it would be subject to regulatory action. The Commission of the European Communities has published and set maximum levels for certain contaminants in foods. Since maize is a major crop often exposed to multiple mycotoxins contamination, the European Commission has established maximum permitted levels for 1250 ppb DON, 2 ppb aflatoxin B_1, 4 ppb total aflatoxins, 5 ppb OTA, 100 ppb ZON and 2,000 ppb limit fumonisins B_1+B_2 (*2*). T-2 and HT-2 limits are currently under discussion by the EU. It has also established 2 ppb OTA maximum level in wine products (red, white, rose, other wines and/or grape must beverages) produced from the 2005 harvest onwards. Other wines refer to fruit wines, sparkling wines excluding liqueur wines and wines with alcohol content greater than 15% by volume (*3*).

The range limits for these toxins are country jurisdiction dependant and may be subject to change with time. To help avoid unpleasant consequences, it is essential to have precise and reliable multiple mycotoxins / multi-toxin analytical methods applicable at regulatory limits for various mycotoxins and commodities for enforcement purposes by food control laboratories. Also, for cost and expediency, it is desirable to integrate several mycotoxins into a common sample preparation and a single LC method. Vicam's prototype mixed-bed immunoaffinity column (IAC), containing a 5-toxin (AflatoxinDONFumonisinOTAZON) monoclonal antibody resin support, abbreviated as "Myco5in1" IAC has antibodies for the simultaneous purification of 10 mycotoxins namely DON, aflatoxins B_1, B_2, G_1, G_2, fumonisins B_1, B_2, B_3, ZON, and OTA. In collaboration with Pickering Laboratories (Mountain View,

CA), we developed a new single LC/Ultraviolet (u.v.)/Fluorescence and post-column derivatization technique using this prototype IAC clean-up to analyze multi-toxin in spiked beer and rice wine. At low spike concentration, triplicate recoveries ranged from 73 to 111% with a majority of the relative standard deviations (RSD) less than 10% (*4*). Vicam Group of Waters Technologies (Watertown, MA) has patents pending for the recently commercialized Myco6in1™ IAC that has additional antibodies for T-2 and HT-2 toxins when compared to the "Myco5in1" column (*5*).

In another collaborative study with Varian Inc., all the 12 relevant mycotoxins, namely DON, aflatoxins B_1, B_2, G_1, G_2, fumonisins B_1, B_2, B_3, OTA, T-2, HT-2, and ZON present in complex wheat beer and rice wine matrices were extracted using multifunctional mixed-bed immunoaffinity columns and were detected using a LC/MS/MS method that achieved good separation and detection (*6*). At medium spike level, triplicate recoveries for all 12 mycotoxins using Myco6in1™, previously also known as "AOFZDT2" IAC ranged from 64 - 127% with RSD < 14% (*6*). The R-Biopharm Rhône Ltd. (Glasgow, Scotland) prototype IAC containing antibodies for ZON and trichothecenes (DON, T-2 and HT-2) was also studied (*6*). At low spike level, triplicate recoveries obtained for DON, T-2, HT-2 and ZON in fortified beer and sake ranged from 96 to 113% with RSD < 5%. The prototype R-Biopharm Rhône IAC recovery results were consistent with the Vicam prototype Myco6in1™ IAC recoveries for DON, T-2, HT-2 and ZON at 96 – 116% with RSD < 5% (*6*). The LC separation was performed using a reverse phase C_{18} column. MS measurements were done both in the positive and negative ESI modes. Multiple reaction monitoring (MRM) analyses were performed for each compound (*6*). A method for LC/MS/MS in ESI mode and Vicam multi-analyte Myco6in1™ IAC clean-up for simultaneously analyzing 11 mycotoxins (aflatoxins B_1, B_2, G_1, G_2, fumonisins B_1, B_2, OTA, DON, HT-2, T-2, and ZON) in maize has also been reported (*7*). All tested mycotoxins recoveries were greater than 79% with RSD less than 13% (*7*). The method performances fulfill the repeatability and recovery criteria established by European Committee for Standardization for acceptance of an analytical method (*7*).

Earlier collaborative research findings with Waters Corporation on the sample preparations using prototype IAC and multi-mycotoxin LC methods have been reported (*8-10*). LC/MS/MS multi-toxin method for determination of 33 mycotoxins after single extraction in various products has been developed (*11*). Rapid multi-mycotoxin analysis using Ultra Performance Liquid Chromatography (UPLC) and tandem MS (*12*) and rapid analysis of mycotoxin contaminants in pistachio, almond and cashew nuts have also been successfully demonstrated (*13*). This chapter describes the preliminary development of two different analytical methods, LC/u.v./Fluorescence and LC/MS/MS to simultaneously determine 10 mycotoxins (DON, aflatoxins B_1, B_2, G_1, G_2, fumonisins B_1, B_2, B_3, OTA, and ZON) in beer after "Myco5in1" IAC clean-up. Mixed-bed immunoaffinity purification of mycotoxins coupled to the modern LC/MS/MS in ESI mode and/or the classical LC/u.v./Fluorescence system with post-column derivatization analyses are cost effective and efficient solutions to the simultaneous detection of multiple mycotoxins present in complex matrices.

Materials and Methods

Materials

Aflatoxins B_1, B_2, G_1, G_2, deoxynivalenol (DON), ochratoxin A (OTA), zearalenone (ZON), fumonisins B_1 and B_2 standards were purchased from Sigma–Aldrich Corporation, Milwaukee, WI. ACS reagent grade sodium tetraborate, 2-mercaptoethanol, and HPLC grade phthaldialdehyde required to prepare the o-phthalaldehyde (OPA) derivatizing reagent were also obtained from Sigma–Aldrich Corporation, Milwaukee, WI. Fumonisin B_3 was acquired from Promec of MRC (Cape Town, South Africa). Analytical-reagent grade phosphoric acid and acetic acid, and HPLC grade methanol and acetonitrile were purchased from Fisher Scientific (Pittsburgh, PA). Water (conductivity of 18 $M\Omega^{-1}$) was deionized by a Millipore Milli-Q system (Bedford, MA). SurfaSil™ siliconizing fluid to silanize laboratory glassware was obtained from Pierce Biotechnology (Rockford, IL). The prototype mixed-bed 5-toxin (AflatoxinDONFumonisinOchraZON) immunoaffinity column (IAC) or "Myco5in1" IAC, and phosphate buffered saline (PBS), 10x concentrate solution were obtained from Vicam (Watertown, MA).

Reagent Preparation

o-phthalaldehyde (OPA) derivatizing reagent: Borate solution was prepared by mixing 9.55 g sodium tetraborate decahydrate with 0.5 L deionized (D.I.) water. This OPA diluent was filtered and degassed with helium for 10 min. A 150-mg OPA portion dissolved in 5 mL methanol and 500 µL 2-mercaptoethanol were added to the 500 mL borate solution. The OPA derivatizing reagent was then mixed well. This prepared reagent could readily oxidize and should be kept under inert gas.

Multi-toxin stock standard solution preparation: Accurately measured amounts of all 10 mycotoxins were transferred into a silanized borosilicate glass volumetric flask. Accompanying organic solvents were dried and reconstituted with 50% (v/v) acetonitrile/D.I. water, and filled to the mark to prepare a known mixed-toxin stock standard solution to be used for multi-toxin standard calibration and sample spiking purposes. The standard was stored at 4°C when not in use.

Two Instrument Systems Set-up

(1) The classical HPLC/u.v./Fluorescence system contained several components that were assembled in series (HPLC injector – analytical column – photodiode array (PDA) detector – Photochemical Reactor for Enhanced Detection (PHRED™) – post-column derivatizer – fluorescence detector – waste). The HPLC set-up consisted of Waters™ 2695 Alliance®

system. A Waters 2475 fluorescence detector, Waters 2996 PDA detector and Waters Post Column Reaction Module (PCRM) from Waters Corporation (Milford, MA) were also used.

(2) The modern HPLC/MS/MS system consisted of a Waters 2695™ Alliance System with Quattro micro API and the Waters MassLynx™ MS software (version 4) with QuanLynx option. The detector is an atmospheric pressure tandem quadrupole mass spectrometer for quantitative applications.

(1) HPLC/u.v./Fluorescence system

Analytical Conditions

The Waters Symmetry® C_{18}, 4.6 x 150 mm, 3.5 μm particle size, and a 5-μm guard column were from Waters Corporation (Milford, MA). Waters Empower software was used for data management. The mobile phase consisted of combinations of three reagents. The HPLC gradient was as follows:

Time	0.1% (v/v) Phosphoric acid/D.I. water (%)	Methanol (%)	Acetonitrile (%)
0.0	85	0	15
4.0	85	0	15
5.0	60	25	15
16.0	60	25	15
17.0	40	30	30
30.01	85	0	15

The flow rate was 1 mL/min with column temperature of 30° C and injection volume of 50 μL. The analysis took 30 min with no equilibration time.

Photochemical Reactor for Enhanced Detection ("PHRED"™)

The PHRED™ unit (Aura Industries, New York, NY) was equipped with a 254 nm low pressure Hg lamp and the PTFE (poly-tetrafluoro-ethylene) knitted reactor coils. The 254-nm u.v. light was able to perform continuous photolytic derivatization to enhance the sensitivity and/or selectivity of fluorescence detection response. The photochemical reactor was placed between the HPLC analytical column and the detector.

Detection

Analyte	Derivatization	Detection	Wavelength
DON	None	PDA	$\lambda = 220$ nm
Aflatoxins	Photolytic (PHRED™)	Fluorescence	$\lambda_{ex} = 365$ nm $\lambda_{em} = 455$ nm
Fumonisins	Post-column (OPA)	Fluorescence	$\lambda_{ex} = 329$ nm $\lambda_{em} = 465$ nm
Zearalenone	None	Fluorescence	$\lambda_{ex} = 276$ nm $\lambda_{em} = 460$ nm
Ochratoxin A	None	Fluorescence	$\lambda_{ex} = 329$ nm $\lambda_{em} = 460$ nm

The wavelength settings on the fluorescence detector were as follows:

Time	λ_{ex}	λ_{em}
0.1	365	455
18.0	329	465
23.0	276	460
27.9	329	460

The detector sensitivity level for aflatoxins, fumonisins and zearalenone was set at gain 10. Gain was set at 100 for ochratoxin A. All gradient and wavelength changes were programmed through the Empower™ software.

Post-column Conditions

The Waters Post Column Reaction Module (PCRM) was equipped with post-column reaction, temperature control, and reagent manager modules from Waters Corporation (Milford, MA). The reactor temperature was set at 60°C. The OPA derivatizing reagent must be freshly prepared prior to analysis. The flow rate was set at 0.5 mL/min. The post-column pump program was activated by turning the pump on at 17.5 min, then off at 22.5 min.

(2) HPLC/MS/MS System

The Waters™ 2695 Alliance System was equipped with a solvent delivery, autosampler system, and connected to Waters Quattro micro API with dual orthogonal API source (ZSpray™) atmospheric pressure ionization interface and the MassLynx MS software version 4, including the QuanLynx option.

MRM and API conditions, including measurements in positive ESI mode were optimized for multi-component LC/MS/MS method to analyze for beer samples spiked with mycotoxins that were cleaned-up using the prototype immunoaffinity columns. Details of the HPLC conditions are as follows:

LC Conditions

Column:	Waters Symmetry® C_{18}, 2.1 x 150 mm, 3.5 µm particle size
Buffer A:	10% acetonitrile / 0.1% acetic acid in D.I. water by volume
Buffer B:	100% acetonitrile / 0.1% acetic acid by volume
Flow:	200 µL / min
Column Temp.:	30°C
LC Gradient:	Hold at 100% A for 2 min Linear Gradient to 100% B over 30 min Hold at 100% B for 5 min Re-equilibrate for 10 min
Injection Vol.:	50 µL

Sample Preparation: Immunoaffinity Column (IAC) Clean-up Protocols

Vicam prototype "Myco5in1" IAC was used to isolate and extract DON, aflatoxins B_1, B_2, G_1, G_2, fumonisins B_1, B_2, B_3, ZON, and ochratoxin A from beers. Sample preparation was performed using the Zymark RapidTrace™ automatic SPE workstation and TurboVap® LV evaporator from Caliper LifeSciences (Hopkinton, MA). In this study, beer samples were evaporated (de-alcoholized) under nitrogen using the Zymark evaporator at room temperature. Once the known original / neat beer sample aliquot reached approximately half the volume (to ensure most of the volatile organic compounds and alcohol were removed), the remaining de-alcoholized beer sample could then be reconstituted to its original volume using one-tenth diluted PBS solution. The PBS addition was to buffer the normally acidic beer sample (typically at pH < 6) to make the beers less acidic before passing them through the immunoaffinity columns which operate optimally at closer to neutral pH.

The following column clean-up protocols were performed by the Zymark SPE workstation. A 5-mL aliquot of beer mixed with one-tenth diluted PBS solution (~1:1 ratio) was then fortified with known amount of multi-toxin standards. The spiked sample was passed through the "Myco5in1" immunoaffinity column. The column was then washed with 2 mL one-tenth diluted PBS solution and then 4 mL deionized water. Target mycotoxins were eluted with 4 mL methanol. The methanol eluate collected in a silanized borosilicate culture tube was dried under nitrogen using the Zymark evaporator. The dried mycotoxins were reconstituted to a known volume with 1:1 methanol/D.I. water, mixed well prior to either LC/u.v./Fluorescence or LC/MS/MS sample injection.

During the multi-toxin enrichment step, if 5 mL spiked sample aliquot was passed through the clean-up column and the eluate was dried and reconstituted in 1 mL of 1:1 methanol/D.I. water for example, then the original multi-toxin concentrations were enriched 5 times. This analyte enrichment step to concentrate trace level of mycotoxin from the sample extract can be modified to more than 5 times the original mycotoxin concentration as needed. The measured sample mycotoxin concentrations should also be corrected for any trace amounts of naturally-occurring mycotoxins that could be present in the neat beer. In this study, the neat beer samples contained no mycotoxins.

Results and Discussion

This study describes a novel LC/u.v./Fluorescence and post-column derivatization unified approach to analyze 10 mycotoxins in a single run. Of the many derivatization methods for aflatoxins found in the literature, three techniques besides pre-column derivatization by trifluoroacetic acid tend to stand out: iodine which is the basis of the official AOAC method, electrochemically generated bromine (*e.g.*, Kobra Cell), and photochemical reactor for enhanced detection (PHRED™). Comparison of the three different aflatoxin derivatization methods detection limits (MDL) in parts per trillion (ppt) calculated per Environmental Protection Agency (EPA) 40 CFR Ch. 1 Pt. 136 Appendix B, seven replicates MDL = (σ x 3.143) resulted in the same order of sensitivity. MDL ranged between 0.63 to 4.08 ppt for aflatoxins (*10*). The σ is the standard deviation of these seven lowest mycotoxin standard replicates and 3.143 is the students' t value appropriate for a 99% confidence level and a standard deviation estimate with n-1 degrees of freedom. EPA defined MDL as the minimum concentration of a substance that can be measured and reported with 99% confidence when the analyte concentration is greater than 0 and is determined from analysis of a sample in a given matrix containing the analyte. While each derivatization method for aflatoxins has its advantages, incorporation into a mixed mycotoxin scheme presents complications. Fumonisins require a post-column addition of o-phthalaldehyde (OPA) derivatizing reagent for derivatization. This reagent interferes with iodine addition for aflatoxins. On the other hand, the KBr - HNO_3 required for electrochemical bromine generation introduce u.v. active species into the eluent which can diminish the response of DON at 220 nm. This leaves photochemical u.v. as a suitable alternative for aflatoxin derivatization in a mixed mycotoxin analysis. Through use of u.v. detector for DON, a PHRED™ unit for aflatoxins and a programmable post-column addition pump for OPA derivatizing reagent for fumonisins, we have been able to combine DON, the four aflatoxins B_1, B_2, G_1 and G_2, three fumonisins B_1, B_2 and B_3, zearalenone and ochratoxin A into a single analysis.

DON does not have fluorescence activity. Aflatoxins B_2, G_2, OTA and ZON are naturally fluorescent, whereas the weakly fluorescent aflatoxins B_1 and G_1 (if detectable) require photolytic sensitization by PHRED™ as described earlier. Fumonisins require an OPA post-column derivatization to make these analytes fluorescent. This multi-analyte method uses a HPLC system with a post-column reaction module, integrating fluorescence detection of photolytically derivatized aflatoxins, post-column derivatized fumonisins, and zearalenone and ochratoxin A that naturally fluorescence. The PDA detector was placed in series before the post-column hardware for the simultaneous u.v. detection and confirmation of DON.

The above approach requires the post-column addition of OPA be initiated and terminated after the elution of aflatoxin B_1 but before the elution of zearalenone, respectively. This is accomplished using the system control capabilities of the Waters Alliance HPLC system and Empower™ software. Post-column derivatizing reagent manager flow is initiated at 17.5 minutes and stopped at 22.5 minutes. The 2475 Fluorescence detector was time-programmed to change excitation and emission wavelengths for analyte response optimization.

Six-point standard calibration curves at appropriate multi-toxin concentration ranges were performed. The multi-toxin standard linear regression equations have greater than 0.999 correlation coefficients. The LC/u.v./Fluorescence method detection limit (MDL) for the multi-toxin technique in ppb is shown below:

DON = 6	Fumonisin B_1 = 2
Aflatoxin G_2 = 0.03	Fumonisin B_2 = 2
Aflatoxin G_1 = 0.04	Fumonisin B_3 = 2
Aflatoxin B_2 = 0.02	Zearalenone = 1
Aflatoxin B_1 = 0.04	Ochratoxin A = 0.03

A prototype "Myco5in1" IAC used for multiple mycotoxins extraction, isolation and enrichment helped maximize throughput and minimize cost. Our multi-toxin method in collaboration with Pickering Laboratories (Mountain View, CA) using the combination of the Vicam prototype IAC, HPLC/u.v./Fluorescence with post-column photochemical and *o*-Phthalaldehyde derivatizations reported limits of detection (LOD) of 20 μg/L for deoxynivalenol, 0.02 μg/L for aflatoxin G_2, 0.08 μg/L for aflatoxin G_1, 0.01 μg/L for aflatoxin B_2, 0.02 μg/L for aflatoxin B_1, 3 μg/L for fumonisin B_1, 3 μg/L for fumonisin B_2, 8 μg/L for fumonisin B_3, 0.14 μg/L for ochratoxin A, and 9 μg/L for zearalenone. The LOD values were obtained by averaging seven injections of the lowest mycotoxin standards determined using estimated standard analyte signal-to-noise ratio, S/N (peak-to-peak) $\approx$ 3 (4).

The Waters 4.6 x 150 mm Symmetry® column was used for the HPLC/u.v./Fluorescence work (system set-up #1) while the 2.1 x 150 mm Symmetry® column was used for the HPLC/MS/MS (system set-up #2). The Pickering Laboratories MycoTox™, C_{18}, 4.6 x 250 mm, 5 μm particle size column and the Waters Symmetry® C_{18}, 4.6 x 150 mm or 2.1 x 150 mm, 3.5 μm particle size analytical column exhibited different column selectivities under described conditions. The mycotoxins order of elution for the former column was DON, aflatoxins G_2, G_1, B_2, B_1, fumonisin B_1, ochratoxin A, zearalenone, fumonisins B_3, and B_2. The latter column's mycotoxins order of elution was DON, aflatoxins G_2, G_1, B_2, B_1, fumonisins B_1, B_3, B_2, zearalenone and ochratoxin A. There are a few advantages to using the Waters Symmetry® C_{18} column since the OPA reagent can be introduced only once to consecutively derivatize all 3 fumonisins at the appropriate LC conditions. Run time was reduced by half from about 60 mins using the previously published method to 30 mins using the described multi-toxin LC method. It must be emphasized that selection of appropriate analytical columns and optimization of LC conditions are normally needed for any laboratory performing single laboratory method validation for multi-residue analyses and/or implementing interlaboratory LC method transfers.

The beer samples were intentionally spiked at low mycotoxin concentrations since beers normally contain trace level of naturally occurring mycotoxins. The EC mycotoxin regulation for beer is not yet established. The multi-toxin data analyzed between the two different LC methods and the spike recovery data obtained were complementary and in acceptable ranges. For example, the percent recoveries of 5 mL spiked beer sample aliquot passed through the "Myco5in1" column in triplicates at 45 : 0.42 : 0.14 : 0.42 : 0.14 : 5 : 5 : 6 : 0.22 : 6 ppb levels of DON : aflatoxin B_1 : aflatoxin B_2 : aflatoxin G_1 : aflatoxin G_2 : fumonisin B_1 : fumonisin B_2 : fumonisin B_3 : OTA : ZON ranged from 73% to 102% (*14*).

Figures 1 and 2 are representative mixed-toxin standard and spiked wheat beer chromatograms of the classical LC/u.v./Fluorescence system (instrument system set-up #1) with at higher and lower multi-toxin concentrations, respectively. The mycotoxin peaks elution order and resolution are clearly depicted.

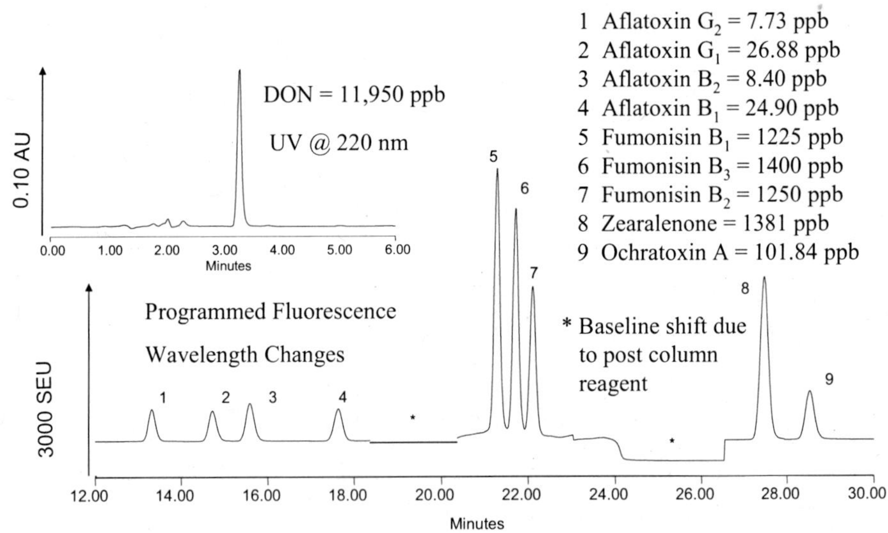

Note: SEU and AU denote standard emission units and absorbance units, respectively.

Figure 1. Multi-toxin standard at higher mycotoxin concentrations analyzed using multi-toxin HPLC/PDA/Fluorescence method - a representative chromatogram.

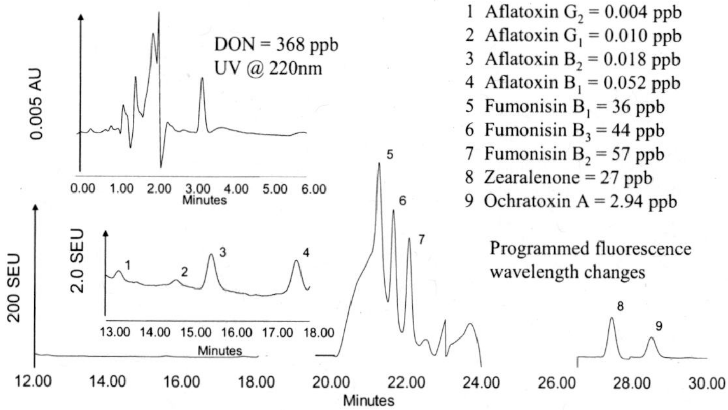

Note: SEU and AU denote standard emission units and absorbance units, respectively.

Figure 2: Fortified wheat beer at lower mycotoxin concentrations analyzed using multi-toxin HPLC/PDA/Fluorescence method - a representative chromatogram.

Figures 3 and 4 are representative mixed-toxin standard and spiked beer MRM chromatograms of the modern LC/MS/MS system (instrument system set-up #2) at lower and higher multi-toxin concentrations, respectively.

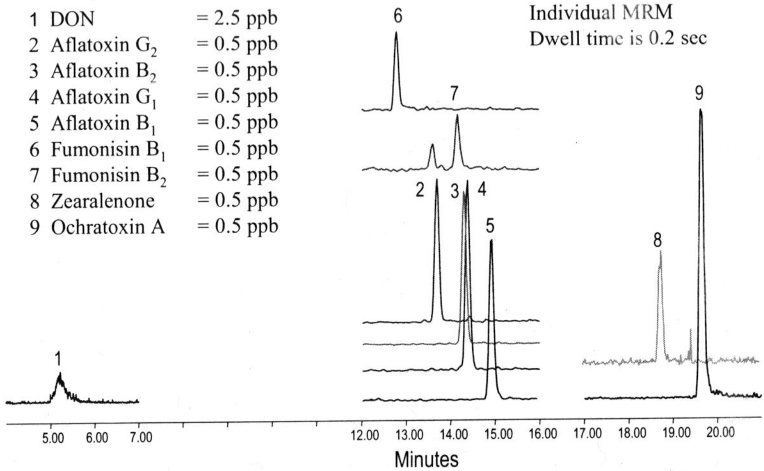

Figure 3. Multi-toxin standard at lower mycotoxin concentrations analyzed using LC/MS/MS – a representative MRM chromatogram.

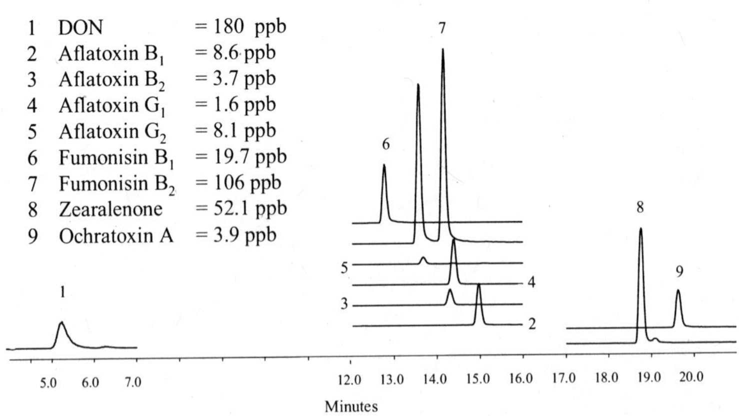

Figure 4. Spiked beer at higher mycotoxin concentrations analyzed using LC/MS/MS – a representative MRM chromatogram.

Table 1 summarized the Waters Quattro micro API Multiple Reaction Monitoring (MRM) ions and conditions.

Table 1. Quattro micro API MRM Ions and Conditions

Analyte	Precursor Ion	Product Ion	Cone	Collision	Dwell
DON	297.25	249.15	12 V	35 eV	.2 sec
Aflatoxin B_1	313.0	241.1	40 V	35 eV	.2 sec
		269.0			
Aflatoxin B_2	315.0	259.1	40 V	28 eV	.2 sec
		287.1			
Aflatoxin G_1	329.0	243.1	42 V	26 eV	.2 sec
		311.0			
Aflatoxin G_2	331.1	245.1	45 V	30 eV	.2 sec
		313.0			
Ochratoxin A	404.2	239.1	20 V	29 eV	.2 sec
Zearalenone	318.8	300.8	20 V	14 eV	.2 sec
		282.9			
		187.0			
Fumonisin B_1	722	334	45 V	30 eV	.2 sec
		352			
Fumonisin B_2/B_3	706	336	45 V	30 eV	.2 sec
		318			

Note: Primary product ion for quantification (regular font style values);
Secondary product ion for confirmation (italic font style values).

The multiple MRM in 3 functions with 50 µL injection of standards prepared in 10% acetonitrile / 0.1% acetic acid in D.I. water produced MDL for LC/MS/MS in ppb:

DON = 2.5
Aflatoxin G_2 = 0.5
Aflatoxin G_1 = 0.5
Aflatoxin B_2 = 0.5
Aflatoxin B_1 = 0.5

Fumonisin B_1 = 0.5
Fumonisin B_2 = 0.5
Zearalenone = 0.5
Ochratoxin A = 0.5

Although there are some chromatographic co-elutions, specific analyte MRM eliminates this problem. There are many variations on the "universal gradient" theme, including the use of methanol or acetonitrile for selectivity differences and eluting power. Volatile buffers, such as formic or acetic acid, or their ammonium salts can aid in ionization. When using the "universal gradient" on a reverse phase column, larger injection volumes could be used provided the sample solution was less than the initial LC solvent conditions, in this method,

10% acetonitrile / 0.1% acetic acid in D.I. water. As the sample solvent became stronger, the early eluting peaks, such as DON could broaden. An initial 2 column volume segment at initial gradient conditions was incorporated to concentrate the analytes for improved efficiency before starting the gradient. With MS technology, samples can be analyzed without sample preparation. However, it is not without its consequences, such as the need for frequent cone cleaning, the possibility of response suppression or enhancement due to chromatographed complex matrix components, and decreased chromatographic and MS performance. The chosen immunoaffinity clean-up column sample preparation was suited to minimize these undesirable issues.

Based upon the MRM standard chromatograms, the MDL in ppb being defined as 3 times S/N in 10% acetonitrile / 0.1% acetic acid in D.I. water using a 50 μL injection is:

DON = 0.9	Fumonisin B_1 = 0.1
Aflatoxin G_2 = 0.05	Fumonisin B_2 = 0.1
Aflatoxin G_1 = 0.05	Zearalenone = 0.2
Aflatoxin B_2 = 0.05	Ochratoxin A = 0.02
Aflatoxin B_1 = 0.03	

In this case, there was no MS response suppression and roughly 10 fold increase in sensitivity between standards in mobile phase versus standards in a blank matrix was observed. There is no universally accepted definition for detection limit. It generally involves the lowest concentration of analyte in the test material that can be unambiguously determined by the method. It is often associated with the "instrument detection limit". This limit can be invariably far lower than that which can be measured in the analytical sample (*15*). Based on the MS response reported, the estimated difference is about 10 times lower. This estimated value will predominantly be reagent matrix dependent. Perhaps a more useful performance characteristic of a method is the limit of determination (*16*) which is analogous to the limit of quantitation. Limit of determination is generally defined as the lowest concentration of a residue that can be identified and quantitatively measured in a test matrix using a validated analytical method at specified accuracy and precision. Limit of quantification or quantitation (LOQ) is the smallest amount of analyte in a test sample which can be quantitatively determined with suitable precision and accuracy under previously established method conditions. It is often estimated to be 10 times the standard deviations of the background noise. When LOQ is calculated, the definition and method used should be stated. LOD and LOQ can be expected to vary considerably since they are determined at the lowest useful ranges of the methods that tend to be regions of poor accuracy and precision (*17*).

The 12 relevant mycotoxins namely DON, aflatoxins B_1, B_2, G_1, G_2, fumonisins B_1, B_2, B_3, OTA, T-2, HT-2 and ZON measured by a LC/MS/MS method using a 100 μL sample injection volume in our collaborative study with Varian Inc. have detection limits of 2, 0.04, 0.04, 0.04, 0.04, 0.4, 0.4, 0.4, 0.18, 0.4, 0.4 and 0.4 ppb, respectively (*6*). The performances of the prototype IAC clean-up and LC methods discussed in this chapter were all satisfactory for the tested multiple mycotoxins at contamination concentrations close to and/or below relevant EU maximum permitted or recommended levels.

Conclusions

Vicam prototype "Myco5in1" and the commercially available Myco6in1™ immunoaffinity columns containing immobilized antibodies have been proven over time to selectively retain mycotoxins to produce cleaner extracts with a minimum level of interfering matrix components. The mixed-bed ("Myco5in1") multifunctional immunoaffinity columns combined with the complementary LC/u.v./Fluorescence and LC/MS/MS techniques in this study, and the previously published Myco6in1™ columns in conjunction with LC/MS/MS analyses, have proven to be effective immunochromatography techniques. Immunochromatography refers to a combined analytical technique of immunoaffinity column clean-up followed by liquid chromatography. The LC/u.v./Fluorescence classical technique is a cost effective and reliable way to perform multi-toxin analysis. Nonetheless, future food safety methods will involve a faster single, multi-toxin LC method as MS instruments become less expensive and accessible by food control laboratories worldwide. Fast performance LC systems capable of maximizing sample throughput while minimizing organic waste, such as the Ultra Performance LC (UPLC), Rapid Resolution LC (RRLC), Rapid Separation LC (RSLC) and Ultra Fast LC (UFLC) equipped with MS have recently gained popularity. In conjunction with new innovations of multifunctional SPE cartridges for sample preparation, high throughput mycotoxin screening of food raw materials and finished products can be accomplished.

Acknowledgements

Darsa Siantar wishes to thank Dr. Mary Trucksess (U.S. FDA, College Park, MD), Nancy Zabe and Dr. Stephen Powers (Vicam Group of Waters Technologies; Watertown, MA) for the valuable discussions and collaborative efforts over the years. He would also like to thank Adiva Sotzsky, Dr. Abdul Mabud and Bill Foster (TTB, Washington, DC) for their supports in this collaborative project. He would like to acknowledge Vicam Group of Waters Technologies as well as Carol Donelly (R-Biopharm Rhône Ltd., Glasgow, Scotland) for supplying the prototype immunoaffinity columns. Jim Krol, currently with Applied Biosystems (Framingham, MA) would like to thank (1) Drs. Janssen and Franken at the Technical Laboratory in Rotterdam, Netherlands for developing a set of chromatographic and MS/MS conditions to quantitate the 9 mycotoxins, (2) Dr. Nico Van Eeckhout from Waters MS Technologies in Belgium for demonstrating and installing this application, and (3) Dr. Martien Spanjer from Inspectorate for Health Protection and Veterinary Public Health in Amsterdam for validating this method for corn and wheat.

References

1 U.S. Food and Drug Administration Office of Regulatory Affairs Compliance Sec. 555.400 Foods - Adulteration with Aflatoxin (CPG 7120.26); http://www.fda.gov/ora/compliance_ref/cpg/cpgfod/cpg555-400.html.

2 European Commission Regulation (EC) No. 1881/2006, Official Journal of the European Union, Brussels, 2006.

3 European Commission Regulation (EC) No. 123/2005, Official Journal of the European Union, Brussels, 2005.

4 Siantar, D.; Ofitserova, M.; Nerkar, S.; Pickering, M.; Cardozo, M.; Peterson, G.; Dugar, S.; "The Simultaneous Analysis of Aflatoxins, Deoxynivalenol, Fumonisins, Ochratoxin A and Zearalenone Using Mixed-bed Immunoaffinity Column and HPLC/UV/Fluorescence with Post-column Derivatization'. In *Mycotoxin Contamination and Control*; Njapau, H., Trujillo, S., Pohland, A. and Park, D., Eds.; AuthorHouse Publishers: Bloomington, IN, 2008, 217-229.

5 Zabe, N.; Basker, C.; *European Patent Application* EP1787699 A1, 2007.

6 Rudrabhatla, M; George III, J.; Hill, N.; Siantar, D.; "Detection of Relevant Mycotoxins in Wheat Beer and Sake by LC-MS/MS Using Prototype Immunoaffinity Coulumn Clean-Up: A Preliminary Study". In *Food Contaminants: Mycotoxins and Food Allergens*; Siantar, D., Trucksess, M., Scott, P. and Herman, E., Eds.; ACS Symposium Series *1001*, ACS Publications: Washington, D.C., 2008, ISBN: 978-0-8412-6954-5, 241-251.

7 Lattanzio, V.; Solfrizzo, M.; Powers, S.; Visconti, A.; *Rapid Commun. Mass Spectrum.*, 2007; *21*, 3253-3261.

8 Benvenuti, M.; Romano, J.; Krol, J.; Siantar, D.; "A Novel Approach for Mixed Mycotoxin Analysis"; Waters Poster Reprint No. 720000605EN.

9 Benvenuti, M.; Romano, J.; Krol, J.; Siantar, D.; "A Unified Method for Mixed Mycotoxin Analysis Using UV and Fluorescence Detection"; Waters Poster Reprint No. 720000908EN and also presented during the XI International IUPAC Symposium on Mycotoxins And Phycotoxins in Bethesda, MD (May 17-21, 2004).

10 Benvenuti, M.; Krol, J.; Romano, J.; "Multi-analyte Mycotoxin Analysis"; Waters Application Note No. 720001050EN.

11 Spanjer, M.; Rensen, P.; Scholten, J.; "LC-MS/MS Multi-method for Mycotoxins after Single Extraction, with Validation Data for Peanut, Pistachio, Wheat, Maize, Cornflakes, Raisins and Figs"; *Food Additives and Contaminants: Part A*, 2008; *25*, 472-489.

12 de Kok, A; Spanjer, M.; Scholten, J.; Rensen, P.; Kearney, G.; "Rapid Multi-Mycotoxin Analysis using ACQUITY UPLC and Quattro Premier XE"; Waters Application Note No. 720001996EN.

13 Morphet, J.; Hancock, P.; de Kok, A.; Rensen, P.; Scholten, J.; Spanjer, M.; "Application of ACQUITY TQD for the Analysis of Mycotoxin Contaminants in Pistachio, Almond and Cashew Nuts"; Waters Application Note No. 720002244EN.

14 Siantar, D.; "The Multi-Step Process for Multiple Mycotoxins Detection"; presented during the Gordon Research Conference Satellite Workshop at

Mount Desert Island Biological Laboratory in Maine (June 17 – 18, 2005); unpublished data.

15 Pohland, A.; "Method Validation Procedures". In *Mycotoxin Contamination and Control*; Njapau, H., Trujillo, S., Pohland, A. and Park, D., Eds.; AuthorHouse Publishers: Bloomington, IN, 2008, 151-166.

16 Garfield, F.; Klesta, E.; Hirsch, J., Eds.; In *Quality Assurance Principles for Analytical Laboratories*; AOAC INTERNATIONAL, Gaithersburg, MD, 2000, 3rd Edition.

17 Horwitz, W., Ed.; In *Official Methods of Analysis of AOAC INTERNATIONAL*; AOAC INTERNATIONAL, Gaithersburg, MD, 2002, 17th Edition, Revision 1, Appendix E: Laboratory Quality Assurance, 1-6.

Chapter 18

Stable Isotope Labeled Mycotoxins as Standards for HPLC-MS/MS Analysis

Review and Evaluation of Published Procedures for the Introduction of Stable Isotopes into Mycotoxins

Benedikt Cramer, Marita Beyer, and Hans-Ulrich Humpf*

Institute of Food Chemistry, University of Muenster, Corrensstr. 45 48149 Muenster, Germany

The production of stable isotope labeled mycotoxins as standards for the quantification of mycotoxins by high performance liquid chromatography with mass spectrometric detection (HPLC-MS/MS) is a crucial point for the application of this technique in routine analysis. Without these standards, laborious calibration procedures are required to overcome matrix effects and the analytical results may posses deficiencies in accuracy and certainty. In the last years four different approaches to stable isotope labeled mycotoxins have been successfully applied. The principles of each strategy are presented in this review and examples for the application of each technique are discussed. As a general consequence, the chemical modification of unlabeled mycotoxins seems to be the most suitable procedure to get access to larger quantities of stable isotope labeled standards with reasonable effort. However, until now the growth of fungal cultures on fully [^{13}C]-labeled media is the only technique that provides commercially available standards of all regulated mycotoxins.

Introduction

Natural food contaminants have been identified as a possible health risk since the early beginning of official food regulations. Among this group, mycotoxins play a major role, as they can be found in nearly every commodity in various toxin combinations. After the first reported cases of aflatoxicosis in the 1960s, the mycotoxins as harmful fungi metabolites are a focus for official authorities (*1*). Since then, further mycotoxins were regulated to improve food safety, so that at the moment limits for aflatoxin B_1, aflatoxin M_1, total aflatoxins, ochratoxin A, patulin, deoxynivalenol, sum of fumonisins B_1 and B_2; and zearalenone are set by the European Union (*2*). Further limits for T-2 toxin and HT-2 toxin are expected for 2009.

The increasing requirement for the analysis of multiple mycotoxins in a single sample is a driving force for the development of multi-mycotoxin methods. They allow a qualitative and quantitative analysis of most mycotoxins in short time and at low cost. Furthermore the strict regulations demand reliable analytical results, especially for compounds that have little or no UV activity and require critical derivatization steps, such as deoxynivalenol, fumonisins B_1 and B_2, and T-2 toxin/HT-2 toxin. The answer to these demands was the introduction of high performance liquid chromatography tandem mass spectrometry (HPLC-MS/MS) with the robust and efficient electrospray ionization (ESI) (*3*) or atmospheric pressure chemical ionization (APCI) (*4*). Both ionization techniques provide a high ion yield despite HPLC-flow rates above 200 µL/min and allow low limits of detection.

Using these techniques, methods for the nonspecific cleanup of food and feed samples were developed and improved. Cartridges containing mixtures of activated charcoal, alumina, celite or silica were introduced for the combined one step cleanup of types A and B trichothecenes (*5, 6*). Deoxynivalenol, 3-acetyl-deoxynivalenol, nivalenol and fusarenone X have been analyzed by HPLC-MS/MS directly after passing through a cleanup cartridge (*7*). Subsequently several groups expanded the multi methods to cover up to 12 major trichothecenes, zearalenone, and fumonisins (*8-11*).

Another, more recent trend in HPLC-MS/MS development is the reduction of sample preparation prior to the extraction process. Due to the enhanced mass spectrometer performance, so-called “dilute and shoot” approaches have been published. These sample preparations simply extract the mycotoxins from commodities with aqueous solutions, dilute the extract and inject it into the HPLC-MS/MS system (*12*). However, up to now, this approach seems not to be suitable for routine analysis, as strong matrix effects (see below), a loss of sensitivity, and more requirements for instrument maintenance can be observed.

With increasing application of mass spectrometric detectors, the difficulties of quantitation became a new focus in method development. Classical fluorescence and UV-detectors show a linear correlation between analyte concentration and signal response over a defined calibration range. The signal is not significantly influenced by coeluting sample matrix, as long as it does not exhibit the same fluorescence or UV-activity. In contrast, mass spectrometric detectors with ESI or APCI interfaces can strongly be influenced by coeluting matrix compounds. This not directly visible matrix effect usually results in signal suppression, but can also cause an enhancement of the analyte response.

The factors responsible for this effect are discussed in several reviews (*13-15*). Generally, this variation of the signal intensities is the result of changes in the ionization efficiency of the target compound caused by the total ion content, buffer salts, analyte polarity and the total amount of non-volatiles. In general the signal suppression is strongly depending on the sample cleanup. Highly specific cleanup procedures such as the use of immunoaffinity chromatography remove most of the coeluting compounds and thus show a significantly lower matrix effect compared to unspecific purification via liquid/liquid extraction or solid phase materials (*16*).

Compensation of Matrix Effects

Several different strategies to achieve a reliable quantification of mycotoxins despite occurring matrix effects are published. A well known basic procedure in a homogenous sample batch is the matrix calibration. Calibration standards are dissolved in blank samples instead of pure solvent, to induce a similar matrix effect to the standard solution, as it can be observed in samples. This procedure is suitable for homogenous solutions, such as in process control in the pharmaceutical and chemical industries. However, for example a matrix calibration applied to blood samples from different individuals might exceed the required limits for accuracy and reproducibility (*17*).

Most of the recently published methods use internal standards to compensate matrix effects. For this purpose, chemical analogs of the target compounds are added to the samples and to standard solutions and a relative response is calculated. In the case of mycotoxins, typical internal standards are neosolaniol, verrucarol and deepoxy-deoxynivalenol for trichothecenes, and zearalanone or zearalanol for zearalenone and the zearalenols (*8-10, 18*). However, the disadvantages of these internal standards are their chemical difference compared to the analyte and the different retention time. As all three factors, retention time, polarity and structure have a strong impact on the matrix effect, a full compensation seems questionable.

Another procedure that compensates for all effects beside the different retention time is the echo peak technique, where a time-delayed second injection of a standard solution is programmed (*19*). This second injection results in the occurrence of double peaks for every analyte, where the first peak corresponds to the analyte and the latter one to a reference of known concentration (echo peak). Based on the relative intensities, the concentration of the analyte can be calculated. Nevertheless, while the compound specific differences are eliminated by the echo peak, the time dependent matrix effect is still a problem that has to be considered. Furthermore, errors related to the second injection in the autosampler may occur and losses during workup are not corrected.

The most suitable way to compensate all matrix effects and to cover all losses during cleanup is the addition of stable isotope labeled standards in an early stage of the sample preparation. These usually [^{13}C]- or [D]- labeled compounds have almost similar chemical and physical properties to the analyte. They elute at nearly identical retention times and are thus affected by similar matrix effects. However, in the case of mycotoxins, the availability of labeled derivatives at reasonable costs is the limiting factor.

Stable Isotope Labeled Mycotoxin Standards

In recent years four different strategies have been used to provide isotope labeled mycotoxins. Each method has its special advantages and disadvantages that are discussed in the following sections.

Total synthesis of labeled mycotoxins

Mycotoxins are complex natural products with often multiple stereogenic centers and reactive groups. As a consequence, the total synthesis of these compounds is a challenging task for natural product chemists. Despite successful total synthesis of some mycotoxins, a synthetic access to larger quantities is probably limited to the substances alternariol, patulin and ochratoxin A. For these compounds synthetic routes with reasonable overall yields have been published (*20-22*). Within the total synthesis, incorporation of [^{13}C]- or [D]- can easily be achieved via commercially available stable isotope labeled precursors. An example for the total synthesis of [^{13}C]-labeled mycotoxins is the total synthesis of [$^{13}C_2$]-ochratoxin α. The synthesis was carried out with modifications of the procedure previously reported by Kraus et al. using relatively inexpensive [$^{13}C_2$]-acetaldehyde as the synthetic precursor to introduce an isotope label (*20*). Using this procedure, [$^{13}C_2$]–ochratoxin α was synthesized with an overall yield of 54 % for the [^{13}C]-labeled component (*23*). A total synthesis of 2,3-[$^{13}C_2$]-patulin, based on the procedure of Bennett et al. is shown in Figure 1 (*21*). Starting from L-arabinose, ethyl bromo-[$^{13}C_2$]-acetate was chosen as a carrier for the isotope label (*24*). In seven steps, [$^{13}C_2$]-patulin was obtained with 2.3 % overall yield.

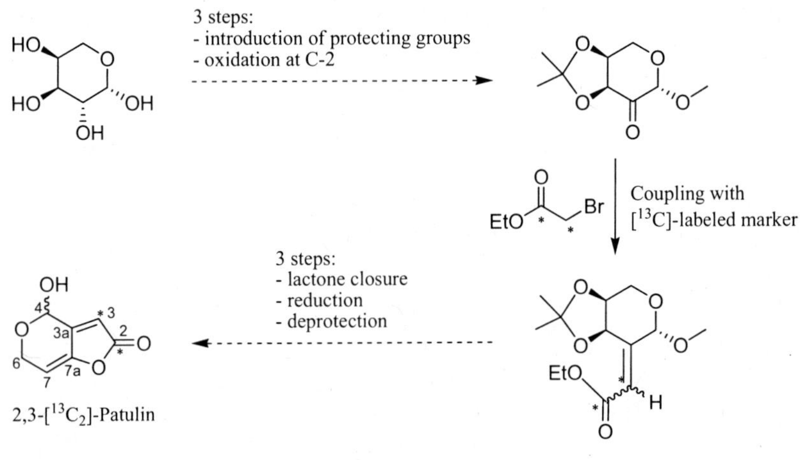

Figure 1: Synthesis of 2,3-[$^{13}C_2$]-patulin via coupling of L-arabinose with ethyl bromo-[$^{13}C_2$]-acetate (24).

Chemical modification of non-labeled mycotoxins

The incorporation of stable isotopes into mycotoxins starting from the non-labeled compound is a convenient method to overcome many synthetic difficulties related to structural complexity. The first approaches of these retro synthetic modifications have been made already in the 70s to introduce a specific radioactivity for metabolic studies into ochratoxin A (*25*). In this experiment, ochratoxin A was hydrolyzed in a first step to ochratoxin α and L-phenylalanine. The former compound was coupled with radioactive labeled L-phenylalanine methyl ester to give ochratoxin A methyl ester, which was subsequently hydrolyzed to yield the free acid. A similar procedure has also been used to synthesize [D_5]-ochratoxin A as a standard for HPLC-MS/MS analysis (*26*).

A further example for the chemical modification of mycotoxins is the large scale synthesis of [D_1]-deoxynivalenol (*27*). The reactive trichothecene structure with three rings, an epoxide function, a conjugated carbonyl and three hydroxyl groups makes chemical modifications a challenging task. However, a selective oxidation of the primary hydroxyl function of the 3-acetyl-deoxynivalenol was successful by the use of the Swern-oxidation (*28*). A subsequent reduction of the aldehyde to an alcohol using sodium borodeuteride allowed a stable introduction of a deuterium label into deoxynivalenol. Due to the characteristic fragmentation of the deoxynivalenol ion in the negative mode, this single labeled compound can easily be used as a stable isotope labeled standard for the quantification by HPLC-MS/MS.

Other modifications of trichothecenes are the synthesis of [$^{13}C_2$]- or [D_3]-derivatives of 3-acetyldeoxynivalenol, 15-acetyldeoxynivalenol, mono- and diacetoxy-scirpenol, T-2 toxin and HT-2 toxin (*28-30*). In these procedures, the isotope label is introduced by acetylation with stable isotope labeled acetyl chloride or acetic anhydride.

The incorporation of deuterium by acid or alkaline catalyzed hydrogen/ deuterium exchange is a common procedure in isotope standard synthesis and was successful applied for the labeling of zearalenone, as shown in Figure 2 (*31*). The hydrogen/deuterium exchange can be carried out unspecifically in the α-positions of the carbonyl function at C-6'and at the aromatic hydrogen at positions 3 and 5 (*32, 33*). In this case, a maximum isotopic purity of less then 40 % for [D_6]-zearalenone can be achieved. Alternatively, 3,5-[D_2]-zearalenone can be synthesized with 95 % isotopic purity in a three step synthesis (*34*). In the latter case the carbonyl function was protected by formation of an 1,3-dioxalane with ethylene glycol before alkali catalyzed hydrogen/deuterium exchange. Cleavage of the acetal under moderate acidic conditions resulted in 3,5-[D_2]-zearalenone (*34*).

The deuteration of double bonds as a possibility for isotope labeling of myco-toxins has recently been carried out by Cervino et al. (*35*). Using this technique, aflatoxin B_1 was converted to [D_2]-aflatoxin B_2 and aflatoxin D_1 converted to [D_2]-aflatoxin D_2.

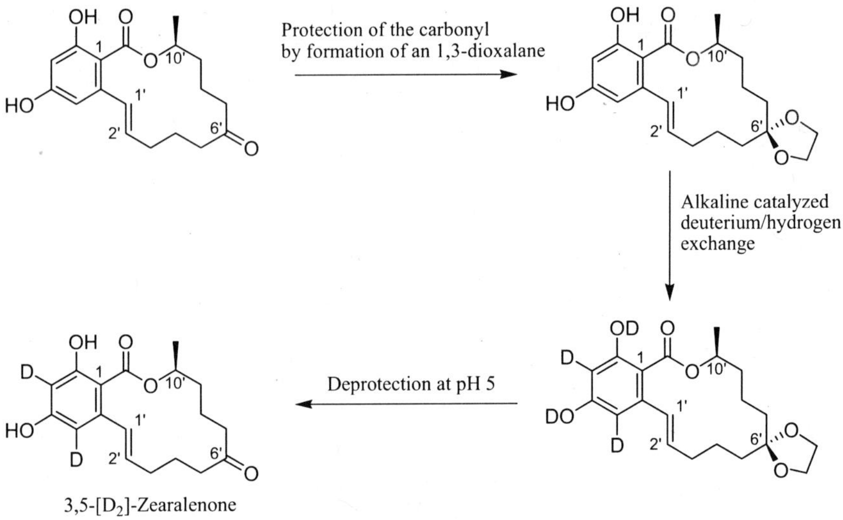

Figure 2: Synthesis of 3,5-[D_2]-zearalenone by alkaline deuterium/hydrogen exchange (33).

Production of labeled mycotoxins in fungal cultures by addition of labeled biosynthetic precursors

The use of [^{13}C]-enriched biosynthetic precursors is a well established strategy for the elucidation of biosynthetic pathways. In these experiments, fungal cultures bioorganisms are fed with selected stable isotope labeled precursor, such as amino acids, sugars or acetate. Afterwards, the incorporation of these molecules into the target compounds can be determined by NMR-studies, where the incorporation is indicated by amplified signal intensities for the respective carbon- or heteroatom signals. As examples, the acetate coupling in ochratoxin A and deoxynivalenol has been investigated by addition of 1-[^{13}C]-acetate or 2-[^{13}C]-acetate to cultures of *A. ochraceus* and *F. graminearum*, resulting in a specific signal enhancement for the coupling carbons in the [^{13}C]-NMR spectra (*36, 37*). The addition of [^{36}Cl]- or [^{13}C]-enriched methionine, glutamic acid and alanine has also been used for biosynthetic studies of fumonisins and ochratoxin A (*38, 39*).

This approach for the identification of biosynthetic pathways can also be used to produce stable isotope labeled standards for mass spectrometry. However, while for the identification of the biosynthetic pathways, only a few percent of incorporation are required, an isotopic purity of more then 90 % is needed for an "easy to handle" mycotoxin standard. To improve the incorporation of the labeled material, three aspects are of special importance. To achieve a selective labeling, the addition of "valuable" precursors like amino acids should be preferred to the addition of principle nutrients like acetate. The latter one will always be in competition with non-labeled carbon sources and thus lead to poor isotopic purities. The occurrence of any other source for the stable isotope labeled precursor supplied must be avoided. In the case of amino acids, this means, that no digested protein mix can be used to improve fungal growth and toxin production. Furthermore, the isotopically labeled supplement must be added in high amounts, such as grams per liter, which makes deuterated precursors a strongly favorable source of labeling, as they are far less expensive compared to [^{13}C]-compounds.

By selection of the appropriate growth medium and precursor compound, the incorporation of the isotopic label into the target can reach levels of > 95%. In the case of mycotoxins, this approach allowed a biosynthetic production of [D_6]-fumonisin B_1, as shown in Figure 3 (*3*). In liquid cultures of *F. verticillioides*, the addition of 1 g/L methyl-[D_3]-L-methionine to the liquid culture broth resulted in an almost complete stopping of the endogen L-methionine biosynthesis and incorporation of the deuterated amino acid to give [D_6]-fumonisin with a final isotopic purity of 96 %.

The addition of ring-[D_5]-L-phenylalanine to a culture of *P. verrucosum* also resulted in a suitable isotope labeled mycotoxin standard. As the final step of the ochratoxin A biosynthesis is the amidation of ochratoxin α with L-phenylalanine by the ochratoxin A synthetase (*40*), the addition of high amounts of ring-[D_5]-L-phenylalanine resulted in a bond formation between the isotopically labeled amino acid with ochratoxin α to [D_5]-ochratoxin A (*41*).

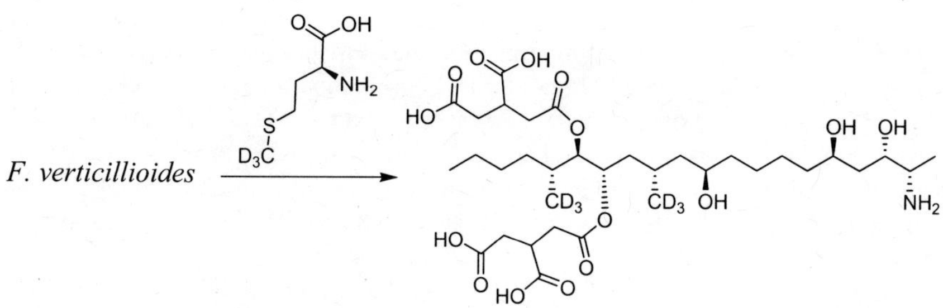

Figure 3: Biosynthetic production of [D_6]-fumonisin B_1 by Fusarium verticillioides in methyl-[D_3]-L-methionine enriched growth medium (3).

Fungal growth on fully [^{13}C]-labeled culture media

A basic approach for the production of [^{13}C]-labeled mycotoxins is the growth of fungal cultures in a fully [^{13}C]-labeled medium. As expected, all fungal metabolites and fungal mycelium incorporates the [^{13}C]-atoms. Simple extraction of the mycotoxins from the growth medium delivers the stable isotope labeled references which consist due to the limited isotopic purity of the starting materials, of different isotopomers. Besides the fully [^{13}C]-labeled mycotoxin, also one and two [^{12}C]–isotopomers are formed, so that for example for deoxynivalenol, a final isotopic purity of about 82 % is reported (*42*). The procedures used for the growth of fungi cultures require efficient fungi cultures and a highly sophisticated extract management to minimize the expenses for the [^{13}C]-labeled starting material, required to produce toxins (*43*).

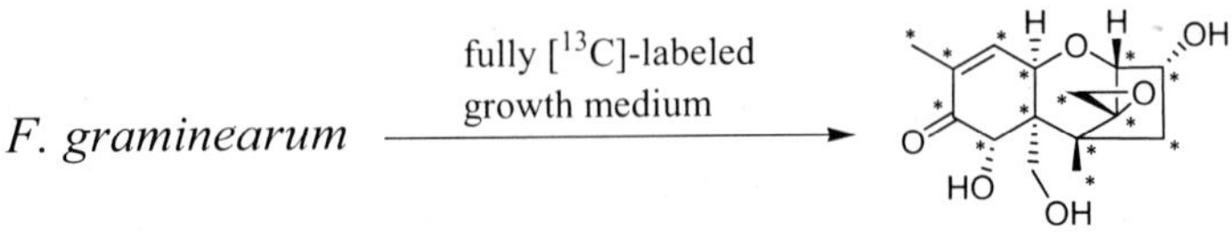

Figure 4: Biosynthetic production of [$^{13}C_{15}$]-deoxynivalenol by growth of Fusarium graminearum on fully [^{13}C]-labeled growth medium (42).

Comparison of the Different Techniques

Mycotoxins cover a wide range of different chemical structures with a variety of reactive groups, which makes the selection of an appropriate method to produce isotope labeled mycotoxins a difficult task. Therefore a general answer which method is the best cannot be given.

However, for those mycotoxins, where different approaches to generate isotopomers are published and their successful application for analysis is reported, a comparison based on the efforts required to produce these standards is possible. One way to compare this is based on the costs of commercially available stable isotope labeled standards. This comparison can be carried out with ochratoxin A and deoxynivalenol, where two types of stable isotope labeled standards with comparable chemical and isotopical purity are commercially available from the same supplier. In the case of deoxynivalenol, the 15-[D_1]-deoxynivalenol and the [$^{13}C_{15}$]-isotopomer are sold, with the latter one being about 20 fold more expensive. In the case of ochratoxin A, the fully [^{13}C]-labeled ochratoxin is about 10 fold more expensive than the [D_5]-derivative.

Even if the price of these compounds does not allow a direct conclusion a factor of 10 and more at least indicates that the synthetic procedures are more efficient. Considering the labeling process itself, it is also clear, that the chemical synthesis or modification of mycotoxins is especially efficient in large batches, as the costs of the labeling compounds are relatively low and the synthetic strategy is nearly independent from batch size. In the biochemical process, the large amount of [^{13}C]-labeled starting material is a cost-intensive factor that is nearly directly correlated to the batch size. Another advantage of synthesized or chemically modified isotope labeled standards is the fact that the isotope label can be selectively introduced at a certain position with only a small difference in the molecular weight (< 5 amu) compared to the unlabeled compound.

Due to the complex chemical structures of mycotoxins, only a few chemically synthesized or modified standards are commercially available. The production of isotope labeled mycotoxins on fully [^{13}C]-labeled media has the advantage that based on known toxin production procedures principally each compounds is available.

Finally, a real breakthrough of high quality HPLC-MS/MS analysis in the mycotoxin field is still missing, as cheap stable isotope labeled standards of all mycotoxins are not available. In this ideal case, a mixture of labeled mycotoxins could be added to a sample already in the extraction process and thus compensate all errors and losses during the whole cleanup and analytical process. However, in such a case, a single analysis based on the procedure of Klötzel et al. (*9*) would require already 2.5 µg of labeled deoxynivalenol, a standard that is, at least fully [^{13}C]-labeled, until now far too expensive.

References

1. Nesbitt, B. F.; O'Kelly, J.; Sargeant, K.; Sheridan, A. *Aspergillus flavus* and turkey X disease: toxic metabolites of *Aspergillus flavus*. *Nature* **1962**, *195*, 1062-1063.
2. European Commission. Commission regulation (EC) No 1881/2006 of 19 December 2006 setting maximum levels for certain contaminants in foodstuffs. *OJEU* **2006**, *L 364*, 5-24.
3. Lukacs, Z.; Schaper, S.; Herderich, M.; Schreier, P.; Humpf, H. U. Identification and determination of fumonisin FB1 and FB2 in corn and corn products by high-performance liquid chromatography - electrospray-ionization tandem mass spectrometry (HPLC-ESI-MS-MS). *Chromatographia* **1996**, *43*, 124-128.
4. Rosenberg, E.; Krska, R.; Wissiack, R.; Kmetov, V.; Josephs, R.; Razzazi, E.; Grasserbauer, M. High-performance liquid chromatography atmospheric-pressure chemical ionization mass spectrometry as a new tool for the determination of the mycotoxin zearalenone in food and feed. *J. Chromatogr. A* **1998**, *819*, 277-288.
5. Romer, T. R. Use of small charcoal/alumina cleanup columns in determination of trichothecene mycotoxins in foods and feeds. *JAOAC Int.* **1986**, *69*, 699-703.
6. Klötzel, M.; Lauber, U.; Humpf, H.-U. A new solid phase extraction clean-up method for the determination of 12 type A and B trichothecenes in cereals and cereal-based food by LC-MS/MS. *Mol. Nutr. Food Res.* **2006**, *50*, 261-269.
7. Lagana, A.; Curini, R.; D'Ascenzo, G.; De Leva, I.; Faberi, A.; Pastorini, E. Liquid chromatography/tandem mass spectrometry for the identification and determination of trichothecenes in maize. *Rapid Commun. Mass Spectrom.* **2003**, *17*, 1037-1043.
8. Royer, D.; Humpf, H. U.; Guy, P. A. Quantitative analysis of *Fusarium* mycotoxins in maize using accelerated solvent extraction before liquid chromatography/atmospheric pressure chemical ionization tandem mass spectrometry. *Food Addit. Contam.* **2004**, *21*, 678-692.
9. Klötzel, M.; Gutsche, B.; Lauber, U.; Humpf, H. U. Determination of 12 type A and B trichothecenes in cereals by liquid chromatography-electrospray ionization tandem mass spectrometry. *J. Agric. Food Chem.* **2005**, *53*, 8904-8910.

10. Biselli, S.; Hummert, C. Development of a multicomponent method for *Fusarium* toxins using LC-MS/MS and its application during a survey for the content of T-2 toxin and deoxynivalenol in various feed and food samples. *Food Addit. Contam.* **2005**, *22*, 752-760.
11. Berger, U.; Oehme, M.; Kuhn, F. Quantitative determination and structure elucidation of type A- and B- trichothecenes by HPLC/ion trap multiple mass spectrometry. *J. Agric. Food Chem.* **1999**, *47*, 4240-4245.
12. Sulyok, M.; Berthiller, F.; Krska, R.; Schuhmacher, R. Development and validation of a liquid chromatography/tandem mass spectrometric method for the determination of 39 mycotoxins in wheat and maize. *Rapid Commun. Mass Spectrom.* **2006**, *20*, 2649-2659.
13. Cech, N. B.; Enke, C. G. Practical implications of some recent studies in electrospray ionization fundamentals. *Mass Spectrom. Rev.* **2001**, *20*, 362-387.
14. King, R.; Bonfiglio, R.; Fernandez-Metzler, C.; Miller-Stein, C.; Olah, T. Mechanistic investigation of ionization suppression in electrospray ionization. *J. Am. Soc. Mass. Spectrom.* **2000**, *11*, 942-950.
15. Taylor, P. J. Matrix effects: The Achilles heel of quantitative high-performance liquid chromatography-electrospray-tandem mass spectrometry. *Clin. Biochem.* **2005**, *38*, 328-334.
16. Mallet, C. R.; Lu, Z.; Mazzeo, J. R. A study of ion suppression effects in electrospray ionization from mobile phase additives and solid-phase extracts. *Rapid Commun. Mass Spectrom.* **2004**, *18*, 49-58.
17. Matuszewski, B. K.; Constanzer, M. L.; Chavez-Eng, C. M. Strategies for the assessment of matrix effect in quantitative bioanalytical methods based on HPLC-MS/MS. *Anal. Chem.* **2003**, *75*, 3019-3030.
18. Pallaroni, L.; von Holst, C.; Eskilsson, C. S.; Bjoerklund, E. Microwave-assisted extraction of zearalenone from wheat and corn. *Anal. Bioanal. Chem.* **2002**, *374*, 161-166.
19. Zrostlikova, J.; Hajslova, J.; Poustka, J.; Begany, P. Alternative calibration approaches to compensate the effect of co-extracted matrix components in liquid chromatography-electrospray ionisation tandem mass spectrometry analysis of pesticide residues in plant materials. *J. Chromatogr. A* **2002**, *973*, 13-26.
20. Kraus, G. A. A facile synthesis of ochratoxin-A. *J. Org. Chem.* **1981**, *46*, 201-202.
21. Bennett, M.; Gill, G. B.; Pattenden, G.; Shuker, A. J.; Stapleton, A. Ylidenebutenolide mycotoxins. Concise syntheses of patulin and neopatulin from carbohydrate precursors. *J. Chem. Soc., Perkin Trans. 1* **1991**, 929-937.
22. Koch, K.; Podlech, J.; Pfeiffer, E.; Metzler, M. Total synthesis of alternariol. *J. Org. Chem.* **2005**, *70*, 3275-3276.
23. Cramer, B.; Nakamura, K.; Uemura, D.; Humpf, H.-U. Total synthesis of 13C-labeled ochratoxin A. *in preparation*.
24. Rychlik, M.; Schieberle, P. Synthesis of 13C-labeled patulin [4-hydroxy-4H-furo[3,2-c]pyran-2(6H)-one] to be used as internal standard in a stable isotope dilution assay. *J. Agric. Food Chem.* **1998**, *46*, 5163-5169.

25. Wei, R. D.; Chu, F. S. Synthesis of ochratoxins T(A) and T(C), analogs of ochratoxins A and C. *Experientia* **1974**, *30*, 174-175.
26. Lindenmeier, M.; Schieberle, P.; Rychlik, M. Quantification of ochratoxin A in foods by a stable isotope dilution assay using high-performance liquid chromatography-tandem mass spectrometry. *J. Chromatogr. A* **2004**, *1023*, 57-66.
27. Bretz, M.; Beyer, M.; Cramer, B.; Humpf, H. U. Stable isotope dilution analysis of the *Fusarium* mycotoxins deoxynivalenol and 3-acetyldeoxynivalenol. *Mol. Nutr. Food Res.* **2006**, *50*, 251-260.
28. Omura, K.; Swern, D. Oxidation of alcohols by "activated" dimethyl sulfoxide. A preparative, steric and mechanistic study. *Tetrahedron* **1978**, *34*, 1651-1659.
29. Bretz, M.; Beyer, M.; Cramer, B.; Humpf, H.-U. Synthesis of stable isotope labeled 3-acetyldeoxynivalenol. *Mol. Nutr. Food Res.* **2005**, *49*, 1151-1153.
30. Asam, S.; Rychlik, M. Synthesis of four carbon-13-labeled type A trichothecene mycotoxins and their application as internal standards in stable isotope dilution assays. *J. Agric. Food Chem.* **2006**, *54*, 6535-6546.
31. Tuck, K. L.; Tan, H.-W.; Hayball, P. J. A simple procedure for the deuteration of phenols. *J. Labelled Compd. Radiopharm.* **2000**, *43*, 817-823.
32. Miles, C. O.; Erasmuson, A. F.; Wilkins, A. L.; Towers, N. R.; Smith, B. L.; Garthwaite, I.; Scahill, B. G.; Hansen, R. P. Ovine metabolism of zearalenone to α-zearalanol (zeranol). *J. Agric. Food Chem.* **1996**, *44*, 3244-3250.
33. Hartmann, N.; Erbs, M.; Wettstein, F. E.; Schwarzenbach, R. P.; Bucheli, T. D. Quantification of estrogenic mycotoxins at the ng/L level in aqueous environmental samples using deuterated internal standards. *J. Chromatogr. A* **2007**, *1138*, 132-140.
34. Cramer, B.; Bretz, M.; Humpf, H. U. Stable isotope dilution analysis of the *Fusarium* mycotoxin zearalenone. *J. Agric. Food Chem.* **2007**, *55*, 8353-8358.
35. Cervino, C.; Asam, S.; Knopp, D.; Rychlik, M.; Niessner, R. Use of isotope-labeled aflatoxins for LC-MS/MS stable isotope dilution analysis of foods. *J. Agric. Food Chem.* **2008**, *56*, 1873-1879.
36. Blackwell, B. A.; Greenhalgh, R.; Bain, A. D. Carbon-13 and proton nuclear magnetic resonance spectral assignments of deoxynivalenol and other mycotoxins from *Fusarium graminearum*. *J. Agric. Food Chem.* **1984**, *32*, 1078-1083.
37. De Jesus, A. E.; Steyn, P. S.; Vleggaar, R.; Wessels, P. L. C-13 nuclear magnetic-resonance assignments and biosynthesis of the mycotoxin ochratoxin-A. *J. Chem. Soc., Perkin Trans. 1* **1980**, 52-54.
38. Wei, R.; Strong, F. M.; Smalley, E. B. Incorporation of chlorine-36 into ochratoxin-A. *Appl. Microbiol.* **1971**, *22*, 276-277.
39. ApSimon, J. W.; Blackwell, B. A.; Edwards, O. E.; Fruchier, A.; Miller, J. D.; Savard, M.; Young, J. C. The chemistry of fumonisins and related compounds. Fumonisins from *Fusarium moniliforme*: chemistry, structure and biosynthesis. *Pure Appl. Chem.* **1994**, *66*, 2315-2318.

40. Ferreira, N. P.; Pitout, M. J. The biosynthesis of ochratoxin. *J. S. Afr. Chem. Inst.* **1969**, *22*, 1-8.
41. Cramer, B.; Königs, M.; Humpf, H.-U. Identification and *in vitro* cytotoxicity of ochratoxin A degradation products formed during coffee roasting. *J. Agric. Food Chem.* **2008**, *56*, 5673-5681.
42. Haubl, G.; Berthiller, F.; Rechthaler, J.; Jaunecker, G.; Binder, E. M.; Krska, R.; Schuhmacher, R. Characterization and application of isotope-substituted (13C15)-deoxynivalenol (DON) as an internal standard for the determination of DON. *Food Addit. Contam.* **2006**, *23*, 1187-1193.
43. Freudenschuss, M.; Haeubl, G.; Krska, R.; Jaunecker, G. Production of highly isotopically labelled, secondary, microbial metabolic products, and corresponding metabolic products. Austrian patent 501,629, 2005.

Chapter 19

Molecularly Imprinted Polymer Nanomaterials for Mycotoxin Extraction

Jorn C.C. Yu[1], Edward P.C. Lai[2]

[1]College of Criminal Justice, Sam Houston State University, Huntsville, TX 77341
[2]Department of Chemistry, Carleton University, Ottawa, ON K1S 5B6, Canada

A new electrochemical imprinting technique, in which pyrrole is used as a functional monomer, has great potential to glue nanomaterials together. The molecularly imprinted polypyrrole (MIPPy), using ochratoxin A (OTA) as the template molecule, is electrochemically deposited onto carbon nanotubes (CNTs) or quantum dots (QDs) to make a solid phase of nanoscale for specific extraction of the mycotoxin. This MIPPy nanomaterial is simply structured within the 0.5-μm pores of a stainless steel frit, or inside a syringe needle, to make a novel micro solid phase extraction (μSPE) device. With the nanostructure of CNTs or QDs, the total number of specific binding sites of MIPPy-on-a-frit or MIPPy-in-a-needle is significantly enhanced for binding of OTA at sub-ppb levels even in the presence of abundant red wine matrix components.

Molecularly Imprinted Polymers

Over the last decade, molecularly imprinted polymers (MIPs) have gained more acceptance regarding their application as synthetic receptors for analytical techniques such as liquid chromatography, capillary electrochromatography, solid phase extraction, immunoassay, and chemical sensors (*1*). MIPs are solid materials that can be synthesized in the presence of a target compound (the template molecule) through polymerization. After removal of template, specific binding cavities with a shape and functional groups complementary to the target compound will be created within the polymer structure. MIPs are synthetic antibodies that mimic the function of natural antibodies which exhibit molecular recognition properties. Many studies concerning MIPs have been reported, such as a quartz-crystal microbalance (QCM) for amino acid sensing (*2*), cholesterol sensor (*3*), colorimetric tests for morphine (*4*), molecularly imprinted quantum dot photoluminescence for caffeine detection (*5*), voltammetric sensor for vanillymandelic acid (*6*), capacitive sensor for amino acid (*7*), just to name a few.

Preparation of Molecularly Imprinted Polymers

To make good MIPs, selection of suitable functional monomers, cross-linkers, porogen solvents, initiators and polymerization procedures require careful consideration. Due to the complexity of such factors as functional monomer-template complexation, solvent effect and cross-linking density that drive the imprinting process, the performance of any new MIP toward the target molecule is rather difficult to predict. The development of MIP for a specific application still relies on empirical optimization. The specificity of a MIP is governed by the factors mentioned above (at the preparation stage) and experimental conditions at the binding stage. Many vinyl monomers and different cross-linkers (polyfunctional acrylics) are available commercially at a low cost (http:// www.sigmaaldrich.com/catalog/search/TablePage/20203289). Free radical polymerization is usually the method of choice for preparing MIPs (*8*). To induce radical polymerization, an appropriate quantity of initiator is required. Macroscopic polymer networks have been most widely synthesized. These MIPs tend to be insoluble materials that provide rigidity and mechanical stability to all imprinted binding sites. Since most MIPs are prepared in the form of a macroporous monolith, the grinding and sieving process is required to yield proper particle sizes for analytical applications. This preparation of MIPs inherits some drawbacks, such as intensive labor, insufficient yield, and potential exposure to hazardous airborne particles when toxic molecules are imprinted.

Molecularly Imprinted Polymers for Solid Phase Extraction

In pharmaceutical, clinical, environmental and food chemistry applications, solid-phase extraction (SPE) has been widely employed for the isolation and pre-concentration of target analytes as well as the clean-up of sample matrices (*9*). However, matrix interference components can be co-extracted with the target analytes when conventional SPE sorbents, such as C18, ion-exchange and size-exclusion materials, are used. The co-extraction and co-elution problems produce undesirable results. For example, detector signals can be suppressed or enhanced when matrix components are co-eluted with the analytes. These effects are particularly problematic in the case of trace analysis.

Many affinity-based sorbent materials have been developed to selectively extract the target analytes, such as boronate, lectin, protein A or G, synthetic dyes, immobilized metal ions, aptamers, peptides, antibodies and molecularly imprinted polymers (MIPs) (*10*, *11*). Among these affinity techniques, both antibodies and MIPs can be used as single analyte- or group-selective sorbents. Particles of an MIP material can be either packed into a micro-column (*12*,*13*,*14*) for selective solid phase extraction (SPE), or glued to the surface of a solid-phase micro-extraction (SPME) fiber (*15*). New fabrication techniques continue to be developed for molecularly imprinted solid phase extraction (MISPE) and molecularly imprinted solid phase micro-extraction (MISPME). In an extraction, the MIP can be washed with an appropriate solvent that is capable of disrupting the non-specific interactions of compounds (other than the analyte) with the polymer matrix. Pulsed elution has been developed for the elimination of non-specific bindings by Mullett *et al* (*16*).

New Development of Molecularly Imprinted Polymers

The interfacing of a suitable transducer to MIPs is still growing and expected to have more significant impact in the field of biochemical sensors. A recent evaluation of MIP films for coulometry used an applied positive potential to induce adsorption of the target molecules (*17*), including morphine (*18*). The resultant sensors showed a high degree of sensitivity, selectivity and a broad linear range. Imprinting a polymer matrix with binding sites located at the surface has been shown to be advantageous as the sensor interface. The binding sites are more accessible, the mass transfer is faster, and the binding kinetics is faster. Such surface imprinted MIPs are potentially suited for the recognition of large-molecular-weight peptides and proteins. However, their preparation is less straightforward and requires specially adapted protocols (*19*).

Chen *et al* observed the enthalpic changes attributed to rebinding of template molecules to the MIP by microcalorimetric studies (*20*). The results suggest that a simple one-point interaction is insufficient to induce selectivity regardless of the strength of this interaction. Selectivity requires molecular recognition based on electrostatic interactions and secondary interactions such as hydrophobicity and macroscopic phase separation. Spivak *et al* have determined that shape selectivity is an important contributing factor to overall MIP selectivity (*21*). It was found that branched-structure templates produce higher selectivity MIPs than their straight-chain counterparts. Shape selectivity, as determined by steric exclusion or optimal fit, maximizes binding interactions.

Techniques for bead formation have been investigated to obtain more homogeneous MIP particles within a narrow size range (*22*, *23*). Precipitation polymerization or emulsion polymerization has gained attention due to their better control of particle sizes and morphologies (*24*). However, polymerization of this type usually requires either the use of special dispersing phases or complicated swelling processes (*25*). Molecular imprinting at the nano-scale to make MIP nanomaterials has gained more and more attention (*26*). MIP nanomaterials can be synthesized by precipitation or emulsion polymerization, to be readily suspended in aqueous media. Ye *et al* have prepared MIP nanomaterials with controllable size in the nano- to micro-meter range (*27*). Varying the composition of the cross-linking monomer allowed the particle size of the MIP to be altered in the range from 130 nm to 2.4 μm, whereas the favorable binding property of remained intact. Applying MIP onto gold nanoparticles used for sensing has been shown to be a promising strategy (*28*).

Electrochemical Preparation of Molecularly Imprinted Polymers

The possibility of inducing selectivity by the presence of the template during polymerization has been tested with several non-crosslinked electrogenerated polymers, such as polypyrrole (PPy), polyphenol, and poly(*o*-phenylenediamine) (*o*-PPD) (*29*). PPy has been used for many successful applications ranging from ion-exchange extraction to hydrophobic extraction, based on the interaction between the polymer and the target analytes (*30*). In addition, tuneable properties of the polymer, such as the oxidation/reduction equilibrium in conductive PPy, can be exploited to control adsorption and desorption (*31*).

Molecularly imprinted polypyrrole (MIPPy) has previously been shown to be a promising material for molecular recognition, as demonstrated by Bachas *et al* (*32*) and Nagaoka *et al* (*33*). Enantioselectivity towards amino acids by MIPPy was studied (*34*). A polypyrrole-based MIP was developed on platinum-black electrode for the detection of bovine leukemia virus glycoproteins (*35*). A caffeine-imprinted polypyrrole was successfully synthesized using galvanostatic electropolymerization directly onto the gold electrode of a quartz crystal microbalance (*36*). Pyrrole rings can contribute π-π and electrostatic interactions with the template molecule (*37*). It has been shown that the pyrrole ring can act as a π-base against the OH group of an alcohol (*38*). The driving force for the stabilization of the complexes is the strong hydrogen bonding interactions localized on the C-3 atom of the pyrrole ring (being the electrostatic interaction between this negatively charged center and the positively polarized H atom of the alcoholic OH group). The interactions between pyrrole and a wide range of potential target analytes make pyrrole an attractive functional monomer. The choice of pyrrole as a functional monomer is gaining more and more attention for the preparation of polypyrrole-based MIPs, as summerized in Table 1-1.

Table 1-1 Development and Applications of Pyrrole-based MIPs

Target Analyte	*Application*	*reference*
L-glutamate	Sensor (quartz crystal microbalance)	*34*
L-lactic acid	Enatioseparation	*39*
1-naphthalenesulfonate	Sensor (quartz crystal microbalance)	*42*
Taurocholate	Sensor (quartz crystal microbalance)	*40*
Bovine leukemia virus (BLV) glycoprotein gp51	Pulsed amperometric detection (PAD)	*43*
L-,D-Tyrosine	Sensor (Electrochemical)	*32*
Caffeine	Sensor design (piezoelectric quartz crystal)	*44*
Glutamic acid	MIPPy nanowires for bioanalytical applicatons	*35*
Ochratoxin A	SPR sensor and micro solid phase extraction	*41,42*

Molecularly Imprinted Polymer Nanomaterials for Solid Phase Micro Extraction

Using MIP for Mycotoxin Extraction

Ochratoxin A (OTA) is a secondary metabolite of the genera of Aspergillus and Penicillium. It has been found as a contaminant in food and feed commodities. OTA exhibits multiple toxicities in animals and mankind, including nephrotoxic, hepatotoxic, immunotoxic, teratogenic and carcinogenic effects, which represent a serious health risk to livestock and the general population. The widespread occurrence of OTA in cereals, maize, rice, beans, nuts, raisins and beverages (such as milk, coffee, grape juice and wine) has prompted health regulation authorities to define maximal tolerable daily intake levels (5.0 ng/kg body weight) (*43*).

Validated methods for OTA determination generally involved liquid–liquid extraction, immunoaffinity column (IAC) clean-up, and high-performance liquid chromatography (HPLC) with fluorescence detection (44). Unfortunately, the IACs are relatively costly and have a short shelf life. The development of a new SPE sorbent for selective OTA preconcentration prior to HPLC analysis is an important issue. An affordable SPE sorbent, compared to IAC columns, will make the screening of foodstuff more frequent by agri-food laboratories in both developed countries and poor countries, thus protecting human health. Conventional preparation of MIP bulks for OTA recognition is challenging, and the several reports found in the literature are all listed in Table 1-2.

Table 1-2 Functional Monomers Used for OTA MIP Preparation from the Literature

Functional Monomer	*Reference*
N-Phenylacrylamide	*13*
Methacrylic acid and acrylamide (dual functional monomer)	*45*
Methacrylic acid	*46*
Racemic, (S)- and (R)-quinuclidin-3-yl methacrylamide (Q-MAA)	*47, 48*

Molecularly imprinted polypyrrole (MIPPy) for OTA recognition has been prepared easily via electrochemical imprinting (*41*). Furthermore, it has been revealed by N1s X-ray photoelectron spectroscopy (XPS) studies that the hydrogen of N-H on the pyrrole ring undergoes protonation acidic conditions. Under alkaline conditions, the N-H will be deprotonated (*49*). The removal of OTA template molecules from MIPPy could be achieved by perturbing the hydrogen bonding between the OTA analyte and the MIPPy binding cavity using a high pH. Uptake of target analyte would be enhanced when the N-H on pyrrole ring is protonated. The particles of an MIP material can be either packed into a micro-column for solid phase extraction (SPE), or possibly glued to the surface of a solid-phase micro-extraction (SPME) fiber. Molecularly imprinting via the electrochemical route has been investigated to prepare polypyrrole-based MIPs (*42*). The resulting MIPPy offerred good selectivity for the target analyte. It has been demonstrated that molecular imprinting by electro-polymerization of pyrrole was an effective way of growing a thin MIP coating directly onto a variety of conductive sensor surfaces. . This new spin-off technology of MIP modified stainless steel frits for selectivity and sensitivity enhancement in food analyses has expanded the use of MIPPy. The molecular recognition properties of the MIPPy-modified frits were competitive to those reported for IAC columns and conventional MIP monoliths specific for OTA.

Molecularly Imprinted Polymer Nanomaterials

Applications of nanomaterials are of great interest in solid phase extraction, due to their high surface-to-volume ratio and their ease of miniaturization. Various nanomaterials, such as polymer nanoparticles, gold nanoparticles, silica nanoparticles, fullerenes and carbon nanotubes, are potentially useful as stationary phases (*50*). Carbon nanotubes (CNTs) represent a new kind of carbon material that has been widely recognized as the quintessential nano-material since their discovery in 1991 (*51*). It was found that CNTs possess a high electro-catalytic effect and a fast electron-transfer rate (*52*). In addition to the electrochemical applications of CNTs, their high conductance, tensile strength, chemical stability, ultra-small size and poor solubility make them potential candidates for development of novel analytical devices.

The desirable merging of CNTs (or MWNTs) and conducting properties of PPy in composite form presents new opportunities to produce superior materials for novel applications. To date, most reports on electrochemically-grown conducting polymer films on CNTs have focused on PPy since this conducting polymer offers a greater degree of electrochemical processing flexibility than most other commercially available conducting polymers (*53*). Nano sized CNTs has been considered as the surface-enhancing component for PPy film. A glucose sensor was developed, using PPy as glucose oxidase immobilization media, into a gold supported nanotube microelectrode array (*54*). PPy can also be synthesized on to CNTs using $(NH_4)_2S_2O_8$ as oxidant (*55*). Compared to PPy, better thermal stability and higher charge delocalization properties were observed in the resulting PPy-CNTs. Nano-porous composite films of multi-walled carbon nanotubes (MWNTs) and either PPy or poly(3-methylthiophene) were grown using an electrochemical polymerization technique in which the nanotubes and conducting polymer were deposited simultaneously (*56*). The electrochemical growth of MWNT-conducting polymer composites offers the ability to produce 3-dimensional nano-structured films that combine the redox pseudo-capacitive charge storage mechanism of conducting polymers with the high surface area and conductivity of MWNTs (*57*). Such films should be very useful in the fabrication of micro-SPE (μSPE) cartridges.

Good combination of new nano-materials with the MIP technology was demonstrated in a unique idea that packed sub-micron interstitial frit spaces. No time-consuming steps of grinding the bulk polymer and removing the template molecules were necessary. This new preparation scheme for MIP is promising for the SPE of other analytes, such as mycotoxins that are difficult to be imprinted with the traditional approach. The ultra-high surface-to-volume ratio and nano-structured surface morphology were utilized to enhance the binding capacity of these new MIPPy/CNT-modified stainless steel frits. The aim was to develop a micro-solid phase preconcentration (μSPP) device for the determination of trace OTA at low levels (0.01-0.1 ppb) in red wine. Due to the high complexity of wine matrices, a on-line μSPP device with minimal dead volume is required for accurate OTA determination by HPLC with fluorescence detection.

MIPPy/CNT-modified Stainless Steel Frits

When growing composite films of CNTs and MIPPy in a frit, a polymerization electrolyte was used with 0.1 M pyrrole monomer and 0.05% by weight (or 0.5 mg/mL) of suspended CNTs in acetonitrile (*58*). There was no need to oxidize the suspended nanotubes using a 3:1 ratio of sulfuric acid and nitric acid refluxed at 130°C for 1 h as a surface functionalization technique. The resultant MIPPy/CNT apparently showed good mechanical stability for μSPP applications. Figure 1 shows a typical MIPPy/CNT structure under a scanning electron microscope.

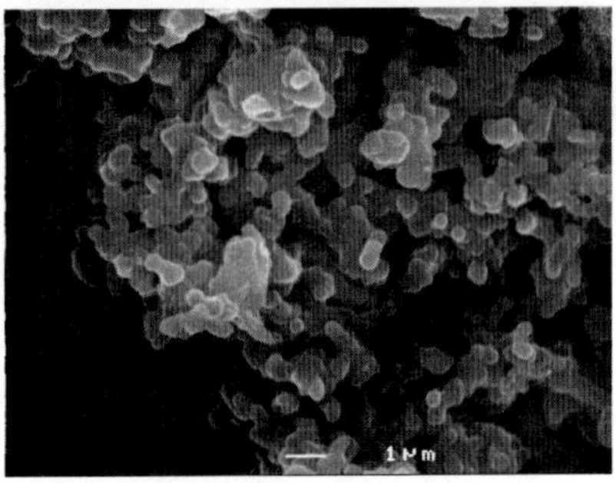

Figure 1. SEM image of of MIPPy/CNTs-on-a-frit.

High Binding Capacity via Surface Area Enhancement

The electrochemical preparation of MIPPy/CNT-modified stainless steel frits was characterized by not only a large absolute surface area, but also a large surface-to-volume ratio, allowing for more efficient μSPP of OTA. It was observed that an increase in the extraction rate and amount were obtained using this frit geometry. The higher surface-to-volume ratio resulted in fast extraction and desorption, all occurring within a very compact size of the frit with an active bed volume of 1.5 μL. A standard solution of OTA (50 ppb) was employed in frontal chromatography to determine the binding capacity of each and every frit. Frontal chromatography is a well-established measurement technique that takes both surface area and sub-surface porosity into account, yielding one single data (*59,60*). The capacity (or the selectivity) provided by MIPPy/CNT can be properly attributed to the CNTs (or the molecular imprinting). In addition, a set of 8 different MIPPy/CNT-modified stainless steel frits yielded a relative standard deviation of 3.6% in their binding capacity measurements (*61*). This indicated a good reproducibility of their electrochemical preparation.

A schematic diagram for MIPPy/CNT-μSPP-HPLC-FD is shown in Fig. 2. The MIPPy/CNT-modified stainless steel frit displayed a good combination of high binding capacity with excellent selectivity for μSPE of OTA in red wine. When the frit was calibrated using standard OTA solutions, it was shown to produce a linear response and constant recovery (nearly 100% binding). Experimental results indicated that the frit was a fast and quantifiable means of trapping OTA in red wine. Using a sample volume of 100 μL red wine for pre-concentration, it was possible to determine OTA (by fluorescence detection) down to a detection limit (S/N=3) of 0.08 ppb, or a quantification limit (S/N=10) of 0.27 ppb. Fast binding kinetics was explored for the cleanup of larger red wine samples (>100 μL) to increase the sensitivity of MIPPy/CNT-μSPP-HPLC analysis. Using a sample volume of 3 mL red wine for pre-concentration, it was possible to determine OTA down to a detection limit of 12 ppt, or a quantification limit of 41 ppt. Note that 3-mL of red wine loaded a total of 3 ng OTA onto the MIPPy/CNT-modified stainless steel frit, special investigations were pursued to determine the extent of the carryover in the extraction phase.

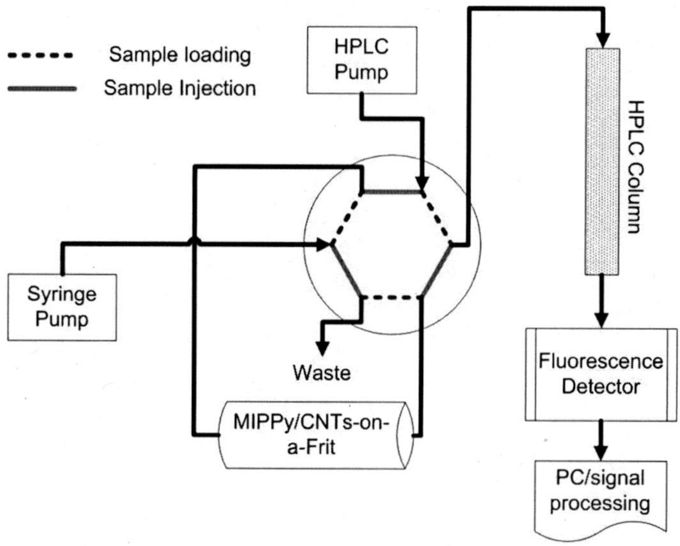

Figure 2. A schematic diagram for MIPPy/CNT-uSPP-HPLC-FD.

Fortunately, the third pulsed elution yielded an OTA peak area as little as 1% of the first pulsed elution. These observations signify that a second pulsed elution must always be made, quickly, to clean out any residual OTA from the MIPPy/CNT-modified stainless steel frit. After that, even without making a third pulsed elution, the extent of carryover would be equivalent to 1% of 1 ppb (= 0.01 ppb) OTA. This especially low analyte concentration is already at the detection limit of MIPPy/CNT-μSPP-HPLC for red wine analysis. The concentration-response profile for the comparison of MIPPy to IAC is shown in Fig. 3, which demonstrates the benefits of the MIPPy/CNT system for μSPP. Comparison of selectivity and linear dynamic ranges, as well as chromatograms showing neat solution, with concentration, and secondary pulsing of the regeneration of MIP can be found in separate reports (41, 42, 58, 61).

A comparison was made with red wine analysis using a MIPPy-modified stainless steel frit (without carbon nanotubes) or commercially available immunoaffinity columns (IACs) for sample cleanup. MIPPy/CNT-μSPP-HPLC and MIPPy-μSPP-HPLC offered the highest sensitivity for OTA determination in red wine. These two methods had almost identical sensitivities, based on the slopes of their regression equations. However, MIPPy/CNT-μSPP-HPLC offered a larger linear dynamic range (up to 15 ppb OTA at least) due to its enhanced binding capacity for OTA.

One of the main findings of our research is the improvement of the capacity of the MIPPy/CNT-modified stainless steel frit. Such capacity is really necessary to determine OTA in red wine samples. From our knowledge, OTA has been detected in red wine at 0.01-0.1 ppb, which means that using a sample volume of 3 mL, a MIP capacity of 0.3 ng would be necessary for the OTA alone. However, red wines may contain a great abundance of matrix components that can readily bind onto the MIPPy-CNT surfaces in a non-specific mechanism. We need a ten-, hundred- or thousand-times higher capacity for OTA analysis in order to accommodate varing matrix effects from different brands of red wines. It is always important to have the extra capacity that ensures quantitative binding of all OTA molecules in the wine sample.

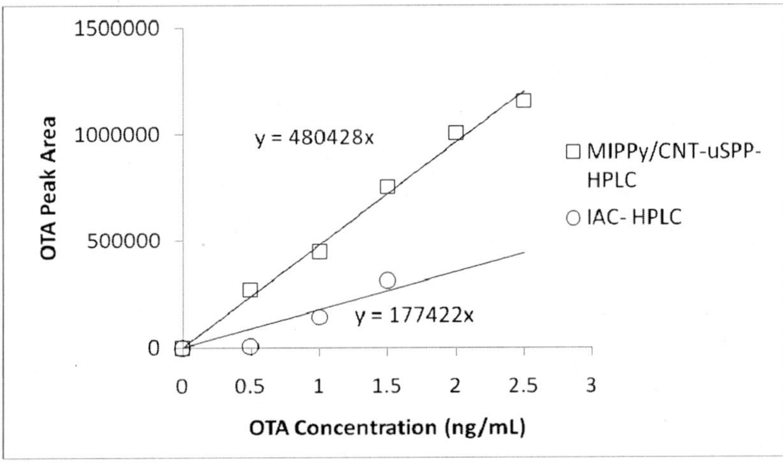

Figure 3. Calibration Curves for MIPPy/CNT-μSPP-HPLC and IAC-HPLC.

Stability of MIPPy/CNT-modified Stainless Steel Frits

The new MIPPy/CNT-modified stainless steel frit seemed to be stable and robust under ordinary lab operation conditions. From the observation, its binding capacity remained relatively constant over the first 15 days, but decreased to about 80% of the original value after one month (used more than 80 times). Good long-term stability resulted from the stability of cross-linking the carbon nanotubes with PPy through EGDMA, thereby encapsulating the CNTs permanently. When not in use, the frit was stored (either dry in air or wet in HPLC mobile phase) at room temperature. Leakage of OTA from the frit was monitored over storage periods ranging from hours to days. In good laboratory practice, a couple of pulsed elutions can readily clean out any OTA leakage prior to using the frit for MIPPy/CNT-μSPP-HPLC analysis of red wine samples (59).

MIPPy/CNT/Quantum Dots in a Needle

A new μSPP device has been developed in which the MIPPy and CNTs were simultaneously deposited along the inner wall of a stainless steel syringe needle to form a three-dimensional network (*62*). The resulting needle exhibited a significant binding capacity for OTA and great effectiveness to remove the interfering components of complex red wine matrices. The MIPPy/CNT-modified needle as a μSPP protocol afforded an enriched sample that is suitable for sensitive HPLC-FD determination of OTA. Compared to the popular Immunoaffinity columns for OTA analysis, the MIPPy/CNT-modified needle is much more robust and cost-effective for repeated use. When compared to conventional C18 SPE columns, the MIPPy/CNT-modified needle offers a lower detection limit (0.04 ppb) and uses a smaller volume of wine sample (0.5 mL). This new preparation technique can be applied to needles in various lengths and shapes. Therefore it can readily handle small sample volumes (several μL) as well as large volumes (several mL), even for autosamplers that are coupled online to HPLC and liquid chromatography–mass spectrometry (LC–MS). Preliminary results indicate that these MIPPy/CNT-modified needles, under the operation of a HPLC autosampler, can afford up to 80–98% recovery of OTA from a 1-ppb standard solution due to their larger total surface areas.

The MIPPy/CNT-modified stainless steel needle seemed to be stable and robust under ordinary lab operation conditions. Our observation showed that its extraction efficiency remained fairly constant over 2 months (when used more than 60 times). MIP combined with nano-technology can lead to the discovery of novel extraction procedures and development of new sample preparation devices. A 22-gauge needle can be modified to introduce a small sample on the microfluidic chip for capillary electrophoresis analysis (*63*). Currently, our group is exploring a new possibility in using quantum dots and carbon nanotubes to build three-dimensional structures inside a stainless steel needle (*64*). New results are verifying that those nano-scale structures can further increase the total surface area of MIPPy film, thus dramatically enhancing the extraction efficiency (or %recovery). The MIPPy/CNTs/QDs-modified needle as a μSPP device affords an enriched sample that is suitable for sensitive determination of OTA by HPLC-FD. This new μSPP procedure can readily handle larger volumes (up to 10 mL) when an autosampler is added to the operation. Automation of the μSPP-HPLC-FD procedure allows for rapid determination of ultratrace OTA. It is everyone's expectation that both the acceptance and the range of potential applications for a new technique with automation can be significantly promoted. In today's modern economy where food products of all sorts are produced and shipped around the world, the safety of these products is of vast importance to the health of everyone. Future work in our labs will be focused on the new application of these μSPP devices in the food industry, clinical tests, forensic detection, and bioterrorism defense.

Conclusions

The molecular imprinting technique has been successfully applied to modify nanomaterials, such as carbon nanotubes and quantum dots. The MIPPy/CNT-modified stainless steel frit is an inexpensive device to fabricate for on-line implementation of μSPP-HPLC. A composite film of MIPPy and CNTs was grown using electrochemical polymerization, in which the conducting polymer and nanotubes were simultaneously deposited. These functionalized nanomaterials can be used for molecularly imprinted micro solid phase extraction (MIMSPE) to selectively concentrate trace amounts of OTA, which is a mycotoxin commonly found in red wine. The resultant frit exhibited a significant pre-concentration capacity to better handle complex red wine matrices. As PPy is an inert extraction phase that is biocompatible, MIPPy/CNT-modified stainless steel frits can potentially be used for pharmacokinetic, clinical, forensic, food safety, and environmental health analyses. Their applications will see better quality and more cost-effective analyses in these industries. MIMSPE in-a-needle is a new MIP technology that can potentially be coupled with electro spray ionization (ESI) to make a rapid MIMSPE-ESI-mass spectroscopy method for the determination of OTA and identification of other mycotoxins.

Acknowledgements

Financial support of the Natural Sciences and Engineering Research Council (NSERC) Canada is gratefully acknowledged.

References

1. Haupt, K.; Mosbach, K. Molecularly imprinted polymers and their use in biomimetic sensors. *Chem. Rev.* **2000**, *100*, 2495-2504.
2. Reddy, P.S.; Kobayashi, T.; Abe, M.; Fujii, N. Molecular imprinted Nylon-6 as a recognition material of amino acids. *Eur. Polym. J.* **2002**, *38*, 521-529.
3. Piletsky, S.A.; Piletskaya, E.V.; Sergeyeva, T.A.; Panasyuk, T.L.; El'skaya, A.V. Molecularly imprinted self-assembled films with specificity to cholesterol. *Sens. Actuators. B* **1999**, *60*, 216-220.
4. Yeh, W.M.; Ho, K.C. Amperometric morphine sensing using a molecularly imprinted polymer-modified electrode. *Anal. Chim. Acta* **2005**, *542*,76-82.
5. Lin, C.I.; Joseph, A.K.; Chang, C.K.; Lee, Y.D. Molecularly imprinted polymeric film on semiconductor nanoparticles Analyte detection by quantum dot photoluminescence. *J. Chromatogr. A* **2004**, *1027*, 259-262.
6. Blanco-Lopez, M.C.; Lobo-Castanon, M.J.; Miranda-Orrdieres, A.J.; Tunon-Blanco, P. Voltammetric sensor for vanillylmandelic acid based on molecularly imprinted polymer-modified electrodes. *Biosens. Bioelectron.* **2003**, *18*, 353-362.
7. Panasyuk, T.L.; Mirsky, V.M.; Piletsky, S.A.; Wolfbeis, O.S. Electropolymerized molecularly imprinted polymers as receptor layers in capacitive chemical sensors. *Anal. Chem.* **1999**, *71*, 4609-4613.
8. Cormack, P.A.G.; Elorza, A.Z. Molecularly imprinted polymers: synthesis and characterization. *J. Chromatogr. B* **2004**, *804*, 173-182.
9. Haginaka J. Selectivity of affinity media in solid-phase extraction of analytes. *Trends Anal. Chem.* **2005**, *24*, 407-415.

10. Luzi, E.; Minunni, M.; Tombelli, S.; Mascini, M. New trends in affinity sensing - aptamers for ligand binding. *Trends Anal. Chem.* **2003**, *22*, 810-818.
11. Naffin, J.L.; Han, Y.; Olivos, H.J.; Reddy, M.M.; Sun, T.; Kodadek, T. Immobilized peptides as high-affinity capture agents for self-associating proteins. *Chem. Biol.* **2003**, *10*, 251-259.
12. Feng, S.Y.; Lai, E.P.C.; Dabek-Zlotorzynska, E.; Sadeghi, S. Molecularly imprinted solid-phase extraction for the screening of antihyperglycemic biguanides. *J. Chromatogr. A* **2004**, *1027*, 155-160.
13. Zhou, S.N.; Lai, E.P.C.; Miller, J.D. Analysis of wheat extracts for ochratoxin A by molecularly imprinted solid-phase extraction and pulsed elution. *Anal. Bioanal. Chem.* **2004**, *378*, 1903-1906.
14. Wu, S.G.; Lai, E.P.C.; Mayer P.M. Molecularly imprinted solid phase extraction - pulsed elution - mass spectrometry for determination of cephalexin and a-aminocephalosporin antibiotics in human serum. *J. Pharm. Biomed. Anal.* **2004**, *36*, 483-490.
15. Liu, M.; Zeng, Z.; Fang, H. Preparation and application of the sol-gel-derived acrylate/silicone co-polymer coatings for headspace solid-phase microextraction of 2-chloroethyl ethyl sulfide in soil. *J. Chromatogr. A* **2005**, *1076*, 16-26.
16. Mullett, W.M.; Lai, E.P.C. Rapid determination of theophylline in serum by selective extraction using a heated molecularly imprinted polymer micro-column with differential pulsed elution. *J. Pharm. Biomed. Anal.* **1999**, *21*, 835-843.
17. Liang, H.J.; Ling, T.R.; Rick, J.F.; Chou, T.C. Molecularly imprinted electrochemical sensor able to enantroselectivly recognize D and L-tyrosine *Anal. Chim. Acta* **2005**, *542*, 83-89.
18. Ho, K.C.; Yeh, W.M.; Tung, T.S.; Li, J.Y. Amperometric detection of morphine based on poly(3,4-ethylenedioxythiophene) immobilized molecularly imprinted polymer particles prepared by precipitation polymerization. *Anal. Chim. Acta* **2005**, *542*, 90-96.
19. Pérez, N.; Whitcombe, M.J.; Vulfson, E.N. Surface imprinting of cholesterol on submicrometer core-shell emulsion particles. *Macromolec.* **2001**, *34*, 830-836.
20. Chen, W.Y.; Chen, C.S.; Lin, F.Y. Molecular recognition in imprinted polymers: thermodynamic investigation of analyte binding using microcalorimetry. *J. Chromatogr. A* **2001**, *923*, 1-6.
21. Spivak, D.A.; Simon, R.; Campbell, J. Evidence for shape selectivity in non-covalently imprinted polymers. *Anal. Chim. Acta.* **2004**, *504*, 23-30.
22. Turiel, E.; Martin-Esteban, A. Molecularly imprinted polymers: towards highly selective stationary phases in liquid chromatography and capillary electrophoresis. *Anal. Bioanal. Chem.* **2004**, *378*, 1876-1886.
23. Liu, Z.H.; Xu, Y.L.; Yan, C.; Gao, R.U. Mechanism of molecular recognition on molecular imprinted monolith by capillary electrochromatography. *J. Chromatogr. A* **2005**, *1087*, 20-28.
24. Ye, L.; Mosbach, K. Molecularly imprinted microspheres as antibody binding mimics. *React. Funct. Polym.* **2001**, *48*, 149-157.

25. Mayes, A.G.; Mosbach, K. Molecularly imprinted polymer beads: suspension polymerization using a liquid perfluorocarbon as the dispersing phase. *Anal. Chem.* **1996**, *68*, 3769-3774.
26. Yang, H.H.; Zhang, S.Q.; Tan, F.; Zhuang, Z.X.; Wang, X.R. Surface molecularly imprinted nanowires for biorecognition. *J. Am. Chem. Soc.* **2005**, *127*, 1378-1379.
27. Yoshimatsu, K.; Reimhult, K.; Krozer, A.; Mosbacha, K.; Sode, K.; Ye, L. Uniform molecularlyimprinted microspheres and nanoparticles prepared by precipitation polymerization: The control of particle size suitablefor different analytical applications. *Anal. Chim. Acta* **2007**, *584*, 112–121.
28. Matsui, J.; Akamatsu, K.; Nishiguchi, S.; Miyoshi, D.; Nawafune, H.; Tamaki, K.; Sugimoto, N. Composite of Au nanoparticles and molecularly imprinted polymer as a sensing material. *Anal. Chem.* **2004**, *76*, 1310-1315.
29. Blanco-Lopez, M.C.; Lobo-Castanon, M.J.; Miranda-Ordieres, A.J.; Tunon-Blanco, P. Electrochemical sensors based on molecularly imprinted polymers. *Trends Anal. Chem.* **2004**, *23-1*, 36-48.
30. Saito, Y.; Jinno, K. Miniaturized sample preparation combined with liquid phase separations. *J. Chromatogr. A* **2003**, *1000*, 53-67.
31. Kim, K.M.; Hur, J.W.; Jung, S.I.; Kang, A.S. Electrochemical characteristics of activated carbon/Ppy electrode combined with P(VdF-co-HFP)/PVP for EDLC. *Electrochim. Acta* **2004**, *50*, 863-872.
32. Ersoz, A.; Gavalas, V.G.; Bachas, L.G. Potentiometric behavior of electrodes based on overoxidized polypyrrole films. *Anal. Bioanal. Chem.* **2002**, *372*, 786-790.
33. Shigi, H.; Kishimoto, M.; Yakabe, H.; Deore, B.; Nagaoka, T. Highly selective molecularly imprinted overoxidized polypyrrole colloids: One-step preparation technique. *Anal. Sci.* **2002**, *18*, 41-44.
34. Deore, B.; Chen, Z.; Nagaoka, T. Potential-induced enantioselective uptake of amino acid into molecularly imprinted overoxidized polypyrrole. *Anal. Chem.* **2000**, *72*, 3989-3994.
35. Ramanaviciene, A.; Ramanavicius, A. Molecularly imprinted polypyrrole-based synthetic receptor for direct detection of bovine leukemia virus glycoproteins. *Biosens. Bioel.* **2004**, *20*, 1076-1082.
36. Ebarvia, B.S.; Cabanilla, S.; Sevilla, F. Biomimetic properties and surface studies of a piezoelectric caffeine sensor based on electro synthesized polypyrrole. *Talanta* **2005**, *66*, 145-152.
37. Stefov, V.; Pejov L.; Optrajanov B. Experimental and quantum chemical study of pyrrole self-association through N–H…π hydrogen bonding. *J. Mol. Struct.* **2003**, *649*, 231-234.
38. Muñoz, M.A.; Galán, M.; Gómez, L.; Carmona, C.; Guardado, P.; Balón, M. The pyrrole ring as hydrogen-bonding π-donor base: an experimental and theoretical study of the interactions of N-methylpyrrole with alcohols. *Chem. Phys.* **2003**, *290*, 69-77.
39. Okuno, H.; Kitano, T.; Yakabe, H.; Kishimoto, M.; Deore, B.A.; Siigi, H.; Nagaoka, T. Characterization of overoxidized polypyrrole colloids imprinted with L-lactate and their application to enantioseparation of amino acids. *Anal. Chem.* **2002**, *74*, 4184-4190.

40. Shiigi, H.; Kijima, D.; Ikenaga, Y.; Hori, K.; Fukazawa, S.; Nagaoka,T. Molecular recognition for bile acids using a molecularly imprinted overoxidized polypyrrole film. *J. Electrochem. Soc.* **2005**, *152*, 129-134
41. Yu, J.C.C.; Lai, E.P.C. Interaction of ochratoxin A with molecularly imprinted polypyrrole film on surface plasmon resonance sensor. *React. Funct. Polym.* **2005**, *63*, 171-176.
42. Yu, J.C.C.; Krushkova, S.; Lai, E.P.C.; Dabek-Zlotorzynska, E. Molecularly imprinted polypyrrole modified stainless steel frits for selective solid phase preconcentration of ochratoxin A. *Anal. Bioanal. Chem.* **2005**, *382*, 1534-1540.
43. Soufleros, E.H.; Tricard, C.; Bouloumpasi. E.C. Occurrence of ochratoxin A in Greek wines. *J. Sci. Food Agric.* **2003**, *83*, 173-179.
44. Visconti, A.; Pascale, M.; Centonze, G. Determination of ochratoxin A in wine by means of immunoaffinity column clean-up and high-performance liquid chromatography. *J. Chromatogr. A* **1999**, *864*, 89-101.
45. Turner, N.W.; Piletska, E.V.; Karim, K.; Whitcombe, M.; Malecha, M.; Magan, N.; Baggiani, C.; Piletsky, S.A. Effect of the solvent on recognition properties of molecularly imprinted polymer specific for ochratoxin A. *Biosens. Bioelec.* **2004**, *20*, 1060-1067.
46. Baggiani, C.; Giraudi, G.; Vanni, A. A molecular imprinted polymer with recognition properties towards the carcinogenic mycotoxin ochratoxin A. *Bioseparation* **2001**, *10*, 389-394.
47. Jodlbauer, J.; Maier, N.; Lindner, W. Towards ochratoxin A selective molecularly imprinted polymers for solid-phase extraction. *J. Chromatogr. A* **2002**, 945, 45-63.
48. Maier N.; Buttinger G.; Welhartizki, S.; Gavioli, E.; Lindner, W. Molecularly imprinted polymer assisted sample clean-up of ochratoxin A from red wine: merits and limitations. *J. Chromatogr. B.* **2004** *804*, 103-111.
49. Zhang, X.; Kang, E. T.; Neoh, K.G.; Tan, K.L.; Kim, D.Y.; Kim, C.Y. Surface studies of pristine and surface-modified polypyrrole films. *J. Appl. Polym. Sci.* **1996**, *60*, 625-636.
50. Nilsson, C.; Birnbaum, S.; Nilsson, S. Use of nanoparticles in capillary and microchip electrochromatography. *J.Chromatogr. A* **2007**, *1168*, 212-224.
51. Iijima, S. Helical microtubules of graphitic carbon. *Nature* **1991**, *354*, 56-58.
52. Gong, K.P.; Zhang, M.N.; Yan, Y.M.; Su, L.; Mao, L.Q.; Xiong, S.X.; Chen, Y. Sol-Gel-Derived ceramic-carbon nanotube nanocomposite electrodes: tunable electrode dimension and potential electrochemical applications. *Anal. Chem.* **2004**, *76*, 6500-6505.
53. Chen, J.H.; Huang, Z.P.; Wang, D.Z.; Yang, S.X.; Li, W.Z.; Wen, J.G.; Ren, Z.F. Electrochemical synthesis of polypyrrole films over each of well-aligned carbon nanotubes. *Synthetic. Met.* **2002**, *125*, 289-294.
54. Gao, M.; Dai, L.; Wallace, G.G. Biosensors based on aligned carbon nanotubes coated with inherently conducting polymers. *Electroanal.* **2003**, *13*, 1089-1094.

55. Fan, J.; Wan, M.; Zhu, D.; Chang, B.; Pan, Z.; Xie, S. Synthesis and properties of carbon nanotube-polypyrrole composites. *Synth. Met.* **1999**, *102*, 1266-1267.
56. Hughes, M.; Chen, G.Z.; Shaffer, M.S.P.; Fray, D.J.; Windle, A.H. Controlling the nanostructure of electrochemically grown nanoporous composites of carbon nanotubes and conducting polymers. *Compos. Sci. Technol.* **2004**, *64*, 2325–2331.
57. Gooding, J.J. Nanostructuring electrodes with carbon nanotubes: A review on electrochemistry and applications for sensing. *Electrochim. Acta* **2005**, *50*, 3049-3060.
58. Yu, J.C.C.; Lai, E.P.C. Molecularly imprinted polypyrrole modified carbon nanotubes on stainless steel frit for selective micro solid phase pre-concentration of ochratoxin A. *React. Funct. Polym.* **2006**, *66-7*, 702-711.
59. Tejeda-Mansir, A.; Montesinos, R.M.; Guzmán, R. Mathematical analysis of frontal affinity chromatography in particle and membrane configurations. *J. Biochem. Biophys. Methods* **2001**, *49*, 1-28.
60. Gutenwik, J.; Nilsson, B.; Axelsson, A. Effect of hindered diffusion on the adsorption of proteins in agarose gel using a pore model. *J. Chromatogr. A* **2004**, *1048*, 161-172.
61. Yu, J.C.C.; Lai, E.P.C. Determination of ochratoxin A in red wines by multiple pulsed elutions from molecularly imprinted polypyrrole. *Food Chem.* **2007**, *105*, 301-310
62. Wei, Y.; Qiu, L.; Yu, J.C.C.; Lai, E.P.C. Molecularly imprinted solid phase extraction in a syringe needle packed with polypyrrole-encapsulated carbon nanotubes for determination of ochratoxin a in red wine. *Food Sci. Technol. Int.* **2007**, *13-5*, 375-380.
63. Footz, T.; Wunsam, S.; Kulak, S.; Crabtree, H.J.; Glerum, D.M.; Backhouse, C.J. Sample purification on a microfluidic device. *Electrophoresis* **2001**, *22*, 3868 – 3875.
64. Wei,Y.; Qiu, L.; Owen, C.; Lai, E.P.C. Encapsulation of quantum dots and carbon nanotubes with polypyrrole in a syringe needle for automated molecularly imprinted solid phase pre-concentration of ochratoxin A in red wine analysis. *Sens. Instrum. Food Qual. Saf.* **2007**, *3*, 133-141.

Chapter 20

A Closer Look at Cyclodextrins in Mycotoxin Analysis

Michael Appell and Chris M. Maragos

Mycotoxin Research Unit, USDA-ARS, National Center for Agricultural Utilization Research, Peoria, IL 61604

Cyclodextrins are a class of cyclic oligosaccharides with a variety of applications, including use as recognition components for low molecular weight molecules in methods of detection. These cycloamyloses are of special interest in mycotoxin analysis for enhancing spectroscopic properties of several mycotoxins under aqueous conditions, including aflatoxins, zearalenone, ochratoxin A, and through chemical derivatization, T-2 toxin. Spectroscopic studies and applications of cyclodextrins are frequently associated with inclusion complex formation. Theoretical modeling studies suggest cyclodextrin-mycotoxin interactions are influenced by the size and nature of the mycotoxin and favorable binding energies. The results provide a better understanding of the effectiveness and limitations of incorporating cyclodextrins into mycotoxin analysis.

Introduction

A broad range of approaches are applied to prevent and control natural mycotoxin contamination of agricultural commodities. The effectiveness of these approaches is determined by accurate analysis of toxin levels and validation of methods of detection (*1-4*). Concern over mycotoxin contamination levels has spurred development of new detection methods to improve accuracy, sensitivity, time, and selectivity. Several materials capable of improving mycotoxin analysis are based on selective recognition of analytes, and these materials have been successful in a variety of detection formats, including ELISAs, selective affinity columns (immunoaffinity), and test strips. In addition, inherent properties of the toxins, such as fluorescence, permit improvements in detection levels and selectivity. Less selective materials, including cyclodextrins, offer a general means to improve detection. The use of cyclodextrins in mycotoxin analysis has focused on spectroscopic phenomena, as well as size exclusion (*5-7*). This manuscript reviews recent developments in the use of cyclodextrins for mycotoxin analysis, and examines the parameters important for application of cyclodextrins in mycotoxin detection.

Cyclodextrins: General Overview

Cyclodextrins are cyclooligosaccharides consisting of six to eight glycopyranose units connected through α-(1,4) linkages (*8*). These cycloamyloses are synthesized enzymatically from degraded starch (*7, 8*). Cyclodextrins are of particular interest as "generic" receptors for the ability to form guest-host complexes with their cavities. Similarities between cyclodextrins with six (α-), seven (β-), and eight (γ-) glucopyranose subunits include a $^4C_1$ conformation, a face possessing all 6-hydroxymethyl substituents, and a large rim with secondary hydroxyl groups in the 2- and 3- positions. Noteworthy differences between α-, β-, and γ- cyclodextrins are solubility and cavity volume, with the larger cyclodextrins possessing larger cavities. It is of interest that higher concentrations of cyclodextrins can form more complex structures (*9*).

Properties of cyclodextrins can be tailored through chemical derivatization, however, this approach is complicated by similar reactivity of dozens of hydroxyl groups. Despite the lack of diverse functionality, moderate selectivity can be achieved by exploiting the difference in chemical reactivity attributed to the three positions of the hydroxyls in the glucopyranose subunit (*10*). The 6-hydroxymethyl substituents can be selectively modified by reactions limited to

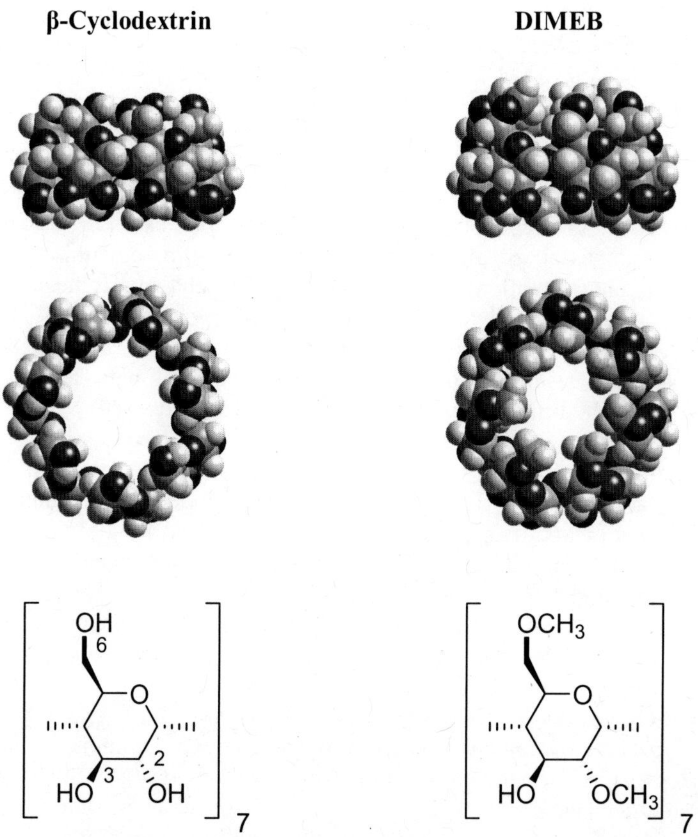

Figure 1. Structure of β-cyclodextrin and 2,6-dimethyl-β-cyclodextrin (DIMEB).

primary hydroxyls. Selectivity between the 2-hydroxyls and 3-hydroxyls can be achieved by the increased acidity of the 2-hydroxyls and the influence of the cooperative hydrogen bond network (*10-12*). Mixtures of cyclodextrin derivatives have found utility in analytical methods. One of the most popular derivatives is heptakis(2,6-di-*O*-methyl)-β-cylodextrin, DIMEB, which occurs as a mixture of cyclodextrins substituted with methoxy groups in various 2- and 6- positions (see Figure 1). The methylation increases solubility compared to β-cyclodextrin, and is capable of greater enhancement of fluorescence intensity for several mycotoxins under certain conditions (*5*, *13*).

Recent Applications of Cyclodextrins in Mycotoxin Analysis

Cyclodextrin-induced fluorescence enhancement of mycotoxins provides a general mechanism for selective detection by utilizing the hydrophobic cavities of cyclodextrins to overcome the fluorescence quenching properties of water (*14*). Applications of cyclodextrins in mycotoxin detection include recognition materials in chromatographic separations, additives to growth media to enhance qualitative assessment of toxin presence, and modulators of quantitative analysis of mycotoxins. These types of mycotoxin-related uses have been reviewed (*5-7*); however, several recent developments are noteworthy.

Aflatoxin production by *Aspergillus* species can be monitored in growth media within the timeframe of three to ten days by cyclodextrin-assisted visual detection (*15, 16*). A fiber optic room temperature phosphorescence method has been developed for aflatoxin detection in growth media using β-cyclodextrin and the surfactant sodium deoxycholate (*17*). The fiber-optic based method reduced the time for detection of aflatoxins in culture medium to within 36 hours. Aflatoxin M_1 contamination of Slovenian milk and cheese has been detected using culture media supplemented with methyl-β-cyclodextrin (*18, 19*). In addition, a method has been developed to detect *Aspergillus* species from Iranian pistachios using PCR-based techniques coupled with methylated-β-cyclodextrin-assisted fluorescence detection in growth media (*20*).

Aflatoxin B_1 in wheat can be detected by use of second order standard addition methods on the cyclodextrin-assisted fluorescence spectra with minimal sample preparation (*21*). The effectiveness of cyclodextrins to improve the detection limits of aflatoxin M_1 in milk has been studied by a rapid screening method using a photomultiplier tube detector (*22*). The instrument is capable of detecting aflatoxin M_1 in milk at 50 ppt without cyclodextrins, and inclusion of cyclodextrins enhanced the detection levels by ~50%. Fundamental studies probing the induced fluorescence enhancement properties of several surfactants, β-cyclodextrin, and calixresorcinarenes have been carried out on aflatoxins B_1, B_2, G_1, and G_2 (*23*).

The inclusion phenomenon of mycotoxin-cyclodextrin complexes has been investigated using computational modeling methods (*24*). Study of the fluorescence enhancement of aflatoxin B_1 by spectroscopy and molecular modeling identified a 1:1 relationship for the guest:host complex (*25*). Interactions of β- and γ- cyclodextrins with aflatoxin B_1, zearalenone, and ochratoxin A were investigated by an analysis developed for hydropathic interactions of proteins and protein-ligand interactions (*26, 27*). Quenching effects of water on aflatoxin B_1 fluorescence were studied using time dependent density functional methods (*28*).

It is clear that the binding interactions of cyclodextrins are more complicated than simple stoichiometric relationships and detection-enhancing phenomena are influenced by the structures of toxin and cyclodextrin. A closer look at the interactions of zearalenone using both fluorescence spectroscopy and molecular modeling can shed light on the merits and limitations of the use of cyclodextrins in mycotoxin analysis. Herein, we report fluorescence spectroscopic studies of the interaction of zearalenone with popular cyclodextrins β-cyclodextrin and DIMEB. Furthermore, to gain insight into the interactions of cyclodextrins and mycotoxins, semi-empirical studies were carried out on complexes of cyclodextrins with zearalenone, zearalenone analog resorcylic acid, aflatoxin B_1, ochratoxin A, and T-2 toxin. The structures for zearalenone **1** and aflatoxin B_1 **2** are shown in Figure 2. The resorcylic acid **3** and coumarin **4** moieties are associated with the fluorescence of these molecules. Fluorescence of ochratoxin A **5** (see Figure 3) is related to the dihydroisocoumarin **7** moiety, and detection of T-2 toxin **6** by fluorescence is possible through chemical derivatization with pyrene 1-carbonyl cyanide, **8**.

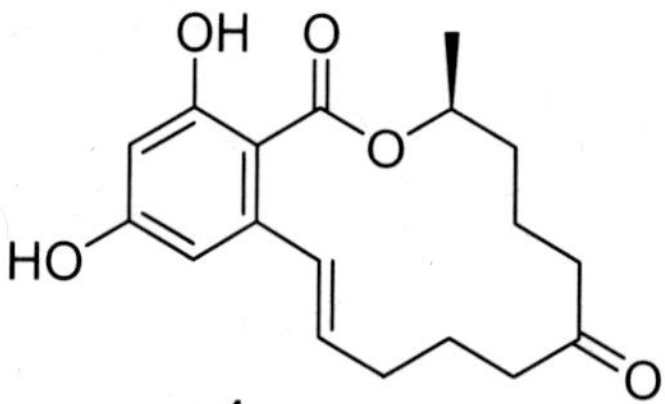

1

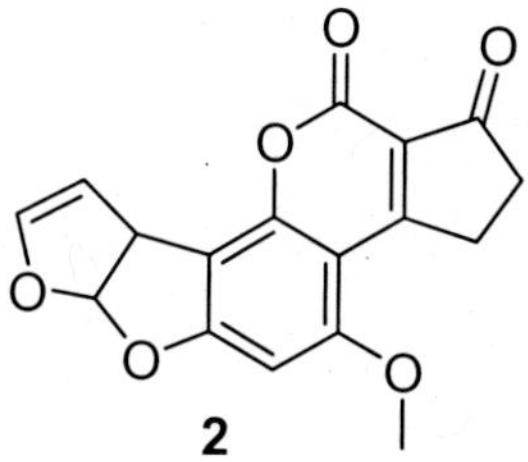

2

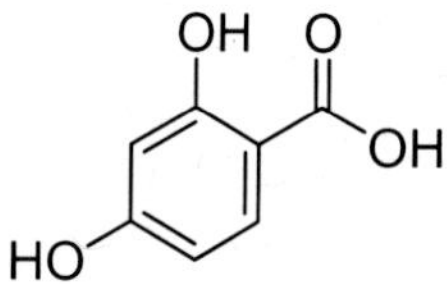

3

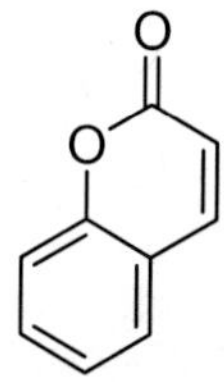

4

Figure 2. Structures for zearalenone (1), aflatoxin B_1 (2), resorcylic acid (3), and coumarin (4).

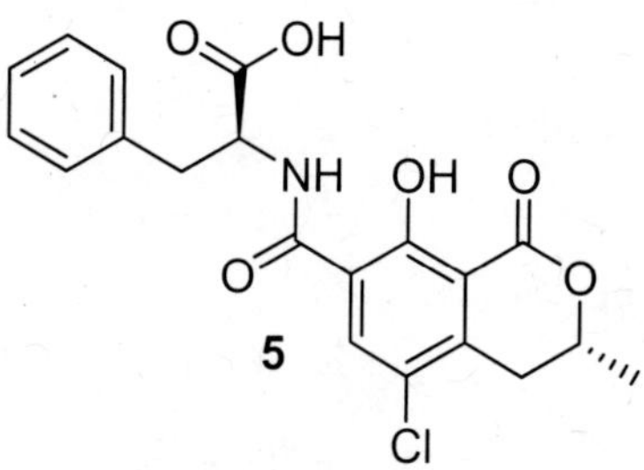

5

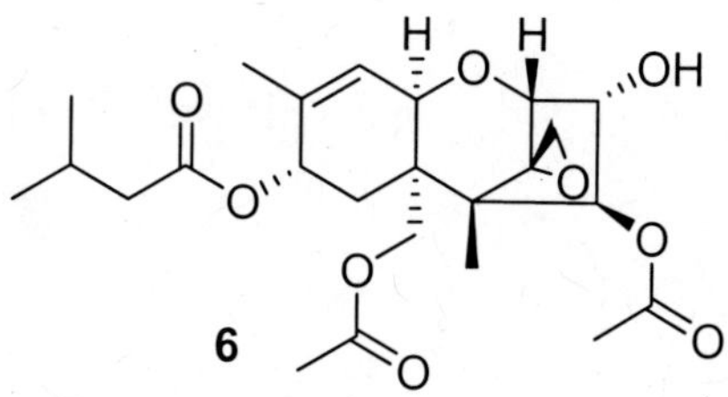

6

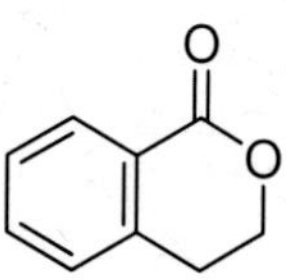

7

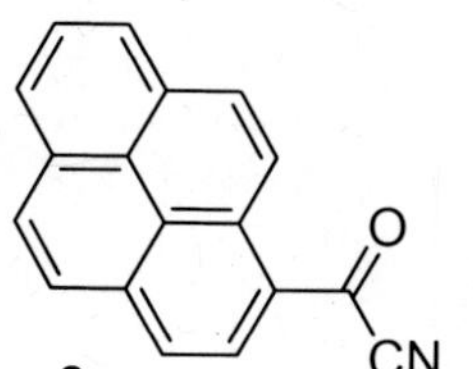

8

Figure 3. Structures for ochratoxin A (5), T-2 toxin (6), dihydroisocoumarin (7), and pyrene 1-carbonyl cyanide (8).

Materials and Methods

All chemicals and reagents were purchased from Sigma-Aldrich Company (St. Louis, MO, U.S.A.). Heptakis(2,6-di-*O*-methyl)-β-cylodextrin (DIMEB) was used as the mixture provided. Deionized water was used for the preparation of all reagents (Nanopure II, Sybron/Barnstead). All solvents were HPLC grade.

Fluorescence Spectroscopy Fluorescence spectra were recorded on a Varian Cary Eclipse (Palo Alto, CA, U.S.A.) instrument in a 10 x 10 mm quartz cell. The zearalenone fluorescence intensity was measured at 460 nm after excitation at 270 nm. All experiments were recorded using a 5 nm slit width. Solutions were incubated for five minutes prior to analysis. Experiments were carried out at room temperature 22-26° C.

Molecular Modeling Calculations were carried out using Parallel Quantum Solutions (Fayetteville, AR, U.S.A.) hardware and software v3.2 (*29*). Initial structures were built using the HyperChem 7.52 program (Gainesville, FL, U.S.A.) and PM3 semi-empirical method (*30*). Geometry optimization was performed on delocalized internal coordinates using the Eigenvector Following Algorithm with the convergence criteria set at 1×10^{-6} Hartree and a gradient of less than 3×10^{-4} a.u. Carbon atoms are displayed in grey, oxygen atoms in black, and hydrogen atoms in white.

Data Analysis Data were analyzed using OriginPro v7.5 SR6 Software OriginLab Corporation (Northampton, MA, U.S.A.). Experimental results were fitted with linear and sigmoidal regression functions.

Results and Discussion

Influence of Cyclodextrins on Fluorescence of Zearalenone

The effect of various cyclodextrins on the fluorescence intensity of zearalenone (10 μM) over pH values of 4 to 10 (100 mM sodium phosphate buffer) is depicted in Figure 4 for α-, β-, γ-cyclodextrins and DIMEB. It is noteworthy that α-cyclodextrin, with its smaller cavity, does not significantly enhance zearalenone fluorescence, and β-cyclodextrin exhibits the greatest enhancement of the unsubstituted cyclodextrins. DIMEB has the most significant fluorescence enhancement at the concentration studied (10 μM

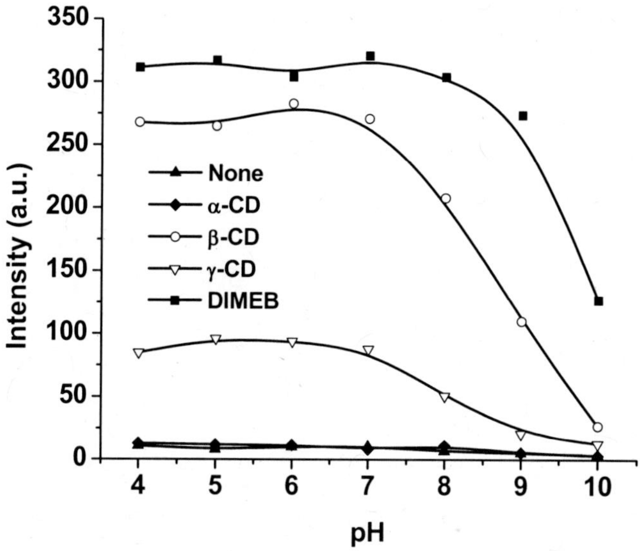

Figure 4. Effect of pH on fluorescence enhancement of zearalenone by cyclodextrins (100 mM sodium phosphate).

zearalenone, 1 mM cyclodextrin). Zearalenone fluorescence is quenched at higher pH. Starting at pH 8, fluorescence intensity is decreased for zearalenone in the presence of all cyclodextrins studied.

The influence of β-cyclodextrin and DIMEB concentration (0.01 mM and 1 mM) on the corrected fluorescence (F-F_0) of zearalenone is displayed in Figure 5. The fluorescence intensity of 10 μM zearalenone is dependent on cyclodextrin structure and concentration for concentrations up to 0.5 mM. However, at cycodextrin concentrations higher than 1 mM, β-cyclodextrin and DIMEB exhibit similar fluorescence-enhancing properties for 10 μM zearalenone. The implications of this dependence of fluorescence intensity enhancement on cyclodextrin concentrations should be an important consideration in the analysis of structure-activity relationships involving zearalenone and cyclodextrins.

Frequently, cyclodextrin structure-activity relationships are characterized as ratios of normalized emission intensities of an analyte at a single cyclodextrin concentration. The normalized emission intensities of zearalenone for several concentrations of cyclodextrins in sodium phosphate buffer (100 mM, pH 7.0) are shown in Table 1. For an equal concentration of cyclodextrin and zearalenone (10 μM), the ratio for normalized emission intensity is 3.6. However, as the concentration of cyclodextrin increases, the effect of the structure-activity relationship decreases for a comparison of DIMEB to β-CD.

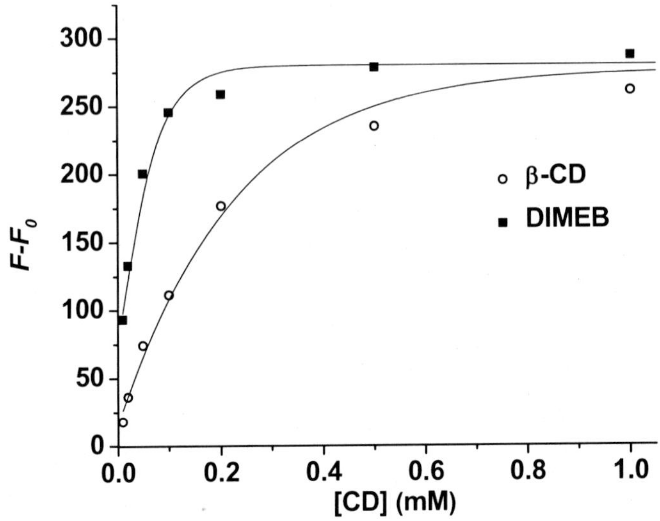

Figure 5. Influence of [CD] on fluorescence enhancement of zearalenone (10 μM zearalenone, pH 7.0).

Table 1. Dependence of normalized emission of 10 μM zearalenone at various concentrations of cyclodextrins.

[Cyclodextrin] (mM)	F/F_0 Zearalenone β-CD	DIMEB	DIMEB/ β-CD
0.01	2.7	9.6	3.6
0.05	7.9	19.6	2.5
0.10	11.7	23.8	2.0
0.5	22.8	26.8	1.2
1.0	23.8	27.5	1.2
3.0	27.9	27.9	1.0

The spectrofluorimetric studies of the interaction of cyclodextrin with zearalenone can be analyzed by Scatchard plot analysis (see Figure 6):

$$\frac{(F - F_0)}{[\mathrm{CD}]} = (F_\infty - F_0)K - (F - F_0)K$$

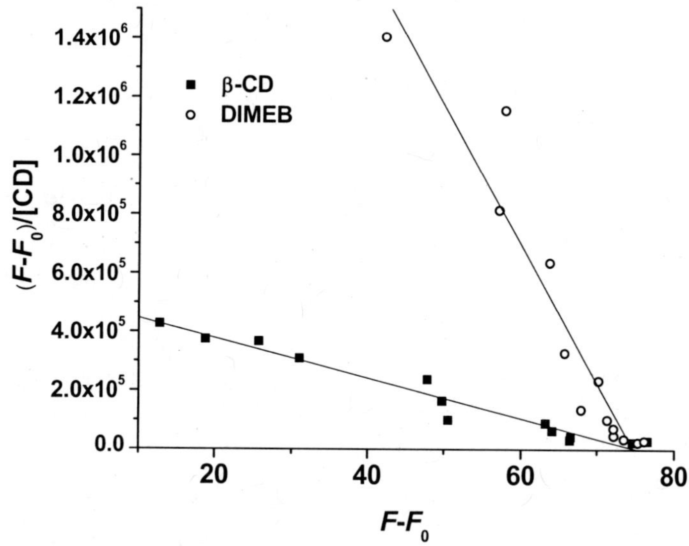

Figure 6. Scatchard plot of $(F\text{-}F_0)/[CD]$ vs. $F\text{-}F_0$ for zearalenone in the presence of β-cyclodextrin and DIMEB (3 μM zearalenone).

Where F is the fluorescence intensity of the zearalenone:cyclodextrin complex, F_0 is the fluorescence intensity of free zearalenone, F_∞ is the fluorescence intensity when all zearalenone is complexed with the cyclodextrin, and K is the association constant (*31-33*). Scatchard plot analysis is a popular method to elucidate binding interactions and applicable to a broad range of binding phenonoma. Furthermore scatchard plot analysis provides similar results to the Benesi-Hildebrand analysis commonly applied to study cyclodextrin-guest inclusion complexes (31, 32). The linearity of the $(F\text{-}F_0)/[CD]$ vs. $F\text{-}F_0$ plot indicates zearalenone forms a 1:1 complex with β-CD and DIMEB. Solving the equation gives a binding constant of 6,800 M^{-1} for β-CD and 47,100 M^{-1} for DIMEB.

Molecular Modeling Studies

Energetics and structural information of mycotoxin:cyclodextrin complexes were carried out using the PM3 semi-empirical method. It should be noted that PM3 calculations are limited in accuracy for certain types of hydrogen bond interactions compared to computationally expensive *ab initio* calculations (*34*). However, PM3 methods have been effective in the study of supramolecular inclusion complexes (*35-37*). Calculated parameters for the mycotoxin-cyclodextrin interactions are given in Table 2. The interaction energies associated with the heat of formation are calculated by:

$$\Delta E = E_{Complex} - (E_{Mycotoxin} + E_{Cyclodextrin})$$

Where $E_{Complex}$ is the heat of formation energy of the mycotoxin:cyclodextrin complex, $E_{Mycotoxin}$ and $E_{Cyclodextrin}$ are the energies for the free molecules, and ΔE is the stabilization energy. The value μ is the dipole moment of the mycotoxin-cyclodextrin bound complex.

Results presented in Table 2 offer some interesting insights into mycotoxin-cyclodextrin interactions. First, complex formation is driven by negative changes in energies for the mycotoxins investigated. Secondly, considerable variation exists in dipole moments, μ, suggesting that the mycotoxin has significant influence over the electronic effects of the bound complexes. Lastly, comparing ΔE of β-CD to DIMEB complexes of zearalenone and aflatoxin B_1, methylation lowers the energies of interactions of complexes suggesting the van der Waals interactions between DIMEB and the mycotoxin are important binding interactions.

Several mycotoxins are capable of forming multiple complexes with β-CD. Two orientations were identified for the zearalenone:cyclodextrin inclusion complex. The most favorable complex for zearalenone and the fluorescent derivative of T-2 toxin is with the fluorophore outside of the cavity, suggesting that binding interactions are more complex than exclusive fluorophore:cyclodextrin complexation. Through fluorescence spectroscopic studies it has been determined that the T-2 toxin pyrene derivative forms a complex with two cyclodextrins (*5*).

Table 2. Calculated parameters of mycotoxin-cyclodextrin interactions

Cyclodextrin Complex	Inclusion Moiety	ΔE (kcal/mol)	μ (Debye)
Zearalenone			
ZEN:β-CD	resorcylic acid	-8.34	2.48
ZEN:β-CD	ketone	-19.44	2.67
ZEN:DIMEB	resorcylic acid	-11.82	2.95
Resorcylic acid			
RA:β-CD	resorcylic acid	-5.71	2.02
RA:DIMEB	resorcylic acid	-7.85	2.84
Aflatoxin B_1			
AFB_1: β-CD	coumarin	-12.77	3.12
AFB_1: β-CD	dihydrofuran	-8.97	4.06
AFB_1: DIMEB	coumarin	-14.84	3.42
T2-pyr			
T2-pyr:DIMEB	T2-toxin	-30.25	3.37
T2-pyr:DIMEB	pyrene	-20.87	2.48
Ochratoxin A			
OTA:β-CD	phenyl	-8.91	4.39

Conclusion

The inclusion complexes of zearalenone and β-CD and DIMEB have been investigated by fluorescence spectroscopy and molecular modeling. Computational results were compared to cyclodextrin complexes of aflatoxin B_1, a pyrene derivative of T-2 toxin, and ochratoxin A. Structure-activity relationships for fluorescence spectroscopic studies of zearalenone with cyclodextrins are dependent on the type of cyclodextrin and pH. Theoretical calculations suggest that mycotoxins can interact with cyclodextrins in multiple binding modes and stabilization of the mycotoxin:cyclodextrin complex is associated with favorable interactions between the toxin and cyclodextrin. Other types of binding forces are likely to contribute to complex formation, including hydrophobic forces, conformational strain relief, and entropically favorable release of water from the cyclodextrin cavity (*24*). These results provide insight into the interaction of mycotoxins and cyclodextrins, and may assist in further applications of cyclomaltoses in methods of detection.

Acknowledgments

The authors thank Lijuan C. Wang for excellent technical assistance. Mention of trade names or commercial products in this article is solely for the purpose of providing specific information and does not imply recommendation or endorsement by the U.S. Department of Agriculture.

References

1 Council for Agricultural Science and Technology (CAST). Mycotoxins: Risks in Plant, Animal and Human Systems. Task Force Report No. 139, CAST, Ames, Iowa, 2003.

2. Maragos C. M. Emerging technologies for mycotoxin detection. *J. Toxicol. Toxin. Rev.* **2004**, *23*, 317-344.

3. Zheng, M. Z.; Richard, J. L.; Binder, J. A review of rapid methods for the analysis of mycotoxins. *Mycopathologia* **2006,** *161*, 261-273.

4. Ueno, A. Review: Fluorescent cyclodextrins for molecule sensing. *Supramol. Sci.***1996**, *3*, 31-36.

5. Maragos, C. M.; Appell, M.; Lippolis, A.; Visconti, A.; Catucci, L.; Pascale, M. Use of cyclodextrins as modifiers of fluorescence in the detection of mycotoxins. *Food Addit.Contam.* **2008**, *25*, 164-171.

6. Galaverna G.; Dall'Asta C.; Corradini, R.; Dossena, A.; Marchelli, R. Cyclodextrins as selectors for mycotoxin recognition. *World Mycotoxin J.* **2008**, *4*, 397-406.

7. Marten Del Valle, E. M. Cyclodextrins and their uses: a review. *Process Biochem.* **2004**, *39*, 1033-1046.

8. Szetjli, J. Introduction and general overview of cyclodextrin chemistry. *Chem. Rev.* **1998**, *98*, 1743-1753.

9. Saenger, W.; Jacob, J.; Gessler, K.; Steiner, T.; Hoffmann, D.; Sanbe, H.; Koizumi, K.; Smith, S. M.; Takaha, T. Structures of the common cyclodextrins and their larger analogues - beyond the doughnut. *Chem. Rev.* **1998**, *98*, 1787-1802.
10. Khan, A. R.; Forgo, P., Stine, K. J.; D'Souza, V. T. Methods for selective modifications of cyclodextrins. *Chem. Rev.* **1998**, *98*, 1977-1996.
11. Rong, D.; D'Souza, V. T. A convenient method for functionalization of the 2-position of cyclodextrins. *Tetrahedron Lett.* **1990**, *31*, 4275-4278.
12. Takahashi, K.; Hattori, K.; Toda, F. Monotosylated α- and β-cyclodextrins prepared in an alkaline aqueous solution. *Tetrahedron Lett.* **1984**, *25*, 3331-3334.
13. Maragos, C. M.; Appell, M. Capillary electrophoresis of the mycotoxin zearalenone using cyclodextrin-enhanced fluorescence. *J. Chromatogr. A* **2007**, *1143*, 252-257.
14. Mallick, A. Purkayastha, P., Chattopadhyay, N. Photoprocess of excited molecules in confined liquid environments: An overview. *J. Photochem. Photobiol.,C* **2007**, *8* 109-127.
15. Rojas-Durán, T. R.; Fenta, C. A.; Váquez, B. I. Franco, C. M. Sanz-Medel, A. Cepeda, A. Study of a room temperature phosphorescence phenomenon to allow the detection of aflatoxigenic strains in culture media. *Int. J. Food Microbiol.* **2007**, *115*, 149-158.
16. Abbas, H. K.; Zablotowicz, R. M.; Weaver, M. A.; Horn, B. W., Xie, W.; Shier, W. T. Comparison of cultural and analytical methods for determination of aflatoxin production by Mississippi Delta *Aspergillus* isolates. *Can. J. Microbiol.* **2004,** *50*, 193-1999.
17. Rojas-Durán, Sánchez-Barragán, I.; Costa-Fernández, J. M.; San-Medel, A. Direct and rapid discrimination of aflatoxigenic strains based on fiber-optic room temperature phosphorescence detection. *Analyst* **2007**, *132*, 307-313.
18. Torkar, K. G. Vengušt, A. The presence of yeasts, moulds and aflatoxin M_1 in raw milk and cheese in Slovenia. *Food Control* **2008**, *19*, 570-577.
19. Torkar, K. G. Vengušt, A. The microbiological quality of raw milk after introducing the two day's milk collecting system. *Acta Agriculturae. Slov.* **2008**, *92*, 61-74.
20. Rahimi, P., Sharifnabi, B., Bahar, M. Detection of aflatoxin in *Aspergillus* species isolated from pistachio in Iran. *J. Phytopathol.* **2008**, *156*, 15-20.
21. Hashemi, J.; Kram, G. A.; Alizadeh, N. Enhanced spectrofluorimetric determination of aflatoxin B_1 in wheat by second-order standard addition method. *Talanta* **2008**, *75*, 1075-1081.
22. Cucci, C. Mignani, A. G., Dall'Asta, C.; Pela, R.; Dossena, A. A portable fluorometer for the rapid screening of M1 aflatoxin. *Sens. Actuators, B* **2007**, *126*, 467-472.
23. Goryacheva, I. Y.; Rusanova, T. Y.; Pankin, K. E. Fluorescent properties of aflatoxins in organized media based on surfactants, cyclodextrins, and calixresorcinarenes. *J. Anal. Chem.* **2008**, *63*, 751-755.
24. Lipowitz, K. B. Applications of computational chemistry to the study of cyclodextrins. *Chem. Rev.* **1998**, *98*, 1829-1873.

25. Aghamohammadi, M.; Alizadeh, N. Fluorescence enhancement of the aflatoxin B_1 by forming inclusion complexes with some cyclodextrins and molecular modeling study. *J. Lumin.* **2007**, *127,* 575-582.
26. Amadasi, A.; Dall'Asta, C.; Ingletto, G.; Pela, R.; Marchelli, R.; Cozzini, P. Explaining cyclodextrin-mycotoxin interactions using a 'natural' force field. *Bioorgan. Med. Chem.* **2007**, *15*, 4585-4594.
27. Cozzini, P.; Ingletto, G.; Singh, R.; Dall'Asta, C. Mycotoxin detection plays "cops and robbers": Cyclodextrin chemosensors as specialized police. *Int. J. Mol. Sci.* **2008**, *9*, 2474-2494.
28. Rameriz-Galicia, G.; Garduno-Jaurez, R.; Vargas, M. G. Effect of water molecules on the fluorescence enhancement of aflatoxin B1 mediated by aflatoxin B1:β-cyclodextrin complexes. A theoretical study. *Photochem. Photobiol. Sci.* **2007**, *6*, 110-118.
29. PQS version 3.2, Parallel Quantum Solutions, 2013 Green Acres Road, Fayetteville, AR 72703 USA
30. Hyperchem, Release 7.52, Hypercube Inc., 1115 NW 4th Street, Gainesville, FL 32601 USA
31. Berzas, J. J.; Alañón, A.; Lázaro, J. A. Cyclodextrin enhanced spectrofluorimetric determination of fluoxetine in pharmaceuticals and biological fluids *Talanta* **2002**, *58*, 301-309.
32. Verrone, R.; Catucci, L.; Cosma, P.; Fini, P.; Agostiano, A.; Lippolis, V.; Pascale, M. Effect of β-cyclodextrin on spectroscopic properties of ochratoxin A in aqueous solution. *J. Incl. Phenom. Macrocycl. Chem.* **2007**, 57, 475-479.
33. Durán-Merás, I.; Muñoz de la Peña, A.; Salinas López, F.; Rodríguez Cáceres. M. I. Complexation study and spectrofluorimetric determination of pipemidic acid with γ-cyclodextrin. *J. Incl. Phenom. Macrocycl. Chem.* **2007**, *51*, 137-142..
34. Dolenc, J.; Koller, J. An improved semiemperical MO PM3 method for hydrogen bond systems. *Acta. Chim. Slov.* **2006**, *53*, 229-237.
35. Castro, R.; Berardi, M. J.; Cordova, E.; do Olza, M. O.; Kaifer, A. E.; Evanseck, J. D. Unexpected roles of guest polarizability and maximum hardness, and of host solvation in supramolecular inclusion complexes: A dual theoretical and experimental study. *J. Am. Chem. Soc.* **1996**, *118*, 10257-10268.
36. Song, L X.; Wang, H. M.; Guo, X, Q.; Bai, L. A comparative study on the binding behaviors of β-cyclodextrin and its two derivatives to four fanlike organic guests. *J. Org. Chem.* **2008**, *73*, 8305-8316.
37. Leclercq, L.; Schmitzer, A. R. Multiple equilibria in the complexation of dibenzylimidazolium bromide salts by cyclodextrins: Toward controlled self assembly. *J. Phys. Chem. B* **2008**, *112*, 11064-11070.

Indexes

Author Index

Abbas, H. K., 37–58
Appell, M., 293–305
Arreguin, I., 167–182

Bandoh, S., 69–76
Benvenuti, M., 247–263
Beyer, M., 265–276
Bhatnagar, D., 93–106
Bianchini, A., 1–16
Binder, E. M., 107–117
Brown, R. L., 157–165
Bullerman, L. B., 1–16
Busman, M., 237–246
Butchko, R. A. E., 167–182

Calvo, A. M., 183–203
Cardozo, S., 247–263
Cary, J. W., 119–142, 183–203
Chen, Z.-Y., 157–165
Cleveland, T. E., 93–106, 157–165
Cramer, B., 265–276

Du, L., 167–182

Fakhoury, A., 157–165

Gerber, R., 167–182
Goto, T., 69–76

Heperkan, D., 77–90
Higashihara, K., 69–76
Hill, N. R., 247–263
Huffman, J., 167–182
Humpf, H.-U., 265–276

Iha, M. H., 59–68

Jaynes, J. M., 119–142

Karbancioglu-Güler, F., 77–90
Kendra, D. F., 237–246
Krol, J., 247–263

Lai, E. P. C., 277–292
Li, L. Q., 167–182
Lin, T., 167–182
Lou, L., 167–182

Mabud, A., 247–263
McCormick, S. P., 143–155
Malone, R. J., 223–236
Maragos, C. M., 293–305
Menkir, A., 157–165

Ohsawa, K., 69–76

Proctor, R. H., 167–182

Rader, J. I., 59–68
Rajasekaran, K., 119–142
Rodrigues, I., 107–117
Romano, J., 247–263

Schatzmayr, G., 107–117
Scott, P. M., 17–35
Shier, W. T., 37–58
Siantar, D. P., 247–263
Szerszen, L., 183–203

Takeuchi, M., 69–76
Trucksess, M. W., 17–35, 59–68, 207–221

Whitaker, T. B., 207–221

Yu, J., 93–106
Yu, J. C. C., 277–292

Zhu, X., 167–182

Subject Index

A

AAL-toxins
- chemical structure, 169*f*
- sphinganine-analog mycotoxins (SAMT), 168
- *See also* Fumonisins

Acetyltransferase gene (*Tri101*), *Fusarium* protection from trichothecenes, 145

Activated charcoal, patulin from apple juice, 24

Activity, synthetic lytic peptides, 129

Aflatoxin
- *Bacillus* inhibiting growth, 2
- chemical analysis, 215–216
- chilies, 27
- coconut products, 22
- collaborative study, 249
- contamination of agricultural products, 93–94
- corn from southern U.S., 40, 41*t*
- detection, 252
- dried figs, 79
- ethanol vs. methanol for extraction in corn, 231–232
- European Commission sampling plan, 211, 212*t*
- extraction studies, 224, 225–226
- heat treatment and, in harvested corn, 46*t*
- metabolic precursors, 184, 186*f*
- microbial degradation, 10
- monitoring production in growth media, 296
- mycotoxin-cyclodextrin interaction parameters, 302*t*
- production on infected corn cobs, 39–40, 43*f*
- regulatory levels, 248
- schematic, 184, 186*f*
- screening for resistance to accumulation, 158–159
- stability in food and food products, 45
- stability with cooking, 26
- storage time and, in ground corn, 46*f*
- structure, 39*f*, 297*f*
- *See also* Mycotoxin; Sampling plans

Aflatoxin biosynthesis
- analysis of *Aspergillus flavus veA* mutant, 188*f*
- development, 187–191
- environmental factors, 191–195
- fungal development, 191–195
- future prospects, 195–196
- gene, 94–95
- gene cluster, 94–95, 184–185
- genomics, 99, 194–195
- global regulatory factors, 187–190
- histone acetylation and toxin biosynthesis, 190–191
- LaeA nuclear generator, 189–190
- pathway, 94
- pathway-specific regulators, 187
- regulation, 95–96, 99
- role of nitrogen, carbon and pH, 191–192
- role of plant metabolites, 192–193
- transcriptional regulation, 187–191
- *veA* gene, 188–189

Aflatoxin extraction efficiency
- procedure, 225–226
- results, 227–229
- solvent comparison for, in corn gluten feed, 229*f*
- solvent comparison for, in corn gluten meal, 228*f*
- solvent comparison for, in cottonseed meal, 227*f*

solvent comparison for, in dried distillers grain, 228*f*
trial conclusions, 233
Aflatoxin pathway-specific regulators, aflatoxin biosynthesis, 187
Agricultural crops, *Aspergillus flavus*, 93
Agricultural products
aflatoxin contamination, 93–94
mycotoxin stability in stored, 45–46
Agricultural raw commodity foodstuffs, microbiological contamination, 248
Agriculture, mycotoxins and, 38, 39*f*
Aldose reductase, resistance-associated proteins (RAPs), 160*t*, 161
Almonds, Codex Alimentarius Commission sample plan, 212, 213*t*
Alternaria alternata
activity of antimicrobial peptide D4E1, 137*t*
transgenic tobacco and resistance to, 136
Alternaria solani, transgenic tomato plants, 136
Amino acids
aflatoxin biosynthesis, 99
hydrophobicity/hydrophilicity, 122, 123*t*
lytic peptides, 126*f*
protein structure and function, 122
volume, 122, 123*t*
Ammoniation, aflatoxin removal, 45
Amphipathy
lytic peptides, 120–121
protein structure and function, 121–125
Antifungal activity
Bacillus, 2
inhibition of *Aspergillus niger* by *Lactobillus rhamnosus*, 4, 5*f*
inhibition of *Penicillium commune* by *L. rhamnosus*, 4, 5*f*
lactic acid bacteria (LAB), 3–7
L. plantarum treatment on cucumber spoilage fungi, 6*t*
propionic acid bacteria, 3
publications for, of LAB, 4*t*
Antifungal peptides, overexpression, 120
Antifungal proteins, maize kernels, 160*t*
Antimicrobial defense system, lytic peptides, 120
Antimicrobial peptides
potential problems, 137–138
See also Lytic peptides
Apples
decay size and patulin accumulation, 75
distribution of patulin in damaged, 73–74
gamma-radiation, 24
heat pasteurization, 72
high-pressure water spraying, 23
patulin contamination, 70
patulin removal from juice, 24
pre-processing reduction of *Penicillium expansum*, 23–24
prevention of patulin, 22–24
storage temperature and patulin contamination, 75
See also Fruit and products; Patulin
Arabidopsis thaliana
comparing *Chlamydomonas* and, bioassays, 151, 152
detoxification of trichothecens, 150
model system, 148–149
toxin library screen, 149
Arkansas
aflatoxin levels in corn by year, 41*t*

fumonisin levels in corn by year, 42*t*
rice samples, 43
Aspergillus section *Flavi*
A. flavus complex, 78
determination of, members, 80
determination of cyclopiazonic acid (CPA) producing, members, 80–81
enumeration, isolation and identification of molds, 79–80
fig samples from contaminated orchards, 81
food fermentations, 184–185
generally recognized as safe (GRAS), 185
morphological properties, 83
mycotoxigenic properties of, from dried figs, 85*t*
See also Figs, dried
Aspergillus species
Bacillus inhibiting growth, 2
botanicals, 24
fungi for agriculture and food, 38
genomics, 194–195
lactic acid bacteria against, 4*t*
monitoring aflatoxin production by, 296
Aspergillus flavus
activity of antimicrobial peptide D4E1, 137*t*
aflatoxin biosynthesis, 94–96
aflatoxin biosynthetic pathway, 94
aflatoxin biosynthetic pathway gene and gene cluster, 94–95
aflatoxins B_1 and B_2, 93
agricultural crops, 93
A. parasiticus with, 78
comparative genomics in fungal evolution, 100
complex, 78
corn contamination, 39–40, 43*f*
Expressed Sequence Tags (EST), 96–97
functional genomics for mechanism of aflatoxin formation, 98–99
genomics, 96–97, 194–195
genomics for identifying genes in aflatoxin formation, 97–98
inducible promoters, 130
regulation of aflatoxin biosynthesis, 95–96
ribosome inactivating protein (RIP), 159
synthetic peptides, 135, 136*f*
zeamatin, 159
See also Figs, dried
Assays, phototoxicity, 145–146
Attacin gene, bacterial pathogen control, 134–135

B

Bacillus species, antifungal activity, 2
Bacillus pumilus, inhibiting aflatoxin production, 7
Banana, disease resistance, 136
Beers
fermentation process, 248
See also Multiple mycotoxins analyses
Beta-conformations, lytical peptides, 127, 128*f*
Binding, mycotoxin, 8–9
Bioassay, *Chlamydomonas*, 146
Bioavailability, food processing-induced changes to fumonisins, 48–50
Biocontrol
future trends, 11
molds, 2
Biodegradation
fumonisins, 112–113
ochratoxin A (OTA), 110–111
zearalenone, 111–112
Biosynthesis
aflatoxin pathway, 94
aflatoxin pathway gene and gene cluster, 94–95

biosynthetic genes of fumonisins, 168–171
Fusarium trichothecenes, 144
histone acetylation and toxin, 190–191
mycotoxins using labeled precursors, 270–271
regulation of aflatoxin, 95–96
See also Fumonisins
Biotransformation, trichothecenes, 109–110
Black shank disease, *Phytophthora parasitica*, 134
Botanicals
changing cultural practices, 27
drying, 25
gamma irradiation, 25–26
harvesting, 25
processing modification, 26
sampling for mycotoxins in, roots, 213–219
storage, 25
treating *Monascus* strains with mutagens, 27
See also Ginger capsule study
Botrytic cinerea, transgenic tobacco and resistance to, 136
Brazil nuts, product sample sizes by US Food and Drug Administration, 210*t*
Bread, lactic acid bacteria and *Aspergillus niger* growth, 6
Breeding resistance, fruit trees, 18

C

Cacopsylla pyricola, synthetic peptides, 135
Candida sake, biocontrol agent for *Penicillium expansum* and patulin, 23
Carbon, aflatoxin biosynthesis, 191–192
Carbon nanotubes (CNTs)
merging with polypyrrole (PPy), 282–283
molecularly imprinted PPy (MIPPy)/CNT-modified stainless steel frits, 283–287
See also Molecularly imprinted polymers (MIPs)
Carbon source, aflatoxin formation, 99
Cecropins-class
antimicrobial peptide, 134
natural peptides, 127, 128*f*
Cercospora kikuchii, activity of antimicrobial peptide D4E1, 137*t*
Cereal, sampling procedures by European Commission, 212*t*
Charge density, protein structure and function, 121–122
Chilies, aflatoxin and ochratoxin A, 27
Chlamydomonas reinhardtii
bioassay, 146
comparing, and *Arabidopsis* bioassays, 151, 152
detoxification of trichothecenes, 150
trichothecenes on, growth, 147*t*, 148*t*
Chloroplast genome, comparison to nuclear genome, 132, 133*t*
Clarification
apple juice, 24
winery practice, 20
Claviceps purpurea, activity of antimicrobial peptide D4E1, 137*t*
Coconut products
aflatoxin, 22
See also Fruit and products
Codex Alimentarius Commission
apple juice recommendations, 22–23
patulin content in apple products, 70
sampling plan for aflatoxins in tree nuts, 212, 213*t*
vineyard recommendations, 19
See also Sampling plans

Cold-regulated protein (ZmCORp), resistance-associated proteins (RAPs), 160*t*, 161, 162
Cold storage, apples with *Penicillium expansum*, 23
Collaborative study. *See* Multiple mycotoxins analyses in beer
Colletotrichum destructivum, activity of antimicrobial peptide D4E1, 135, 137*t*
Color scale, hydrophobicity/hydrophilicity of amino acids, 125
Comparative genomics, fungal evolution, 100
Cooking, aflatoxin stability, 26
Corn (maize)
  aflatoxin and fumonisin production, 39–40, 43*f*
  aflatoxin levels by year, 40, 41*t*
  cobs infected with fungi, 43*f*
  ethanol vs. methanol extraction of aflatoxins, 231–232
  ethanol vs. methanol extraction of fumonisins, 233
  ethanol vs. methanol extraction of zearalenone, 232
  fumonisin levels by year, 40, 42*t*
  heat treatment and aflatoxin contamination, 46*t*
  mycotoxin contamination, 39–42
  near-isogenic maize pairs varying aflatoxin levels, 161*t*
  nixtamalization, 51–52
  ochratoxin A (OTA) extraction, 230, 231*f*
  oxylipin biosynthetic gene, 193
  product sample sizes by US Food and Drug Administration, 210*t*
  resistance-associated proteins (RAPs), 159–161
  screening resistance to aflatoxin, 158–159
  storage time and aflatoxin level, 46*f*
  storage time and fumonisin, 47*f*
  susceptibility to mycotoxins, 38
  toxicity and food processing changes, 51–53
  use in ethanol facilities, 238
  See also Ethanol production from corn
Corn gluten feed, aflatoxin extraction efficiency, 228–229
Corn gluten meal, aflatoxin extraction efficiency, 228
Cottonseed
  aflatoxin extraction efficiency of, meal, 227
  product sample sizes by US Food and Drug Administration, 210*t*
Coumarin, structure, 297*f*
Cucumbers, lactic acid bacteria prolonging, 5–6
Cultivated crops, susceptibility to plant diseases, 120
Cultivation, vineyards, 19
Cyclodextrins
  applications in mycotoxin analysis, 295–297
  concentration of, and fluorescence enhancement of zearalenone, 300*f*
  dependence of emission of zearalenone by, concentration, 300*t*
  fluorescence spectroscopy, 298
  influence on fluorescence of zearalenone, 298–301
  materials and methods, 298
  molecular modeling, 298, 301–302
  mycotoxin:cyclodextrin interaction parameters, 302*t*
  overview, 294–295
  pH and fluorescence enhancement of zearalenone by, 299*f*
  scatchard plot analysis, 300, 301*f*
  structures of -cyclodextrin and 2,6-dimethyl- -cyclodextrin (DIMEB), 295*f*

structures of mycotoxins, 297*f*
Cyclopiazonic acid (CPA)
determining CPA-producing *Aspergillus* section *Flavi* members, 80–81
production in fig samples, 82*t*, 84*t*
structure, 39*f*
toxic metabolite of *A. flavus*, 78
See also Figs, dried
Cytotoxicity, food processing-induced changes to fumonisins, 51–53

D

Defensins
fungal resistance, 134
lytical peptides, 127, 128*f*
Degradation
fumonisins, 112–113
See also Biodegradation
Deoxynivalenol (DON)
animals, 108
Chlamydomonas growth, 147, 148*t*
collaborative study, 249
conversion to 3-keto-deoxynivalenol, 150
detection, 252
organisms for metabolizing, 10
production of ^{13}C-labeled, 271, 272*f*
regulatory levels, 248
transformation, 109–110
See also Vomitoxin (DON)
Design parameters, lytic peptide, 126–129
Detoxification, trichothecenes, 150
Deuteration, chemical modification of non-labeled mycotoxins, 269, 270*f*
4,15-Diacetoxyscirpenol (DAS)
O-methyl derivatives, 150
phytoxic trichothecene, 149
structure, 145*f*
Dihydroisocoumarin, structure, 297*f*
Dilution plating, mold isolation in dried figs, 82*t*, 83
Direct plating, mold isolation in dried figs, 82*t*, 83
Disease resistance
genetic engineering, 133–137
peptides of non-plant origin, 134–135
peptides of plant-origin, 133–134
synthetic peptides, 135–137
DNA (deoxyribonucleic acid), biocontrol trend, 11
Dried distillers grain, aflatoxin extraction efficiency, 227, 228*f*
Dried fruit
product sample sizes by US Food and Drug Administration, 210*t*
sampling procedures by European Commission, 212*t*
Dried vine fruit
preventing mycotoxins in dried figs, 21–22
reducing ochratoxin A (OTA) contamination, 21
See also Figs, dried; Fruit and products
Dry heating, aflatoxin, 45
Drying
chilies, 27
figs, 21–22
ginseng roots, 25
senna pods, 25

E

Edible seeds, sample sizes by US Food and Drug Administration, 210*t*
Environmental factors, aflatoxin biosynthesis, 191–195
Enzymes, mycotoxin detoxification, 9–11
Ergot-alkaloids, medieval times, 107–108
Erwinia amylovora, synthetic peptides, 135

Ethanol
comparison with methanol for mycotoxin extraction, 231–233
extraction of mycotoxins, 226
Ethanol production from corn
analysis of prefractionation components, 242*t*
byproducts distiller dried grains (DDG) and distiller dried grains with solubles (DDGS), 238
chemical analysis of mycotoxins, 239
chemical modification of toxins by yeasts, 243
DDGS fumonisin spike experiments, 242–243
expansion in U.S., 243
extracted ion chromatogram for detection of fumonisin B_1 (FB_1) from DDG sample, 241*f*
FB_1 recovery from spiked DDGS samples, 241*t*
fumonisins accumulating in U.S. corn, 242
fumonisin spike, 240
liquid chromatography–mass spectrometry (LC–MS), 240
method, 239–240
mycotoxin determination in grains, 238–239
prefractionation processes, 243
spike recovery experiments, 241–242
European Commission (EC)
established maximum levels, 248
maximum aflatoxin levels in foods, 211*t*
sampling plan for aflatoxins, 211
sampling procedures by, 212*t*
See also Sampling plans
Expressed Sequenced Tags (EST), *Aspergillus flavus*, 96–97
Extraction. *See* Aflatoxin extraction efficiency; Mycotoxin extraction; Ochratoxin A (OTA) extraction efficiency

F

Fermentation
apple juice, 24
winery practice, 20
Figs, dried
aflatoxigenic *Aspergillus flavus* isolates, 86
cyclopiazonic acid (CPA) production, 78, 86, 87
determining *Aspergillus* section *Flavi* members, 80
determining CPA producing *A.* section *Flavi* members, 80–81
drying, 21–22
enumeration, isolation and identification of molds, 79–80
materials and methods, 79–81
mold counts, method of isolation and properties of molds, 82*t*, 84*t*
morphological properties of *A.* section *Flavi*, 83
mycotoxigenic properties of *A.* section *Flavi*, 85*t*
orchard contamination with *A.* section *Flavi*, 81
orchard management, 21
processing, 21–22
product sample sizes by US Food and Drug Administration, 210*t*
sampling procedures by European Commission, 212*t*
sclerotia, 81, 82*t*
storage, 21–22
See also Fruit and products
Flavobacterium aurantiacum
aflatoxin removal, 10
antifungal activity, 2
Fluorescence

cyclodextrins and, of zearalenone, 298–301
See also Cyclodextrins
Food
accurate estimates of mycotoxin concentration, 208
persistence of mycotoxins, 38
Food and Drug Administration (FDA)
guidance for food and feed industry, 248
product sample sizes, 210*t*
sampling plans for aflatoxins, 209
See also Sampling plans
Food crops
welfare of humanity, 120
See also Genetic engineering of food crops
Food processing
bioavailability and changes to fumonisins, 48–50
induced changes in fumonisins, 48–53
mycotoxin content, 47
proposed mechanisms for formation of *N*-fatty acid hydrolyzed fumonisin, 52–53, 54*f*
toxicity and changes to fumonisins, 51–53
Food safety
contaminants and health risk, 266
food processing and toxicity, 53
mycotoxins and, 38, 39*f*
Fresh vegetables, lactic acid bacteria prolonging, 5–6
Fruit and products
aflatoxin in coconut products, 22
breeding resistance in fruit trees, 18
mycotoxin prevention in dried figs, 21–22
ochratoxin A (OTA), 18–21
OTA occurrence, 18
patulin prevention, 22–24
preventing mycotoxin contamination, 18
prevention of OTA in grape and products, 18–20
reducing OTA contamination in dried vine fruit, 21
See also Apples; Figs, dried
Frying, food processing and toxicity, 52–53
Fumonisin B_1 (FB_1)
biosynthetic production of $[D_6]$-FB_1, 271*f*
structure, 39*f*
Fumonisin biosynthetic genes (FUM)
cluster, 168, 169–171
complementation of biosynthesis, 173*f*
polyketide chain assembly, 171–172
polyketide chain release, 172–173
tailoring polyketide chain, 173–176
See also Fumonisins
Fumonisin paradox, oral absorption, 48–49
Fumonisins
bioavailability and food processing-induced changes, 48–50
biodegradation, 112–113
biosynthetic genes, 168–171
biosynthetic mechanism, 171–176
chemical structure of, and AAL-toxins, 169*f*
collaborative study, 249
comparison of structure of lysyl-fumonisin B_1 with drugs, 50, 51*f*
corn from southern U.S., 40, 42*t*
degradation ability of isolates in complex media, 113*t*
detection, 252
ethanol vs. methanol for extraction in corn, 233

food processing-induced changes, 48–53
"fumonisin paradox," 48, 49
polyketide chain assembly, 171–172
polyketide chain release, 172–173
production on infected corn cobs, 39–40, 43*f*
proposed biosynthetic pathway for carbon backbone modifications, 176, 177*f*
proposed mechanism for covalent binding of tritium labeled, 49, 50*f*
proposed mechanisms for formation of *N*-fatty acid hydrolyzed, 52–53, 54*f*
regulatory levels, 248
sphinganine-analog mycotoxins (SAMT), 168
storage time and, in ground corn, 47*f*
structure, 168–171, 175
tailoring polyketide chain, 173–176
toxicity and food processing-induced changes, 51–53
See also Ethanol production from corn
Fungal development, aflatoxin biosynthesis, 191–195
Fungal evolution, comparative genomics, 100
Fungal resistance, defensin gene, 134
Fusarium species
Bacillus inhibiting growth, 2
botanicals, 24
fungi for agriculture and food, 38
lactic acid bacteria against, 4*t*
rice, 44
trichothecene biosynthesis, 144
trichothecene protection, 145
trichothecenes, 150
Fusarium graminearum
activity of antimicrobial peptide D4E1, 137*t*
phytotoxicity of, mycotoxins, 147, 148*t*
scab disease, 44
Fusarium moniliforme
activity of antimicrobial peptide D4E1, 137*t*
fungal resistance, 134
Fusarium oxysporum
activity of antimicrobial peptide D4E1, 137*t*
synthetic peptides, 135
transgenic banana plants, 136
Fusarium verticillioides
corn contamination, 39–40, 43*f*
production of [D_6]-fumonisin B_1 by, 271*f*
radiolabeled fumonisin B_1, 49
synthetic peptides, 135

G

Gamma irradiation
botanicals, 25–26
licorice root, 26
mycotoxins in fruits, 24
Gas cooking, aflatoxin stability, 26
Gene constructs
nuclear vector, 130, 131*f*
plant transformation, 129–132
plastic vector, 130, 131*f*
Genes
aflatoxin and sterigmatocystin biosynthetic, clusters, 184, 186*f*
aflatoxin biosynthesis, 94–95
biosynthetic, of fumonisins, 168–171
expression studies for aflatoxin formation, 98–99
identifying, in aflatoxin formation, 97–98
metagenome, 11
mycotoxin detoxification, 9
Gene silencing, plant transformation, 132
Genetic engineering of food crops

Aspergillus flavus spores, 136*f*
binary vector for nuclear transformation, 131*f*
broad spectrum antimicrobial activity of D4E1 *in vitro*, 137*t*
comparing chloroplast and nuclear genetic engineering, 132, 133*t*
disease resistance, 133–137
gene constructs for plant transformation, 129–132
gene silencing, 132
genetic plasmid vectors for nuclear and plastic transformation, 131*f*
hybrid peptides, 138
inducible promoters, 130
leaf extracts from T_1 generation plants, 137
nuclear transformation, 132
peptides of non-plant origin, 134–135
peptides of plant-origin, 133–134
plastic transformation, 132
plastid vector, 131*f*
potential problems of antimicrobial peptides, 137–138
promoters and transformation vectors, 132
rational design of substitution analogs, 138
synthetic peptide combinatorial library, 138
synthetic peptides, 135–137
tissue-specific promoters, 130
See also Lytic peptides

Genomics
aflatoxin biosynthesis, 194–195
Aspergillus flavus, 96–97
Aspergillus flavus, for genes in aflatoxin formation, 97–98
comparative, in studying fungal evolution, 100
determining genetics of organism, 96
functional, for aflatoxin formation, 98–99
gene expression studies, 98–99
regulation of aflatoxin biosynthesis, 99
See also Aspergillus flavus

Georgia
aflatoxin levels in corn by year, 41*t*
fumonisin levels in corn by year, 42*t*

Ginger capsule study
aflatoxin analytical variances, 217*t*
chemical analysis of aflatoxins and ochratoxin A (OTA), 215–216
design, 213–214
material and sampling design, 214, 215*f*
measuring variability, 216
OTA analytical variances, 218*t*
variances in aflatoxin measurement, 216–217
variances in OTA measurement, 218, 218–219
See also Sampling plans

Ginger root, product sample sizes by US Food and Drug Administration, 210*t*

Ginseng roots, drying, 25

Gliocladium roseum, detoxification of zearalenone, 10

Global regulatory factors
aflatoxin biosynthesis, 187–190
LaeA nuclear regulator, 189–190
veA gene, 188–189
See also Aflatoxin biosynthesis

Globulins, maize kernels, 160*t*

Glyoxalase I, resistance-associated proteins (RAPs), 160*t*, 161, 162

Good Agricultural Practice (GAP)
apple juice recommendations, 22–23
harvest, 18

Good Manufacturing Practice (GMP)

apple juice recommendations, 22–23
harvest, 18
Grains, sample sizes by US Food and Drug Administration, 210*t*
Grape and grape products
cultivation practices, 19
dried fruit and ochratoxin A (OTA) contamination, 21
harvest practices, 19–20
prevention of OTA, 18–20
treatment at winery, 20

H

Harvesting
botanicals, 25
mycotoxins, 38
preventing mycotoxicoses, 108–109
Harvest practices, ochratoxin A in vine-based products, 19–20
Hazelnuts, Codex Alimentarius Commission sample plan, 212, 213*t*
Heat pasteurization, apples and patulin, 72
Heat treatment, aflatoxin in harvested corn kernels, 46*t*
-Helical classes, lytic peptides, 126, 127*f*
Herbal plants, preventing mycotoxins, 24–27
High performance liquid chromatography–tandem mass spectrometry (HPLC–MS/MS)
development, 266, 273
"dilute and shoot" approaches, 266
See also Stable isotope labeled mycotoxins
Histone acetylation, toxin biosynthesis, 190–191
HT-2 toxin, collaborative study, 249
Humanity, food and feed crops, 120
Hybrid peptides
potential problems, 138
See also Peptides
Hybrid poplar, bacterial disease resistance, 136
Hydrophobicity
amino acids, 122, 123*t*
protein structure and function, 121–125
Hydroxylation, phytotoxicity, 146–147, 149–150
-Hydroxypropionaldehyde (reuterin), antifungal activity, 6

I

Illinois
aflatoxin levels in corn by year, 41*t*
fumonisin levels in corn by year, 42*t*
Immunoaffinity column (IAC)
calibration curve for IAC/HPLC, 286*f*
See also Multiple mycotoxins analyses
Inducible promoters, plant transformation, 130
Inhibitory activity, lactic acid bacteria (LAB), 6
International Institute of Tropical Agriculture (IITA–Nigeria), aflatoxin-resistant maize, 158
In vivo experiments, confirming in vitro findings, 113–114
Irradiation, apple juice with patulin, 24
Isotope labeled mycotoxins. *See* Stable isotope labeled mycotoxins
Isotrichodermin, structure, 145*f*
Isotrichodermol
O-methyl derivatives, 150
structure, 145*f*
Isotrichotriol
phytoxic trichothecene, 149
structure, 145*f*

K

Kernel-screening assay (KSA)
aflatoxin production, 158–159
near-isogenic maize pairs
varying aflatoxin levels by, 161*t*

L

Labeled mycotoxins. *See* Stable isotope labeled mycotoxins
Lactic acid bacteria (LAB)
antifungal activity, 3–7
mycotoxin binding, 8–9
prolonging cucumber shelf–life, 5–6
publications, 4*t*
Lactobacillus species
antimycotoxigenic activity of *L. casei*, 7, 8*t*
inhibiting aflatoxin production, 7
inhibition of *Aspergillus niger* by *L. rhamnosus*, 4, 5*f*
inhibition of *Penicillium commune* by *L. rhamnosus*, 4, 5*f*
lactic acid bacteria, 3
mycotoxin binding, 8–9
mycotoxin production and *L. rhamnosus*, 7
publications on antifungal activity, 4*t*
Lactococcus species, lactic acid bacteria, 3
LaeA nuclear regulator, aflatoxin production, 189–190
Late embryogenesis abundant proteins (LEA), maize kernels, 160*t*
Leuconostoc species
lactic acid bacteria (LAB), 3
mycotoxin binding by *L. mesenteroides*, 9
Licorice root
gamma irradiation, 26
processing, 26
Licorice tea, ochratoxin and processing, 26
Lipoxygenase (LOX) pathways, plant metabolites, 192–193
Livestock animals, confirming in vitro findings with in vivo experiments, 113–114
Louisiana
aflatoxin levels in corn by year, 41*t*
fumonisin levels in corn by year, 42*t*
Lytic peptides
activity, 129
amino acids, 122, 123*t*, 127*f*
amino-terminus of peptide, 127
amphipathy, 120–121, 121–125
antimicrobial defense systems, 120
-conformation, 127, 128*f*
-sheets without disulfide linkages (D4E1), 127, 128*f*
carboxy-terminus, 127
charge density, 121–122
color scale, 125
D4E1 antimicrobial activity, 137*t*
design parameters, 126–129
differences and similarity in protein structure, 124–125
-helical, classes, 127*f*
hydrophobicity, 121–125
hydrophobicity/hydrophilicity of amino acids, 122, 123*t*
Molly motif, 124, 127
natural peptides, 127, 128*f*
protein functionality, 124
structural clues, 124
volume of amino acids, 122, 123*t*
See also Genetic engineering of food crops

M

Macadamia integrifolia, antimicrobial peptide, 134

Magainins-class, natural peptides, 127, 128*f*

Magnaporthe grisea, resistance to rice blast, 134

Maize. *See* Corn (maize)

Matrix effects, mycotoxins, 267

Maturing wine, winery practice, 20

Melittin-class, natural peptides, 127, 128*f*

Metabolic precursors, aflatoxin, 184, 186*f*

Metagenome, biocontrol trend, 11

Methanol
- comparison with ethanol for mycotoxin extraction, 231–233
- extraction of mycotoxins, 226

Microbial degradation, mycotoxin, 9–11

Microwave cooking, aflatoxin stability, 26

Milk, sample sizes by US Food and Drug Administration, 210*t*

Mississippi
- aflatoxin levels in corn by year, 41*t*
- fumonisin levels in corn by year, 42*t*
- rice samples, 43

Molds
- biocontrol, 2
- dried fig samples, 82*t*, 84*t*
- future control trends, 11
- *See also* Figs, dried

Molecularly imprinted polymers (MIPs)
- applications of nanomaterials, 282–283
- carbon nanotubes (CNTs) merging with polypyrrole (PPy), 282–283
- development and applications of pyrrole-based, 281
- electrochemical preparation, 280, 281*t*
- high binding capacity via surface area enhancement, 284–286
- MIP nanomaterials for solid phase micro extraction, 281–287
- MIPPy/CNT– SPP–HPLC and IAC/HPLC calibration, 286*f*
- MIPPy/CNT-modified stainless steel frits, 283–286
- MIPPy/CNT/quantum dots in needle, 286–287
- new development, 279–280
- preparation, 278
- scanning electron microscopy (SEM) image of MIPPy/CNTs-on-a-frit, 284*f*
- schematic of MIPPy/CNT–HPLC–fluorescence detector, 285*f*
- solid phase extraction, 278–280
- stability of MIPPy/CNT-modified stainless steel frits, 286
- use in analysis, 277–278
- using MIP for mycotoxin extraction, 281–282

Molecular modeling, mycotoxin:cyclodextrin complexes, 298, 301–302

Molly motif, protein structure, 124, 127

Monascus, treatment with mutagens, 27

Monilia spp., lactic acid bacteria against, 4*t*

Multiple mycotoxins analyses
- aflatoxin derivatization methods, 254
- collaborative study, 249–250
- detection limits, 260
- experimental, 250–254
- fluorescent activity, 254–255
- fortified wheat beer analysis, 257*f*
- fumonisins derivatization, 254
- HPLC/MS/MS system, 252–253
- HPLC/UV/fluorescence system, 251–252
- HPLC/UV/fluorescence work, 255–256

immunoaffinity column (IAC) clean-up protocols, 253–254
instrument detection limit, 260
ions and conditions of Quattro micro API Multiple Reaction Monitoring (MRM), 259*t*
modern LC/MS/MS analysis of spiked beer, 257, 258*f*
multi-toxin stock solution preparation, 250
o-phthalaldehyde (OPA) derivatizing reagent, 250
photochemical reactor for enhanced detection (PHRED™), 251
post-column conditions, 252
prototype "Myco5in1" IAC, 248–249, 255, 261
spiked beer samples, 256
two instrument systems set-up, 250–253
universal gradient, 259–260
See also Stable isotope labeled mycotoxins

Musa spp., disease resistance, 136
Mycosphaerella musicola, transgenic banana plants, 136
Mycotoxicoses, approaches to preventing, 108–109
Mycotoxin extraction
comparing ethanol vs. methanol, 231–233
extraction of ochratoxin A in wheat, 231*f*
extraction of total aflatoxins in corn, 232*f*
extraction of total fumonisins in corn, 233*f*
extraction of zearalenone in corn, 232*f*
molecularly imprinted polymers for, 281–282
procedure, 226
trial conclusions, 234

Mycotoxin resistance, rice, 43–44
Mycotoxins
binding, 8–9
biotransformation and biodegradation, 109–113
collaborative study, 249
comparing analysis techniques, 272–273
concentration in food, 208
confirming in vitro findings with in vivo experiments, 113–114
contamination in U.S. grown corn, 39–40, 43*f*
extraction studies, 224, 226
food processing and, content, 47
future control trends, 11
inhibition of production, 7–8
matrix effects, 267
microbial degradation, 9–11
phytotoxicity of *Fusarium graminearum*, 147, 148*t*
pre- and post-harvest, 38
prevention in botanicals, 24–27
prevention in dried figs, 21–22
rapid multi-, analysis, 249
research projects on sampling, 213–219
rice contamination in southern U.S., 43–44
secondary metabolites, 108
stability in stored agricultural products, 45–46
structures, 39*f*
worldwide occurrence, 108
See also Cyclodextrins; Ginger capsule study; Multiple mycotoxins analyses; Ochratoxin A (OTA); Sampling plans; Stable isotope labeled mycotoxins

Myrothecium, trichothecene-producing fungi, 144

N

Nanomaterials
molecularly imprinted polymers, 282–283
See also Molecularly imprinted polymers (MIPs)

Natural peptides

classification system, 127, 128*f*
See also Lytic peptides
Nicotiana tabacum, 135, 136
Nitrogen, aflatoxin biosynthesis, 191–192
Nivalenol
Chlamydomonas growth, 147, 148*t*
structure, 145*f*
Nixtamalization, food processing, 51–52
Non-plant origin peptides, disease resistance, 134–135
North Carolina, fumonisin levels in corn by year, 42*t*
Nuclear genome, comparison to chloroplast genome, 132, 133*t*
Nuclear vector, gene constructs, 130, 131*f*
Nuts, sampling procedures by European Commission, 212*t*

O

Ochratoxin A (OTA)
analysis method, 64–65, 215–216
biodegradation and non-toxicity of new metabolite, 110–111
chemical modification of non-labeled, 269
chilies, 27
detection, 252
distribution during tofu processing, 66, 67*t*
distribution in liquid and residual soybeans after processing, 66*t*
extraction by molecularly imprinted polypyrrole (MIPPy), 281–282
extraction studies, 224, 226
licorice processing, 26
liquid chromatogram (LC) of OTA, soymilk and tofu, 65, 66*f*
major mycotoxin, 59–60
metabolism by *Aspergillus niger*, 10
mycotoxin:cyclodextrin interaction parameters, 302*t*
occurrence in fruit and products, 18
okara, 60
prevention in grape and products, 18–20
prevention vs. removal, 20
processing soybean samples, 62, 63*f*
reducing contamination of dried vine fruit, 21
regulatory levels, 248
soybeans, 60
soy milk and tofu production, 62, 64*f*
structure, 39*f*, 297*f*
tofu, 60
Ochratoxin A (OTA) extraction efficiency
high level OTA in corn high extraction solvent comparison, 231*f*
high level OTA in wheat extraction solvent comparison, 230*f*
low level OTA in corn extraction solvent comparison, 230*f*
OTA in wheat extraction solvent comparison, 229*f*
procedure, 226
results, 229–230
trial conclusions, 234
Oilseed meals, product sample sizes by US Food and Drug Administration, 210*t*
Okara
analysis, 65
nutritional value, 60
ochratoxin A (OTA) distribution, 66, 67*t*
See also Ochratoxin A (OTA)
Orchard management, figs, 21
Organic acid production, inhibitory activity of lactic acid bacteria, 6

Organisation Internationale de la Vigne et du Vin (OIV), recommendations, 19
Overexpression, antifungal peptides, 120
Oxygenation, phytotoxicity, 146–147
Oxylipins, plant metabolites, 193

P

Pantoea agglomerans, biocontrol agent for *Penicillium expansum* and patulin, 23
Pathogenesis related protein 10 (PR-10), resistance-associated proteins (RAPs), 160*t*, 161, 162
Patulin
- analysis, 71
- apple product contamination, 70
- decay size and, accumulation on apple, 75
- distribution in damaged apple, 73–74
- extraction from apple juice, 71
- heat pasteurization and, in apple juice, 72
- liquid chromatography (LC) and LC–MS analysis, 72, 73*f*
- materials and methods, 70–71
- mold damaged apple sample preparation, 70–71
- pre-processing reduction of *Penicillium expansum*, 23–24
- prevention in fruit products, 22–24
- production and decay by *P. expansum*, 75*t*
- removal from apple juice, 22, 24
- storage temperature effects, 75
- structure, 39*f*
- synthesis of labeled, 268
- trimming and, reduction, 73–74
- *See also* Apples; Fruit and products

Peanut butter, sample sizes by US Food and Drug Administration, 210*t*
Peanuts, sample sizes by US Food and Drug Administration, 210*t*
Pediococcus species, lactic acid bacteria, 3
Penicillium species
- *Bacillus* inhibiting growth, 2
- botanicals, 24
- fungi for agriculture and food, 38
- lactic acid bacteria against, 4*t*

Penicillium expansum, pre-processing reduction, 23–24
Penicillium italicum, activity of antimicrobial peptide D4E1, 137*t*
Peptides
- hybrid, 138
- non-plant origin, and disease resistance, 134–135
- plant-origin, and disease resistance, 133–134
- potential problems of antimicrobial, 137–138
- synthetic, and disease resistance, 134–135
- *See also* Lytic peptides

Peroxiredoxin antioxidant (PER1), resistance-associated proteins (RAPs), 160*t*, 161, 162
pH
- aflatoxin biosynthesis, 191–192
- fluorescence of zearalenone by, 299*f*

Phytopathogens, activity of antimicrobial peptide D4E1, 136, 137*t*
Phytophthora cinnamomi, activity of antimicrobial peptide D4E1, 137*t*
Phytophthora infestans, transgenic tomato plants, 136
Phytophthora parasitica
- activity of antimicrobial peptide D4E1, 137*t*
- black shank disease, 134
- fungal resistance, 134

Phytotoxicity
- assays, 145–146

Fusarium graminearum
mycotoxins, 147, 148*t*
See also Trichothecenes
Pistachio nuts
Codex Alimentarius Commission sample plan, 212, 213*t*
sample sizes by US Food and Drug Administration, 210*t*
Plant diseases, susceptibility of cultivated crops, 120
Plant metabolites, aflatoxin biosynthesis, 192–193
Plant-origin peptides, disease resistance, 133–134
Plant transformation
gene constructs for, 129–132
gene silencing, 132
Plastid transformation, transgene product, 132
Plastid vector, gene constructs, 130, 131*f*
Polyketide chain
assembly, 171–172
biosynthetic mechanism, 171–176, 177*f*
release, 172–173
tailoring, 173–176
See also Fumonisins
Polypyrrole (PPy)
molecularly imprinted polymers, 280, 281*t*
See also Molecularly imprinted polymers (MIPs)
Populus tremula, bacterial disease resistance, 136
Post-harvest
mycotoxins, 38
preventing mycotoxicoses, 108–109
Pre-fermentation, winery practice, 20
Pre-harvest
mycotoxins, 38
preventing mycotoxicoses, 108
Processing
dried figs, 21–22
food, on mycotoxin content, 47
Propionibacterium species
antifungal activity, 3
mycotoxin binding, 9
Propionic acid bacteria, antifungal activity, 3
Protegrins, lytical peptides, 127, 128*f*
Proteinaceous compounds, lactic acid bacteria activity, 6–7
Proteins
differences and similarities in structure, 124–125
properties unifying structure and function, 121–125
structural clues, 124
See also Lytic peptides
Proteomics, resistance-associated proteins (RAPs), 160–161
Pseudomonas syringae
activity of antimicrobial peptide D4E1, 137*t*
transgenic tomato plants, 136
Pyrene 1-carbonyl cyanide, structure, 297*f*
Pyrrole-based molecularly imprinted polymer
development and applications, 281*t*
electrochemical preparation, 280
See also Molecularly imprinted polymers (MIPs)
Pyrus communis, synthetic peptides, 135
Pythium ultimum, activity of antimicrobial peptide D4E1, 137*t*

Q

Quantum dots
molecularly imprinted polypyrrole/carbon nanotubes/QDs in needle, 286–287
See also Molecularly imprinted polymers (MIPs)

R

Raisins, reducing ochratoxin A (OTA) contamination, 21
Red fermented rice (RFR), fermentation product, 27
Regulation
  aflatoxin biosynthesis, 95–96, 99
  aflatoxin pathway-specific regulators, 187
  global regulatory factors, 187–190
  mechanism of aflatoxin formation, 98–99
  transcriptional, of aflatoxin biosynthesis, 187–191
Resistance
  genetic engineering for disease, 133–137
  rice to mycotoxin, 43–44
  screening for, to aflatoxin accumulation, 158–159
  targets for, 151
  trichothecenes, 144–145
Resistance-associated proteins (RAPs)
  aldose reductase, 160*t*, 161
  characterization, 161–162
  cold-regulated protein (ZmCORp), 160*t*, 161, 162
  comparative proteomics to identify, 160–161
  glyoxalase I, 160*t*, 161
  identification of, 159–161
  maize kernels using proteomics, 160*t*
  near-isogenic maize pairs varying aflatoxin levels, 161*t*
  pathogenesis related protein 10 (PR-10), 160*t*, 161, 162
  peroxiredoxin antioxidant (PER1), 160*t*, 161, 162
  trypsin inhibitor-10kDa (ZmTIp), 160*t*, 161, 162
Resorcylic acid
  mycotoxin:cyclodextrin interaction parameters, 302*t*
  structure, 297*f*
Reuterin, antifungal activity, 6
Rhizoctonia solani, activity of antimicrobial peptide D4E1, 137*t*
Rhizopus stolonifer, lactic acid bacteria against, 4*t*
Ribosome inactivating protein (RIP)
  inhibiting *Aspergillus flavus*, 159
  maize kernels, 160*t*
Rice
  Monascus, 27
  mycotoxin contamination in southern U.S., 43–44
Roasting, aflatoxin stability, 45

S

Saccharomyces cerevisiae, apple juice fermentation, 24
Safety. *See* Food safety
Sampling plans
  aflatoxins and ochratoxin A (OTA) analysis, 215–216
  Codex Alimentarius Commission, for aflatoxins in tree nuts, 212, 213*t*
  definition, 208
  European Commission (EC), for aflatoxins, 211, 212*t*
  material and sampling design, 214, 215*f*
  research projects for mycotoxins in botanical roots, 213–219
  U.S. Food and Drug Administration (FDA), for aflatoxins, 209, 210*t*
  using variance and distribution, 208–209
  variances for aflatoxins, 216–217
  variances for OTA, 218, 218–219
  See also Ginger capsule study

Scab disease, *Fusarium graminearum*, 44
Scatchard plot analysis, mycotoxin:cyclodextrin complexes, 300, 301*f*
Sclerotia, dried figs, 81, 82*t*
Sclerotinia sclerotiorum, transgenic tobacco and resistance to, 136
Screening, resistance to aflatoxin accumulation, 158–159
Senna pods, drying, 25
Solid phase extraction
- micro, using molecularly imprinted polymers (MIPs), 281–287
- MIP for, 278–280
- *See also* Molecularly imprinted polymers (MIPs)

Southern Regional Research Center (SRRC), aflatoxin-resistant maize, 158
Southern United States. *See* Corn (maize)
Soybeans
- analysis, 65
- nutritional value, 60
- ochratoxin A (OTA) distribution during tofu processing, 66, 67*t*
- OTA distribution after processing, 65, 66*t*
- processing samples, 62
- production, 62, 63*f*
- *See also* Ochratoxin A (OTA)

Soy milk
- liquid chromatogram (LC) of ochratoxin A (OTA), 65, 66*f*
- OTA distribution, 66, 67*t*
- production, 62, 64*f*
- *See also* Ochratoxin A (OTA)

Stable isotope labeled mycotoxins
- chemical modification of non-labeled mycotoxins, 269, 270*f*
- compensation of matrix effects, 267
- fungal growth of fully [^{13}C]-labeled culture media, 271–272
- HPLC–MS/MS analysis, 266, 273
- production of, in fungal cultures, 270–271
- total synthesis, 268
- *See also* Multiple mycotoxins analyses

Sterigmatocystin
- precursor to aflatoxin, 183, 184–185
- role of pH in production, 192
- schematic of biosynthetic gene clusters, 184, 186*f*

Storage
- aflatoxin levels in ground corn and time, 46*f*
- agricultural products, 45–46
- botanicals, 25
- cold, of apples, 23
- dried figs, 21–22
- fumonisin levels in ground corn and time, 47*f*
- maize kernels, 160*t*
- mycotoxin stability in stored agricultural products, 45–46
- patulin and apple juice, 24
- patulin contamination on apple and, temperature, 75

Stress-responsive proteins, maize kernels, 160*t*
Structural clues, protein functionality, 124
Synthetic peptides, disease resistance, 135–137

T

T-2 toxin
- collaborative study, 249
- mycotoxin:cyclodextrin interaction parameters, 302*t*
- phytoxic trichothecene, 149
- structure, 145*f*, 297*f*

Temperature

aflatoxin production, 99
aflatoxin stability, 45
Tennessee
aflatoxin levels in corn by year, 41*t*
fumonisin levels in corn by year, 42*t*
Texas
aflatoxin levels in corn by year, 41*t*
fumonisin levels in corn by year, 42*t*
Thielaviopsis basicola
activity of antimicrobial peptide D4E1, 137*t*
synthetic peptides, 135
Tissue-specific promoters, plant transformation, 130
Tobacco, disease resistance, 136
Tofu
analysis, 65
liquid chromatogram (LC) of ochratoxin A (OTA), 65, 66*f*
nutritional value, 60
OTA distribution, 66, 67*t*
production, 62, 64*f*
See also Ochratoxin A (OTA)
Toxicity, food processing-induced changes to fumonisins, 51–53
Toxin biosynthesis, histone acetylation and, 190–191
Toxin library, *Arabidopsis thaliana* screen of, 149
Transgenic crops. *See* Genetic engineering of food crops
Tree nuts
Codex Alimentarius Commission sample plan, 212, 213*t*
sample sizes by US Food and Drug Administration, 210*t*
Trichodiene, structure, 145*f*
Trichodiene derivatives, phytoxic trichothecene, 149
Trichothecenes
acetyltransferase gene (*Tri101*), 145
Arabidopsis thaliana model system, 148–149
A. thaliana screen of toxin library, 149
biotransformation and non-toxicity of new metabolite, 109–110
C-3 oxygenation, 150
C-7 and C-8 oxygenation, 146–147
chemical modification, 269
Chlamydomonas bioassay, 146
collaborative study, 249
comparing *Chlamydomonas* and *Arabidopsis* bioassays, 151, 152
detoxification of, 150
Fusarium, biosynthesis, 144
general structure, 144*f*
growth of *C. reinhardtii* after 6 days, 147*t*, 148*t*
hydroxylation, 149–150
phytotoxicity assays, 145–146
phytotoxicity of *F. graminearum* mycotoxins, 147, 148*t*
resistance, 144–145
structures of, and precursors trichodiene and isotrichotriol, 145*f*
targets for resistance, 151
Trichothecium
trichothecene metabolites, 150
trichothecene-producing fungi, 144
Trypsin inhibitor-10kDa (ZmTIp), resistance-associated proteins (RAPs), 160*t*, 161, 162
Trypsin inhibitor protein
maize kernels, 160*t*
resistant maize lines, 159
Tryptophan, aflatoxin biosynthesis, 99
Turkey-X disease, England, 108
Tyrosine, aflatoxin biosynthesis, 99

U

United States
mycotoxin contamination of corn grown in southern, 39–40, 43*f*
See also Corn (maize); Food and Drug Administration (FDA)

V

Variances
aflatoxin measurement, 216–217
ochratoxin A (OTA) measurement, 218
See also Ginger capsule study; Sampling plans
VeA gene, regulator of aflatoxin production, 188–189
Vegetables, lactic acid bacteria prolonging, 5–6
Verticillium dahliae
activity of antimicrobial peptide D4E1, 137*t*
synthetic peptides, 135
Vine fruit (dried)
reducing ochratoxin A (OTA) contamination, 21
See also Fruit and products
Volume, amino acids, 122, 123*t*
Vomitoxin (DON)
rice samples, 44
structure, 39*f*
See also Deoxynivalenol (DON)

W

Washing, high pressure water, of apples, 23
Wheat
ethanol vs. methanol extraction of ochratoxin A (OTA), 231
ochratoxin A (OTA) extraction, 229, 230*f*
Winery, grape treatment, 20
Wines, fermentation process, 248
Wounding, fungus access to seed, 158

X

Xanthomonas campestris, activity of antimicrobial peptide D4E1, 137*t*

Y

Yeasts, lactic acid bacteria against, 4*t*

Z

Zeamatin
inhibiting *Aspergillus flavus*, 159
maize kernels, 160*t*
Zea mays. *See* Corn (maize)
Zearalenone (ZON)
biodegradation and non-toxicity of new metabolite, 111–112
chemical modification of non-labeled, 269, 270*f*
collaborative study, 249
cyclodextrin concentration and fluorescence enhancement of, 300*f*
cyclodextrins and fluorescence of, 298–301
detection, 252
detoxification by *Gliocladium roseum*, 10
emission of, by cyclodextrin concentrations, 300*t*
E-screen assay, 112*f*
ethanol vs. methanol for extraction in corn, 232
mycotoxin:cyclodextrin interaction parameters, 302*t*
regulatory levels, 248
rice samples, 44
scatchard plot analysis of ZON:cyclodextrin complexes, 301
structure, 39*f*, 297*f*